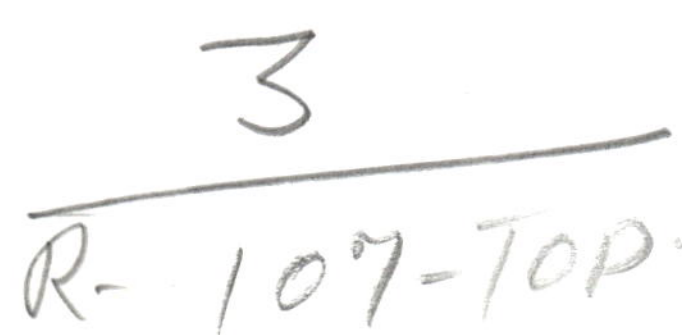
3
R- 107-TOP.

AF522426

THERMODYNAMICS

THERMODYNAMICS

[For B.Sc. & M.Sc. Students of Indian Universities]

By

M.S. YADAV

M.Sc., Ph.D.

ANMOL PUBLICATIONS PVT. LTD.

NEW DELHI - 110 002 (INDIA)

ANMOL PUBLICATIONS PVT. LTD.
Regd. Office: 4360/4, Ansari Road, Daryaganj,
New Delhi-110002 (India)
Tel.: 23278000, 23261597, 23286875, 23255577
Fax: 91-11-23280289
Email: anmolpub@gmail.com
Visit us at: www.anmolpublications.com

Branch Office: No. 1015, Ist Main Road, BSK IIIrd Stage
IIIrd Phase, IIIrd Block, Bengaluru-560 085 (India)
Tel.: 080-41723429 • Fax: 080-26723604
Email: anmolpublicationsbangalore@gmail.com

Thermodynamics

PRINTED IN INDIA

Printed at Balaji Offset, Delhi

Contents

Preface to the Second Edition

With the changes in syllabus and wise suggestions of teachers, this thoroughly revised and enlarged edition of the book is brought out.

Two new chapters are added. New additions are done to give emphatic treatment to the subject. Various errors of first edition are rectified.

We hope, this edition will prove more useful and informative for the students and teachers alike.

I remain thankful to my colleagues who have assisted me in one way or other. I am equally thankful to Mr. J.L.Kumar, Managing Director, Anmol Publications Pvt. Ltd., New Delhi for encouragement in my creative ventures.

Constructive comments and suggestions from the readers are welcome which will be incorporated in the next edition of the book.

M.S. Yadav

Preface to the First Edition

The book "Thermodynamics" includes thirteen chapters. This division of book into chapters has been done to present the subject matter in a comprehensive but easily understandable manner. It has also helped to avoid repetition of the subject matter. Most of the topics included in this book form a part of syllabus for B.Sc. and M.Sc. students of the Indian Universities. The topics included are laws of thermodynamics, thermochemistry, thermodynamic relations, entropy, free energy, chemical potential, Nernst heat theorem, partial molar properties, activity, chemical affinity, specific heats of solid's and gases, etc. Some of these are of interest to a general reader.

An attempt has been made to present the subject matter in such a form that it is liked by students and teachers alike.

I am thankful to a number of authors whose works have been freely consulted while preparing this book. I am especially thankful to Dr. R.C. Dhankhar and Mr. J. Mittra for giving their invaluable suggestions on some of the difficult topics included in the book.

I am thankful to M/s Anmol Publications Pvt. Ltd. for bringing out the book promptly.

I shall thankfully accept the suggestions for modification and improvement in the book.

M.S. Yadav

1

Introduction to Thermodynamics

Thermodynamics

One of the fundamental manifestations in nature in the *energy* that accompanies all changes and tranformations. The most common form in which energy appears, and the form to which all others tend, is *heat.*

Besides heat, we have *mechanical energy,* involved in the motion of machines; *electrical energy, chemical energy, radiant energy* etc. As diversified and distinct as these various forms may at first glance appear, they are, never-the-less, related to one another. *A quantitative relationship of various forms of energy in a system constitutes the subject of thermodynamics.* It is thus an exact science that *describes quantitative inter-relationship of various forms of energy accompanying the physical and chemical changes.* It is not concerned with total energy of a body; it concerns itself only with *energy changes* accompanying a given process or transformation.

The science which deals with the macroscopic properties of matter is known as *classical thermodynamics.* Here we can carry out the entire formulation without the knolwedge that matter consists of atoms and molecules. The science of classical thermodynamics has been developed without an enquiry into the structure of the matter. It is concerned only with the average characteristics of large aggregation of molecules and is not concerned with the average characteristics of individual molecules. In *statistical thermodynamics* we deal with the calculations of *thermodynamic properties* of matter from the classical or quantum mechanical behaviour of a large congregation of atoms or molecules. The statistical thermodynamics considers the microscopic structure of matter and adopts the laws of mechanics as the statistical analysis of the individual particles.

Experimental criteria for equilibrium or for the spontaneity of the process are readily established with the help of thermodynamic principles. These can be used to indicate the direction the system will take to achieve equilibrium. *Time* is not a thermodynamic variable and so thermodynamics cannot give any information about the length of the time that would be required for the completion of the process.

Historically, thermodynamics first dealt with the conversion of heat into mechanical motion, such as we find when heated steam is used to move the piston of a steam engine. Later thermodynamics broadened its scope and now-a-days it deals with the role of heat and work in all types of changes.

Thermo (heat) Dynamics (work) is the study of those interactions among various materials which involve the transfer of heat and the performance of work.

Chemical Thermodynamics

It is that branch of thermodynamics which deals with the study of the processes in which only chemical energy is involved, e.g., neutralization of an acid by an alkali, combustion of coal, discharge of storage battery etc.

Scope and Objectives. The influence of thermodynamics in modern physics, chemistry, biology, biochemistry, space science and engineering can not be over stressed. Though it embraces many areas of science and technology, the use of thermodynamic principles in chemistry are most significant and have been successfully employed. All chemical processes are accompanied by energy changes and such changes in many instances can be measured experimentally. The value of these quantitative measurements lies in the use we can make of such information in understanding the processes by means of which chemical reactions come about and the way we can use the same information to predict the course of reactions about which we have little or inadequate experimental knowledge.

Thermodynamics is an exact mathematical and deductive science and is based on three concise statements known as *three laws of thermodynamics*. These are based on experience and are justified by the fact that their consequences, especially their predictive consequences are consistent with observations.

These laws can be used to predict which chemical or physical reactions may occur and which chemical or physical reactions cannot occur. The basic objective of thermodynamics is to predict the *feasibility* of a given transformtion which may be physical or chemical.

Some facts discussed below illustrate the importance of thermodynamics.

(i) *Fundamental laws*. No exception to the thermodynamic laws has been reported usually, if a conflict exists between theory and experiment, the error is ascribed to the theory in most of the cases. But in thermodynamics we are so sure that if a conflict arises between the theory and an experiment, the validity of the experiment is questioned. Thus the laws of thermodynamics are the most certain aspects of chemistry.

(ii) *Generalization of experimental results*. Generalisation of Physical Chemistry such as Gas Equation, Laws of chemical equilibrium, Phase rule, Distribution Law, Vant Hoff's law of dilute solutions, Raoult's law of lowering of vapour pressure can all be deducted from the laws of thermodynamics.

Equations co-relating the variations of certain physical properties with temperature and pressure have also been derived from these laws.

(iii) *Direction of a reaction*. The laws of thermodynamics can be used to predict the direction in which a process would proceed. If a reaction is feasible on thermodynamic considerations then we are on a safe ground to try to find the proper experimental conditions to make the reaction proceed, e.g., combination of H_2 and O_2 to give water is thermodynamically possible. Nevertheless, both gases can coexist without combining for a long time. They do not react at room temperature and a mixture of H_2 and O_2 will not produce H_2O. However if the mixture is sparked, water is produced with explosive violence, thereby proving the prediction made on the basis of thermodynamic considerations.

(iv) *Prediction of Relationship between Directly Observable Properties.* It is possible with the help of laws of thermodynamics to predict the relationship between directly observable properties of substances quantitatively. Relationship between quantities which can be derived from the other observable properties can also be predicted. For example, it is possible to derive the relationship between pressure, temperature and volume of a gas thermodynamically.

It is to be noted that the great contribution of thermodynamics lies in that it systematises the information obtained from various experiments or measurements and allows us to draw conclusions about other aspects, than that measured, of the behaviour of the system.

Limitations of Thermodynamics

The science of thermodynamics which has proved of great help and importance to the chemist has certain limitations. The most serious weaknesses are as follows:

(i) It deals with matter on the macroscopic scale so these laws are applicable to an assembly of large number of molecules and not to individual molecules and thus independent of atomic and molecular structure of matter. Thus thermodynamics does not make any hypothesis as to the structure of matter.

(ii) Though it gives the relationship between various properties experimentally observed, it is unable to give the actual values of these properties and offer no explanation as to why these properties arise in a system.

(iii) The thermodynamic approach is totally different from that of the approach in atomic constitution of matter or kinetic molecular approach. In kinetic theory the pressure of a gas is considered as a measure of the impact of gaseous molecules on the walls of the container per second, while in thermodynamics it is regarded as the property of the state of a system regardless of its inherent constitution, based on the observation of a body as a whole. Similarly, temperature is a measure of the average kinetic energy according to kinetic theory but from the view point of thermodynamics it is a simple property of state.

(iv) Thermodynamics provides no information regarding the time taken to reach equilibrium and so it fails to give any information about the time taken for the completion of a change and the rate at which a change takes place.

(v) It fails to tell us the behaviour of a system away from equilibrium. It deals with substances in stationary state or equilibrium state.

(vi) It fails to tell us the way the system undergoes a change and is also silent about the path the change follows.

(vii) It gives no information regarding the conditions under which the change takes place.

(viii) Thermodynamics deals with interconversion of various forms of energy; *matter finds no consideration except as a carrier of energy.*

(ix) Mass is also a form of energy as the two are related by the Einstein equation, $E = mc^2$. In thermodynamic processes there is no measurable conversion of mass into energy. Hence, the *fact of mass being a form of energy is left out of consideration in thermodynamical problems.*

Basic Definitions

The definitions of some terms are given below:

Energy

It is a mathematical abstraction and has no existence apart from its relationship to other variables which have a physical interpretation and can be measured, e.g., the kinetic energy of a given mass of material is a function of its velocity and it has no other reality.

First law of thermodynamics depends on the concept of energy and *energy* is an essential thermodynamic function because it allows the formulation of the first law.

System

A system can be any object, any quantity of matter, any region of space etc. selected for study and set apart (mentally) from everything else.

A system is thus defined "*as any portion of universe (any portion of matter under investigation) separated from the rest (called the surroundings) by real or imaginary boundaries in order to study the effect of various variables such as temperature, pressure, volume etc., upon its contents.*

Surroundings

All other matter that can interact with the system are considered as surroundings.

It may be defined as *the portion of the universe excluded from the system which may be capable of exchanging energy and matter with it.* It may consist of one or more substances.

Illustration of terms. A cylinder containing a gas (or liquid) is placed in a thermostatic bath which is maintained at a constant temperature, say, 25°C. This is shown in Fig. 1.1. In this case the gas (or liquid) constitutes the system whereas the constant temperature bath is the surroundings. The walls of the cylinder constitute the boundaries.

As we come across various types of systems as well as walls or boundaries, it becomes essential to define some important types of walls. These are:

(i) *A rigid wall.* This is the wall whose position and shape are

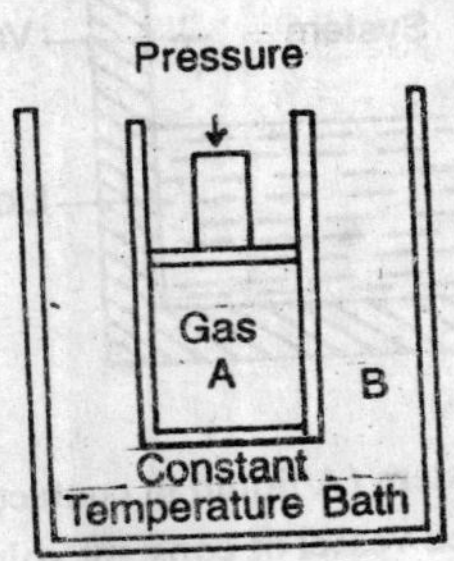

Fig. 1.1

fixed.

(ii) *An impermeable wall.* This is the wall which does not permit the passage of matter.

(iii) *A permeable wall.* It is that type of wall which permits the passage of matter and consequently, also of energy.

(iv) *An adiabatic wall.* It is that type of wall which when held rigid will not permit the passage of matter or energy provided no external force fields are present. An interesting example is the thermos bottle.

(v) *A diathermal or diathermic wall.* It is that type of wall which when held rigid will not permit the passage of matter but allow the passage of energy in the absence of any external force field. It means that a diathermal wall is impermeable but not adiabatic. An aluminium can is an example of a container whose walls are diathermal.

Types of Systems

Depending on the interactions between system and its

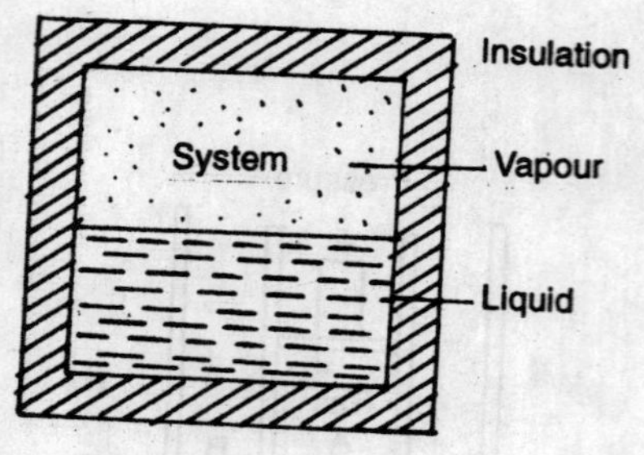

Fig. 1.2 An isolated system (A liquid in contact with its vapour in an insulated vessel of constant volume).

surroundings,three kinds of systems are distinguished *isolated system, closed system* and *open system.*

(i) *Isolated system.* A system which neither exchanges energy nor matter with its surrounds is known as an *isolated system* e.g., a liquid in contact with its vapour in an insulated closed vessel (Fig. 1.2), is an *isolated system.* As the vessel is closed (vapour or liquid), can neither enter nor leave the vessel, i.e., no exchange of matter from within the vessel is possible with the surroundings. Further the vessel is insulated. The system is not in a position to gain heat from or lose heat to its surroundings, i.e., it cannot exchange energy with its surroundings. Hence it is an isolated system.

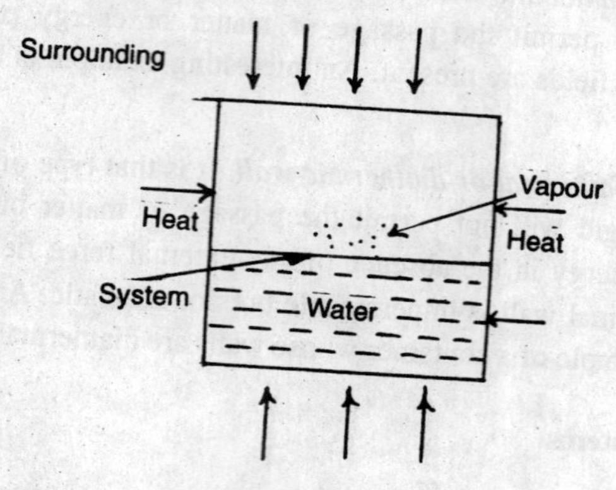

Fig. 1.3

Process: When any property of a system charges there occurs a change of state and the system is said to undergo a process.

It is the path long which a change of state takes place. The process can occur under a variety of conditions which must be defined as many things are dependent on the nature of the process.

Types of process. We came across the following types of processes.

(i) *Isothermal process.* A process is said to be isothermal if the temperature of the system remains constant throughout the process. However, the system can exchange heat energy with the surroundings to keep the temperature constant. Thus the heat content of the system may change in such a process. *An isothermal process occurs under constant temperature condition.*

(ii) *Adiabatic process.* In such a type of process heat can neither be added to nor removed from the system. Thus the total heat content (enthalpy) of the system remains constant. The temperature of the system may rise or fall in these cases. The adiabatic process occurs under constant enthalpy conditions.

(iii) *Isobaric processes.* The process in which the pressure of the system remains constant during each step of the change in the state of

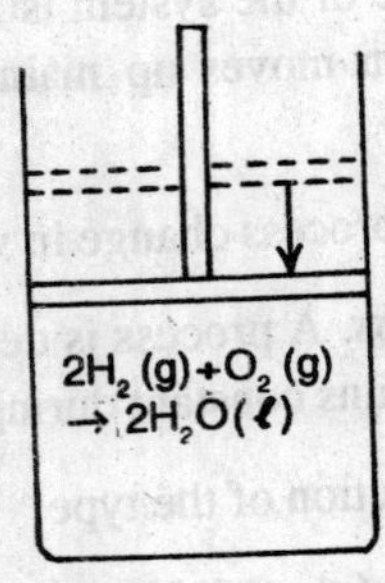

Fig. 1.6 Isobaric process followed by decrease in volume.

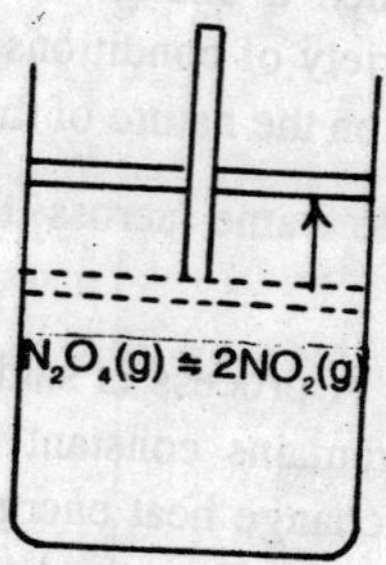

Fig. 1.7. Isobaric process followed by increase in volume.

the system is known as isobaric process. This occurs under constant pressure conditions.

Let us consider the following examples of isobaric processes:

(a) Consider a mixturee of oxygen and hydrogen gases in the ratio 1:2 contained in a cylinder closed by an airtight, weightless and frictionless piston. This is the case of a closed system. When an electric spark is passed at room temperature, the reaction proceeds followed by an appreciable fall in volume of the system. The piston moves down maintaining the pressure of the system constant, Fig. 1.6.

(b) If the reaction $N_2O_4(g)$, $\rightarrow$ $2NO_2(g)$, which is accompanied by the increase in volume of the system is allowed to take place in the above cylinder, the piston moves up maintaining the pressure of the system constant, Fig. 1.7.

Thus in an isobaric process change in volume may take place.

(iv) *Isochoric process.* A process is defined to be isochoric of the volume of the system remains constant during each step of the process.

Let us consider a reaction of the type

$$N_2O_4(g) \rightarrow 2NO_2(g)$$

For the dissociation of every one mole of N_2O_4, two moles of

NO_2 gas are produced and, thus, the volume of the system will tend to increase. In order to keep the volume fixed, weights are to be kept over the piston. Thus in such a case the volume of the system remains constant whereas the pressure of the gaseous system within the cylinder will increase.

Let us consider the reaction between hydrogen and oxygen. On passing an electric spark the gases will react to form liquid water resulting in considerable decrease in volume. But if the piston is fixed, i.e., not allowed to move down, the volume of the system will remain constant and there will be considerable fall in the pressure of the system.

From the above examples, it follows that in isochoric processes, change of pressure may occur.

(v) *Cyclic Process.* A process in which a system undergoes a series of changes and finally returns to the initial state is called a cyclic process.

A cyclic process carried out isothermally is known as isothermal cycle. If these changes are carried out reversibly it is known as a reversible cycle. Carnot's cycle is a reversible cycle.

(vi) *Reversible and Irreversible processes.* Various types of processes given above can be carried out either reversibly or irreversibly.

Reversible Process (Quasistatic Process)

If the process is carried out in such a way that its properties at any instant remain practically uniform, it is called a *reversible process.* Such a process proceeds through a succession of infinitesimal steps and the process proceeds very slowly. The small changes should be carried out in such a way that, throughout the transition, the system must be maintaining virtual thermodynamic equilibrium, at each of the small changes. It should be possible to retrace the path back to the initial conditions in case of reversible process. So the reversible process may be defined as.

"It is one consisting of a series of equilibrium stages which, subsequent to the completion of the process, can be brought back to the initial state without producing any permanent effect in the adjacent surroundings."

Alternatively, *"It may be defined as one in which the process can be made to go in either direction."*

From the above discussion it follows that in a step where heat is absorbed in the direct process, it is given out in the reverse process and vice versa. Similarly, in a step where work is done by the working substance in the direct process an equal amount of work is done on the working substance in the reverse process. Some examples of reversible processes are given below:

(1) An ideal example for a reversible process is the working of a Galvanic cell (Daniell cell) which consist of a zinc and a copper electrode, each dipping in the solution of its sulphate and separated by a porous pot. The cell has a definite potential. The electrodes of this cell are connected to an opposing source of E.M.F. by a battery through a rheostat. If the E.M.F. of the cell is equal to the opposing E.M.F., no current will flow in either direction in the ammeter. If, however, the opposing E.M.F. is made infinitesimally smaller than the E.M.F. of the cell, then an infinitesimally small amount of current will start flowing through the ammeter. The reaction in the Galvanic cell that takes place is

$$Zn + Cu^{2+} \rightarrow Zn^{2+} + Cu$$

The cell under these conditions is said to be working in a thermodynamically reversible manner.

If the potential of the opposing source is infinitesimally larger than the E.M.F. of the cell, the ammeter reading will indicate the flow of the current in the opposite direction and the chemical reaction takes place in the opposite direction.

$$Zn^{+2} + Cu \rightarrow Zn + Cu^{+2}$$

The flow of current in this way is again a case of thermodynamically reversible process.

If the opposing E.M.F. is made appreciably smaller or appreciably larger than the E.M.F. of the cell. In that case an appreciable amount of current will be flowing in one direction or the other. Current flowing in this manner is an example of irreversible process.

Characteristics of a Reversible Process

(1) *At any time during the reversible process, the driving force is only infinitesimally greater than the opposing force. Such a process can be reversed by an infinitesimal increase in the opposing force.*

(2) *At all times the system is in virtual equilibrium with the surroundings, therefore, at any time, the initial state can be regained by infinitesimal small changes in the state variables.*

(3) *Reversible changes are very slow.*

(4) *Work obtained in a reversible process is maximum.*

(b) *Irreversible Process.* When in a process the conditions of reversibility are not followed it is called an *irreversible process.*

An Irreversible Process can be defined *as a process in which the successive steps of the direct process cannot be retraced and an infinitesimal alteration in the conditions governing the equilibrium cause a spontaneous and complete change in the equilibrium.*

For example, quick evaporation by sudden decrease in external pressure is an irreversible process. In this case the temperature and pressure within the system change and disturb the equilibrium.

If the pressure on the piston (external pressure) of the cylinder containing gas is suddenly lowered considerably so that it becomes much smaller than the pressure of the gas within the cylinder, the gas expands rapidly and the piston is pushed up suddenly. The expansion of the gas, in this case is taking place irreversibly.

It is of interest to note that most of the spontaneous processes are irreversible in nature, e.g., water flowing down a hill, gases expanding from regions of high pressure to region of low pressure, the use of a battery or cell directly, heat flowing of its own accord from hot bodies to cooler bodies, etc. These spontaneous processes happen unidirectionally and certainly not in successive stages of equilibrium in minute quantities of transformation. Natural or spontaneous processes are, therefore, irreversible.

From this study we should not conclude that an irreversible process takes place at a very high speed, it should mean only that the direction of the process is spontaneous.

Some more examples of the irreversible processes are given below:

(i) Example for this is the time which is associated with an irreversible process of biological aging.

(ii) Work done against friction is irreversible. It is because in reversing the direction of motion again work has to be done against friction.

(iii) Joule Kelvin effect is irreversible because reversing the flow of gas, a similar cooling or heating effect is not observed. Similarly, diffusion, conduction and radiation are also irreversible.

(iv) Quick evaporation by sudden decrease in external pressure is an irreversible process. In this case the temperature and pressure within the system change and disturb the equilibrium.

(v) Most of the spontaneous processes are irreversible in nature. Some examples are like water flowing down a hill, the heat flowing of its own accord from hot bodies to cooler bodies, etc. All these processes take place in one direction only.

A Comparative Study of the Reversible and Irreversible Processes

(i) A reversible process is very slow through a continuous series of equilibrium states as the driving force is vanishingly small. But an irreversible process goes with a measuring speed towards the final state.

(ii) A reversible process can be made to go in either direction but an irreversible process can proceed only in one direction.

(iii) Another important difference is that the work done by a reversible process is greater than the corresponding irreversible process in between two states.

(iv) A reversible process can be brought back to the initial state without producing any permanent effect in the adjacent surroundings. But an irreversible process cannot be traced to the initial state by avoiding a permanent change in surroundings.

Thermodynamic Equilibrium

A system is said to be in a *state of equilibrium* of a change of state cannot take place while the system is subjected to interaction with the environment. For any system in equilibrium state there must be no balanced potential which tends to promote a change of state. The unbalanced potential may be mechanical, thermal, chemical or electrical etc. Thus a system is said to be in thermodynamic equilibrium if, it is in

(i) Thermal equilibrium

(ii) Mechanical equilibrium

(iii) Chemical equilibrium

(iv) Electrical equilibrium.

When there is *no temperature gradient* within a system it is said to be in a state of *thermal equilibrium.*

When there is no *unbalanced force* within the system, it is said to be in a state of *mechanical equilibrium.*

When a system has no tendency to undergo either a chemical reaction or a process such as diffusion or solution, the system is said to be in a state of *chemical equilibrium.*

When there is no *electrical potential gradient* within a system, it is said to be in a state of *electrical equilibrium.*

When any of the above conditions for a complete thermodynamic equilibrium is not satisfied the system is said to be in a *non-equilibrium state.*

Classical thermodynamics deals only with systems in equilibrium states.

A homogeneous system when it is in its equilibrium state can be represented by a single point on a set of thermodynamic coordinates. When the system is not in equilibrium, some of its macroscopic coordinates vary from one point to the other point and so it is not possible to locate a single point on a state diagram for such a system i.e. a system that is in non-equilibrium state.

Mathematical Background

In thermodynamics, generally a change in thermodynamic property is studied with a change of some independent variable. To clearly understand such changes in mathematical forms in which these are generally expressed a basic background of mathematics is essential. The most commonly used mathematical terms and operations are the concepts of exact differentials, inexact differentials integrals etc. These are discussed here.

Perfect differentials. If some valued function *Z* depends on two or more independents variables *x, y, p, q,* ... For the sake of convenience let us consider that *Z* is a single valued function of two independent variables *x, y.*

Mathematically, it can be expressed as

$$Z = f(x, y)$$

If out of the two independent variables on which *Z* depends one is kept constant then *Z* becomes a function of other variable alone.

dZ the perfect differential or the total differential of Z(*x,y*) is given by

$$dz = \left(\frac{\partial z}{\partial x}\right)_y dx + \left(\frac{\partial z}{\partial y}\right)_x dy \qquad (1)$$

where $\left(\frac{\partial z}{\partial x}\right)_y$ and $\left(\frac{\partial z}{\partial y}\right)_x$ are the partial derivatives and are obtained when changes in only one variable are considered. These are given by following mathematical relations.

$$\left[\frac{\partial z}{\partial x}\right]_y = \underset{\Delta x \to 0}{\text{Lt}} \frac{z(x+\Delta x, y) - z(x,y)}{\Delta x}$$

where Δx is a small increment in value of x and y remains constant.

$$\left(\frac{\partial z}{\partial y}\right)_y = \underset{\Delta y \to 0}{\text{Lt}} \frac{z(x, y+\Delta y) - z(x, y)}{\Delta y}$$

where Δx is a small increment in value of x and y remains constant. dz the perfect differential is obtained for a simultaneous change in both the functions x and y.

It has been assumed that the system undergoes an infinitesimal process from an initial state of equilibrium to another. Every *infinitesimal in thermodynamics must satisfy the requirement that it represents a change in quantity which is small with respect to the quantity itself and large in comparison with the effect produced by the behaviour of a few molecules.* This is due to the fact that thermodynamic coordinates such as volume, pressure, temperature have no meaning when applied to a few molecules. In this way, we emphasize that thermodynamic coordinates are macroscopic coordinates.

The differentials of the actual functions are called *exact differentials* such as dP, dV, dθ etc.

An infinitesimal that is not the differential of an actual function is known as *inexact differential.*

The inexact differential can not be expressed by the equation (1).

Another important condition for dz in equation (1) to be a perfect differential is

$$\frac{\partial^2 z}{\partial x\, \partial y} = \frac{\partial^2 z}{\partial y\, \partial x}$$

It can be obtained as follows:

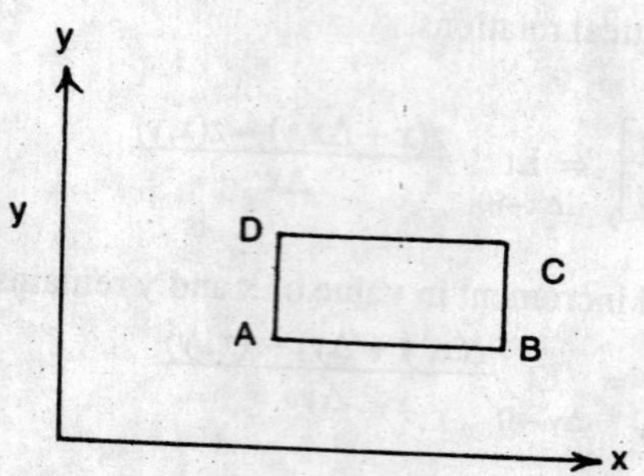

Fig. 1.8

Let A *b* the initial state and C the final state. An infinite number of paths may be adopted to go from *A* to *C*. If we choose the path ABC then the value of quantity is *Z* at *A*, its value at *B* will be $Z+\frac{\partial z}{\partial y}$. dy and at *C* will be

$$\left(Z+\frac{\partial z}{\partial y}dy\right)+\frac{\partial}{\partial x}\left(Z+\frac{\partial z}{\partial y}dy\right)dx$$

Again if we choose the path ADC the value of *Z* at *D* is given by $\left(Z+\frac{\partial z}{\partial x}\,.\,dx\right)$ and at C it will be

$$\left(Z+\frac{\partial z}{\partial x}dx\right)+\frac{\partial}{\partial y}\left(Z+\frac{\partial z}{\partial x}dx\right)dy$$

If *dz* is a perfect differential the value of Z is independent of the path chosen and it should be the same irrespective of the path adopted.

$$\therefore \quad \left(Z+\frac{\partial z}{\partial y}dy\right)+\frac{\partial}{\partial x}\left(Z+\frac{\partial z}{\partial y}dy\right)\mathrm{dx}=\left(Z+\frac{\partial z}{\partial y}dy\right)+\frac{\partial}{\partial x}\left(Z+\frac{\partial z}{\partial y}dy\right)dx$$

or

$$\frac{\partial^2 z}{\partial x\,\partial y}=\frac{\partial^2 z}{\partial y\,\partial x}.$$

Cyclic Rules

To derive the cyclic rule we can either assume *P* and *V* to be state functions, i.e., they depend only on the state of the system and are independent of the path adopted or we can first prove *P* and *V* to be state functions.

(1) (V) Volume of an Ideal Gas is a State Function

To prove it consider the ideal gas equation PV = RT for one mole of an ideal gas

$$V = \frac{RT}{P}$$

$$\therefore \qquad \left(\frac{\partial V}{\partial T}\right)_P = \frac{R}{P} \quad \text{and} \quad \left(\frac{\partial V}{\partial P}\right)_T = \frac{RT}{P^2}$$

Again, exact differential of V which depends on P and T is given by

$$dV = \left(\frac{\partial V}{\partial P}\right)_T dP + \left(\frac{\partial V}{\partial T}\right)_P dT$$

Substituting the values of $\left(\frac{\partial V}{\partial P}\right)_T$ and $\left(\frac{\partial V}{\partial T}\right)_P$, we get

$$dV = -\frac{RT}{P^2}\,dP + \frac{R}{P}dT$$

If *V* is a state function *dV* should be an exact differential and for exact differential, we have

$$\frac{\partial}{\partial T}\left[\left(\frac{\partial V}{\partial P}\right)_T\right]_P = \frac{\partial}{\partial P}\left[\left(\frac{\partial V}{\partial T}\right)_P\right]_T$$

solving it, we get

$$\frac{\partial}{\partial T}\left[\left(\frac{\partial V}{\partial P}\right)_T\right]_P = \frac{\partial}{\partial P}\left(-\frac{RT}{P^2}\right)_P = -\frac{R}{P^2}$$

and

$$\frac{\partial}{\partial P}\left[\left(\frac{\partial V}{\partial T}\right)_P\right]_T = \frac{\partial}{\partial P}\left(\frac{R}{P}\right)_T = -\frac{R}{P^2}$$

These are equal so it is an exact differential. Hence the volume of a gas is a state function.

Similarly, it can be shown that pressure of a gas is a state function.

In this case it can be shown that $\frac{\partial^2 P}{\partial T \partial V} = \frac{\partial^2 P}{\partial V \partial T}$ and thus it is a state function.

To derive the cyclic rules.

Since $\quad P = f(V, T)$

$$\therefore \quad dP = \left(\frac{\partial P}{\partial T}\right)_V dT + \left(\frac{\partial P}{\partial V}\right)_T dV$$

Since P is a state function for a cyclic process $dP = 0$ (the initial and final states are same).

Thus $\quad \left(\frac{\partial P}{\partial T}\right)_V dT + \left(\frac{\partial P}{\partial V}\right) dV = 0$

Dividing by $(\partial V)_P$,

$$\left(\frac{\partial P}{\partial T}\right)_V \left(\frac{\partial T}{\partial V}\right)_P + \left(\frac{\partial P}{\partial V}\right)_T = 0$$

or

$$\left(\frac{\partial P}{\partial T}\right)_V \left(\frac{\partial T}{\partial V}\right)_P = -\left(\frac{\partial P}{\partial V}\right)_T$$

or

$$\left(\frac{\partial P}{\partial T}\right)_V \left(\frac{\partial T}{\partial V}\right)_P \left(\frac{\partial V}{\partial P}\right)_T + 1 = 0.$$

This expression is known as cyclic rule.

Another expression can be drawn for volume. In such a case dV=0 (cyclic process).

Since $V = f(P, T)$

and V is a state function

$$\therefore\ dV = \left(\frac{\partial V}{\partial T}\right)_P dT + \left(\frac{\partial V}{\partial P}\right)_T dP$$

When $dV = 0$

This equation becomes

$$\left(\frac{\partial V}{\partial T}\right)_P (\partial T)_V + \left(\frac{\partial V}{\partial P}\right)_T (\partial P)_V = 0$$

Dividing by $(\partial P)_V$,

$$\left(\frac{\partial V}{\partial T}\right)_P \left(\frac{\partial T}{\partial P}\right)_V + \left(\frac{\partial V}{\partial P}\right)_T = 0$$

Multiplying by $\left(\frac{\partial P}{\partial V}\right)_T$

$$\left(\frac{\partial V}{\partial T}\right)_P \left(\frac{\partial T}{\partial P}\right)_V \left(\frac{\partial P}{\partial V}\right) + 1 = 0.$$

This is another expression for cyclic rule.

The cyclic rule can easily be tested for the equation,

$$\left(P + \frac{a}{V^2}\right)V = RT.$$

Simple systems. Such systems which can be expressed in mass, volume, temperature and pressure only are known as simple systems.

(i) *Thermal expansivity* (α). It is the rate of change of volume with temperature, pressure remaining constant. For a volum V, we have

$$\alpha = \frac{1}{V}\left(\frac{\partial V}{\partial T}\right)_P$$

It has the dimension T^{-1}

(ii) *Compressibility coefficient* (β). It is the rate of change of volume with pressure, temperature remaining constant

$$\beta = -\frac{1}{V}\left(\frac{\partial V}{\partial P}\right)_T$$

It has the dimension P^{-1}

α, β are related as

$$\frac{\alpha}{\beta} = \left(\frac{\partial P}{\partial T}\right)_V$$

This relation can be derived as under

Since $\quad V = f(T, P)$

$$\therefore \quad dV = \left(\frac{\partial V}{\partial T}\right)_T dT + \left(\frac{\partial V}{\partial P}\right) dP$$

From the definitions of α, β.

$$dV = \alpha\, VdT - \beta VdP$$

For $\quad dV = 0 \qquad$ [Condition of volume constant]

We get

$$\alpha V\ (\partial T)_V - \beta V(\partial P)_V = 0$$

or

$$\left(\frac{\partial P}{\partial T}\right)_V = \frac{\alpha}{\beta}.$$

Example. (i) Calculate α and β for an ideal gas obeying gas equation PV = RT (one mole).

(ii) Calculate α and β for a gas obeying Van der Walls equation.

Solution.

For an ideal gas,

$$PV = RT$$

Differentiating,

$$PdV + VdP = RdT$$

Divide by dT, keeping P constant

$$P\left(\frac{\partial V}{\partial T}\right)_P = R$$

By definition, $\alpha = \frac{1}{V}\left(\frac{\partial V}{\partial T}\right)_P$

Substituting value of $\left(\frac{\partial V}{\partial T}\right)_P$

We get $\alpha = \frac{R}{VP} = \frac{1}{T}$

Again, divide the equation

$$PdV + VdP = RdT$$

By dP, keeping T constant

$$P\left(\frac{\partial V}{\partial P}\right)_T + V = 0$$

$$\left(\frac{\partial V}{\partial P}\right)_T = -\frac{V}{P}$$

By definition,

$$\beta = -\frac{1}{T}\left(\frac{\partial V}{\partial P}\right)$$

Substituting the value of $\left(\frac{\partial V}{\partial P}\right)_T$

$$\beta = \left(-\frac{1}{V}\right)\left(-\frac{V}{P}\right)$$

$$= \frac{1}{P}$$

(ii) For a Van der Walls gas, we have

$$\left(P + \frac{a}{V^2}\right)(V - b) = RT$$

or $$PV + \frac{a}{V} - Pb - \frac{ab}{V^2} = RT$$

or $$PV^3 + aV - PbV^2 - ab = V^2RT$$

Differentiating, we get

$$V^3\,dP + 3V^2\,PdV + a\,dV - V^2\,bdP - 2P\,VbdV$$
$$= 2RVT\,dV + V^2RdT$$

Dividing by dT, keeping P constant

$$3V^2P\left(\frac{\partial V}{\partial T}\right)_P + a\left(\frac{\partial V}{\partial T}\right)_P - 2PVb\left(\frac{\partial V}{\partial T}\right)_P$$

$$= 2RVT\left(\frac{\partial V}{\partial T}\right)_P + V^2R$$

or $$\left(\frac{\partial V}{\partial T}\right)_P = \frac{V^2R}{3\,V^2P + a - 2\,PVb - 2\,RVT}$$

since $$\alpha = \frac{1}{V}\left(\frac{\partial V}{\partial T}\right)_P$$

$\therefore$ $$\alpha = \frac{R}{3\,PV + \frac{a}{V} - 2\,Pb - 2\,RT}$$

$$= \frac{R}{3\,PV - \frac{a}{V} + \frac{2ab}{V^2}}$$

On differentiating

$$PV^3 - PbV^2 + aV - ab + RTV^2 = 0$$

With respect to P at constant T

We get

$$V^3 + 3PV^2\left(\frac{\partial V}{\partial P}\right)_T + a\left(\frac{\partial V}{\partial P}\right)_T - bV^2 - 2PbV\left(\frac{\partial V}{\partial P}\right)_T$$

$$= 2\,RVT\left(\frac{\partial V}{\partial P}\right)_T$$

$$\therefore \quad \left(\frac{\partial V}{\partial P}\right)_T = \frac{V^2\ (b-V)}{3PV^2 - 2PVb + a - 2RVT}$$

By definition, $\beta = -\frac{1}{V}\left(\frac{\partial V}{\partial P}\right)_T$

$$\therefore \quad \beta = \frac{V^2 - bV}{3\,PV^2 + a + 2\,PVb - 2\,RVT}$$

$$= \frac{V-b}{3\,PV + \frac{a}{V} - 2\,Pb - 2\,RT}$$

$$= \frac{V-b}{PV - \frac{a}{V} + \frac{2\,ab}{V^2}}$$

2

Zeroth and First Laws of Thermodynamics

Zeroth Law of Thermodynamics

It is assumed that thermal equilibrium is governed by temperature. The zeroth law of thermodynamics, also known as the *law of thermal equilibrium* was put forward much after the establishment of first and second laws of thermodynamics. This law provides a logical basis for the concept of temperature of a system. The law can be stated as

"If two systems are in thermal equilibrium with a third system, then they must be in thermal equilibrium with each other."

Or

"If three or more systems are kept in thermal contact by means of diathermal walls and are in thermal equilibrium together, then any two systems taken separately will be in thermal equilibrium with one another"

Let us consider three fluids A, B and C. Suppose P_A, V_A represent the pressure and volume of A, P_B, V_B, the pressure and volume of B, and P_C, V_C, the pressure and volume of C.

Suppose A and B are in thermal equilibrium with one another.

Then, one can write as

$$\phi_1 (P_A,\ V_A) = \phi_2 (P_B,\ V_B)$$

or $$F_1 [P_A, V_A, P_B, V_B] = 0 \quad (1)$$

On solving equation (1), we get

$$P_B = f_1 [P_A, V_A, V_B] \quad (2)$$

Suppose B and C are in thermal equilibrium. Then, one can write as:

$$\phi_2\ (P_B,\ V_B) = \phi_3 (P_C,\ V_C)$$

or $$F_2\ [P_B, V_B, P_C, V_C] = 0$$

Also $$P_B = f_2 [V_B, P_C, V_C] \quad (3)$$

On equating equations (2) and (3), we get equation (4) for A and C to be in thermal equilibrium separately.

$$f_1 (P_A, V_A, V_B) = f_2 (V_B, P_C, V_C) \quad (4)$$

As *A* and *C* are in thermal equilibrium with *B* separately, then *A* and *C*, in accordance with the zeroth law, are also in thermal equilibrium with one another. It means that

$$F_3 [P_A, V_A\ P_C,\ V_C] = 0 \quad (5)$$

As equation (4) has a variable V_B whereas equation (5) does not have variable V_B, it means that

$$\phi_1\ (P_A,\ V_A) = \phi_3 (P_C,\ V_C) \quad (6)$$

In general,

$$\phi_1\ (P_A, V_A) = \phi_2 (P_B, V_B) = \phi_3 = (P_C, V_C) \quad (7)$$

From equation (7) it follows that all the three functions have the same numerical value but the parameters (P, V) of each are different. This numerical value is termed as the temperature (T) of the body, i.e.,

$$\phi\ (P, V) = T \quad (8)$$

Therefore, the temperature of a system can be defined as

"Temperature is the property which determines whether the body is in thermal equilibrium with the neighbouring systems or not."

From the above definition it is obvious that if two systems are not in thermal equilibrium, they are at different temperatures.

Heat and Work

Heat: When two systems at different temperatures are placed in thermal contact with each other and isolated form of all other systems, a change in the temperature of other thermodynamic property is both systems will always occur due to energy transfer.

The energy transfer that occurs from one system to another solely because of temperature difference between the two systems is called *Heat*. Thus heat can be identified only as it crosses the boundary of a system, hence heat is a form of energy in transit. *Heat* is therefore not an energy contained in a system and thus *it is not a property of the system.* When heat enters a system, it gets converted into potential or kinetic energy of the molecules, atom etc. of the system.

Work: Work is done by a system of the role effect on the surroundings can be reduced to the lifting of a weight. The magnitude of work is the product of the weight and the distance it could be lifted.

Like heat work is an interaction across the boundary of a system and is therefore a form of energy in transit.

Neither heat nor work is an energy contained in the system and so both of these are not the properties of the system.

Work done, dw is given by

$$dw = Pdv.$$

Where υ is the specific volume.

The work is a *path process.*

First Law of Thermodynamics

The first law of thermodynamics is not derived from any

assumptions rather it summarizes our experience with material process. It is law of human experience, and has no theoretical proof. This law was first stated in 1840 by Mayer and Helmholtz in Germany, Joule in England and Colding in Denmark. This law has not been contradicted by any experimental result so far. It is simply the law of *conservation of energy* which states that "*energy can neither be created nor destroyed although it may be converted from one from to the other.*" The law may also be stated in the following ways:

(i) The total energy of an isolated system remains constant although it can undergo a change from one form to the other.

(ii) Whenever a quantity of some form of energy disappears, an exactly equivalent quantity of some other form of energy makes its appearance.

(iii) It is impossible to construct a perpetual motion machine which would produce work without consuming energy.

(iv) For a system in contact with its surroundings, the sum of the energies of the system and the surroundings remain constant, howsoever, differently it may be shared between the two.

This principle of conservation of energy as applied to heat and mechanical energies states "*A certain quantity of heat energy is equivalent to the same quantity of* mechanical *energy.*

The units of energy are same as those of work: *Joule* in *S.I.*, erg in *C.G.S.* system. The units of heat are *calories* in *C.G.S.* system and *kilo calories* in *M.K.S.* system. There is definite conversion factor between the two units which is determined accurately by experiment and is written as *J*, the *mechanical equivalent of heat. J* is equal to 4.2×10^7 ergs per calorie or 4.2 Joules per calorie or 4.2×10^3 joules per kilo calorie. Heat energy can also be measured in work units, Joule or erg.

In 1905, Einstein established the fact that matter and energy are interconvertible according to the relation, $E = mc^2$, where E is the energy produced, m is the mass of the matter destroyed and c is the velocity of light. This is not in contradiction with the first law of thermodynamics since mass in its generalised form may be considered as some kind of potential energy. The law of conservation of mass and the law of conservation of energy are combined together to give the law of conservation of mass and energy. The modified law can be stated as under:

The total mass and energy of an isolated system remains unchanged.

Before we take up the mathematical formulation of this law, let us study briefly the term internal energy, to be used therein.

Internal Energy

Energy may be defined as the capacity to do work. It may appear in the form of the *kinetic* or as *potential energy.* In chemical thermodynamics, we are not concerned with such two forms of energy. We are concerned with an energy which a chemical substance possess because of the arrangement of atoms or electrons within its molecule or which a system possesses because of its temperature, volume, pressure and chemical composition. This is known as the *internal energy* or *intrinsic energy* and is designated by the letter 'E'. Internal energy of an ideal gas is a function of temperature only.

It may be noted that each system can be imagined to have an absolute value of the internal energy. But it is neither necessary nor possible to ascertain this absolute amount of energy. However, the change in the internal energy of a system, accompanying changes in temperature, volume, pressure and composition, is a measurable quantity and can be determined experimentally. Thus is E_1 and E_2 are the internal energies of a system in the initial state and the final state respectively, then the change in the internal energy (ΔE) is given by:

$$\Delta E = E_2 - E_1$$

The internal energy of a system is determined by the state of the system and has nothing to do with the manner or method by which it has been obtained.

The energy change of a system depends upon the initial and the final state of the system and is independent of the path by which this change is brought about.

Let us consider a system in state 'A' (Fig. 2.1) having definite value of pressure, volume and energy content. Let us suppose that the system changes from state A to state B by the path I. At state 'B' the system has definite value of pressure, volume and energy. Let the increase in the internal energy of the system be ΔE.

Further suppose that the system returns to the state A by the Path II and the increase in the internal energy is $\Delta E'$. Let $\Delta E'$ be greater than ΔE, Thus the system will return to its original state but by a gain of energy equal to $\Delta E' - \Delta E$. Thus by the repetition of this process, the energy is continuously produced without the disappearance of an equivalent amount in some other form. In other words, a perpetual motion machine will be possible. This is against the first law of thermodynamics. Hence $\Delta E = \Delta E'$.

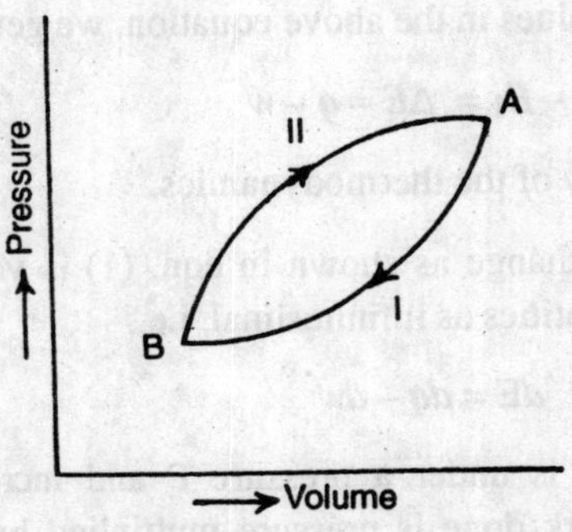

Fig. 2.1

Thus the energy change of a system from one *thermodynamic state of the other depends only on its initial and final states and is independent of the path adopted to bring about the change.*

In *C.G.S.* system, the unit of energy is '*erg*' which is defined as the amount of work done by force of one dyne acting through a distance of one cm.

Mathematical Formulation of the First Law of Thermodynamics (i.e. Relation between internal energy, work and heat). Suppose we put some amount of heat to a system and heat and energy are the only modes of energy exchange between the system and the surroundings. Since the heat energy cannot be lost, it must be either wholly stored within the system as its internal energy or it must be wholly used up by the system in doing mechanical work or it may be partly stored as the internal energy and partly may be used up by the system in doing work. In general, if the heat absorbed goes both to increase the internal energy and to produce some mechanical external work, we must have

Heat absorbed = Increase of internal energy + External work done by the system

or $\left.\begin{matrix}\text{Increase of}\\\text{Internal energy}\end{matrix}\right\} = \text{Heat absorbed} - \left\{\begin{matrix}\text{Work done by}\\\text{the system}\end{matrix}\right.$

If the initial and the final energies of the above system are E_1 and E_2 respectively, then the increase in internal energy is $E_2 - E_1 = \Delta E$. If the heat *absorbed* is q and the work done by the system is w then by substituting these values in the above equation, we get

$$E_2 - E_1 = \Delta E = q - w \qquad (1)$$

which is the first law of the thermodynamics.

If the above change as shown in eqn. (1) is very small, we can write the above quantities as infinitesimal, i.e.,

$$dE = dq - dw \qquad (2)$$

If the system is under a pressure P and increases by a small volume dV, the work done is pressure multiplied by volume change, i.e., PdV and, therefore, the above equation becomes

$$dE = dq - PdV \qquad (3)$$

This is the mathematical expression for the first law for infinitesimal changes where the work is exclusively the work of expansion or contraction only.

It may be noted here that dE is a definite quantity as the change in energy is independent of the path by which the change is brought about and depends only on the initial and final state of the system undergoing a change. In mathematical terminology dE is a *complete or exact differential* in any infinitesimal process. Same is not true of dq and dw as their magnitude (or values) change with the change in path by which the change is brought about and are not functions of the state as work and heat are not the properties of a system or surroundings and are only a mode of exchange of energy between the system and its surroundings. *Thus dq and dw are not complete or exact differentials* in mathematical terms.

The equation (3) may thus be written as

$$dE = \delta q - \delta w \tag{4}$$

Note: In all energy transformations, q is given a **positive sign** when heat is absorbed by a system.

q is given a **negative sign** when heat is evolved by a system.

Work w is given a **positive sign** when work is done by the system. This is the case when the process involves expansion (i.e., increase in volume) of the system as the work is done by the system on the surroundings, work w is given a **negative sign,** when *work is done on the system.* This is the case when the process involves contraction (i.e. decrease in volume) of the system as the work is done on the system by the surrounding.

Equation (4) represents the mathematical form of first law of thermodynamics. In this equation, the differential is distinguished in notation by d from δ. This can be understood from the following discussion:

(A) *Change in internal energy is independent of the path adopted.* It has been already discussed that the change in internal energy is independent of path by which it is brought about and the previous history of the system. So in mathematical terms, an exact differential is always denoted by a notation d e.g. dE in equation [2].

(B) *Nature of* δq and δw. It can be said that 'q' and 'w' are not state functions because changes in their magnitudes for a process depend upon.

(i) the path by which the change is accomplished, and also

(ii) the previous history of the system.

The work done is going from state A to state B by four different paths has been shown in this Fig. 2.2. The vlaue of w will be equal to the area under the curve in a P – V diagram which is represented by shaded regions in Fig. (2.2) for the various paths. It can be easily followed that the work done in passing from A to B by different paths is not same. It means that w is not a definite quantity. Let us again consider the equation,

$$\Delta E = q - w \qquad (1)$$

As ΔE is a definite quantity and w not a definite quantity, it easily follows that q cannot be a definite quantity.

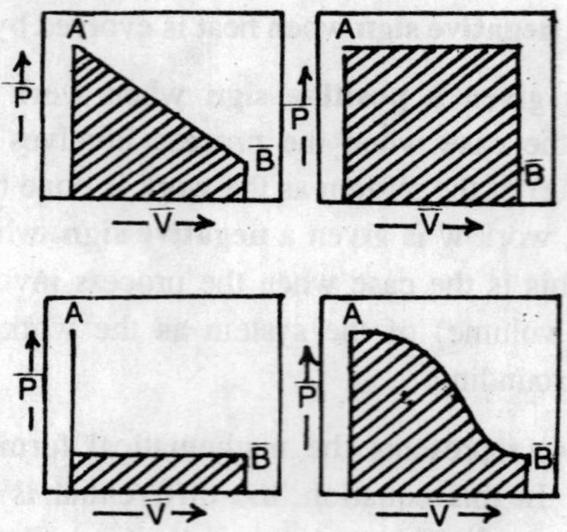

Fig. 2.2

From the above discussion, it follows that w and q are not definite quantities. Mathematically, infinitesimal changes in q and w are not exact differentials but are inexact differentials.

In mathematical language, we always write the inexact differentials by δq , δw, etc.

Let us again consider equation.

i.e., $$dE = \delta q - \delta w \qquad (2)$$

where dE is an *exact differential* and δq and δw are *Inexact differentials.* From equation (2), three cases may arise:

(i) For any cyclic process, the change in the internal energy of the system is zero because the system is brought back to the original state. Hence for a cyclic process $\int dE = 0$.

Therefore, from equation (2)

$$\int \delta q = 0 + \int \delta w$$

or
$$\int \delta q = \int \delta w$$

i.e., total work obtained is equal to the net heat supplied.

(ii) In an isolated system, there is no heat exchange with the surroundings, i.e. $\delta q = 0$. Therefore, equation (2) becomes as

$$dE + \delta w = 0 \qquad \text{or} \qquad \delta\, w = -\, dE$$

It means that in an isolated system, the work performed will be equal to the internal energy decrease.

(iii) In a non-isolated system, the system gains heat (δq) and performs work (δw). For the system

$$d\text{E} = \delta q - \delta w$$

In we consider the surroundings, it loses δq amount of heat, but receives δw units of work. The internal energy change of the surroundings will then be given by

$$dE' = \delta w - \delta q$$

Thus
$$dE = -dE' \text{ or } dE + dE' = 0$$

It means that the net change in the internal energies of the system and the surroundings taken together would be nil.

Significance of First Law of Thermodynamics

This law establishes an exact relation between heat and work. It tells us that a certain quantity of heat will produce a definite amount of work and vice-versa. The first law of thermodynamics denies that work or energy can be produced from nothing. Work cannot appear without disappearance of heat.

Limitations of the First Law of Thermodynamics

According to the first law of thermodynamics, there is an exact equivalence between various forms of energy. This law also tells us that if a process takes place, energy is converted into equivalent amount of work. But *the first law does not give any information regarding the conditions and to what extent it is possible to convert one form of energy into another.*

Illustration. This can be illustrated that if water is taken from a lower level to a higher level, unless heat is supplied from outside, there will be a fall in temperature, the decrease in energy of the system being equivalent to the work done against gravity. But the first law does not tell us under what conditions, water can be taken from a lower to a higher level. This is answered by the second law of thermodynamics.

Enthalpy

Processes at constant pressure are more common in chemistry because most of the operations are carried out in open vessel under conditions of constant pressure the gas expands so that work of expansion will be

$$W = P\Delta V \qquad (1)$$

where ΔV is the increase in volume P is the constant pressure. Putting this value in eqn. (1) of the first law of thermodynamics (viz., $\Delta E = q-w$), we get

$$q_p = \Delta E_p + P\Delta V$$

Subscript p stands for the process at constant pressure.

When the system absorbs q_p calories of the heat, its internal energy increases from E_1 to E_2 and its volume increases from V_1 to V_2. Then we have

$$\Delta E = E_2 - E_1 \text{ and } \Delta V = V_2 - V_1$$

$$q_p = (E_2 - E_1) + P(V_2 - V_1)$$

$$q_p = (E_2 + PV_2) - (E_1 + PV_1) \qquad (3)$$

Since P and V are properties of the state of the system only, it follows that the quantity E + PV, like the internal energy E, is dependent only on the thermodynamic state and not on its previous history. *The thermodynamic property E + PV is called heat content or enthalpy of the system and is represented by the symbol H, i.e.,* by definition.

$$H = E + PV \qquad (4)$$

Thus if H_2 is enthalpy of the system in the final state and H_1 the enthalpy in the initial state, then eqn. (3), becomes

$$q_p = H_2 - H_1 = \Delta H \qquad (5)$$

where $\Delta H = H_2 - H_1$ is the change in enthalpy. Thus *change in enthalpy of a system is equal to the heat absorbed at constant pressure.*

Further combining eqns. (2) and (5), we get

$$\Delta H = \Delta E + P\Delta V$$

i.e., *the change in enthalpy at constant pressure is equal to the increase in internal energy plus external work of expansion.*

From above discussion, we have

(i) at constant volume $\Delta E = q_v$

and (ii) at constant pressure $\Delta H = q_p$.

It follows that *q* under the above two conditions is a state function which shows that Hess's law of constant heat summation is a special case of the first law of thermodynamics.

Enthalpy function is quite signficant since most of the processes in the laboratory are carried out at constant pressure. For example reactions in an open beaker or flask take place under atmospheric pressure, i.e., at constant pressure. Absolute value of enthalpy like energy cannot be known. We are concerned only with the changes in enthalpy for most of the processes rather than its absolute values at various states. The changes in enthalpy may be measured by taking any suitable reference state of the elements. The processes in which heat is absorbed by the system are endothermic whereas the processes in which heat is evolved are exothermic reactions. In Exo. the ΔH is negative and in Endo. the ΔH is positive.

Enthalpy change accompanied by a certain process is called the enthalpy of that process. Thus enthalpy of evaporation means the change in the enthalpy involved when one mole of the substance gets converted from the liquid state into the vapour state. The enthalpies of fusion and sublimation have similar meanings. The enthalpy of a reaction is the difference between the enthalpies of the products and of the reactants as represented by the chemical equation. Enthalpies of some typical processes are listed below in the Table (2.1).

(i) ***Enthalpy of vaporisation.*** A liquid on absorbing heat from the surroundings evaporates and this process is accompanied by the increase in enthalpy. When one mole of water evaporates at 25°C, increase in enthalpy is 10.51 kcal, which may be expressed in the form of a thermo-chemical equation as:

$$H_2O(l) \rightarrow H_2O(g), \Delta H = +10.51 \text{ kcal.}$$

Table 2.1. Enthalpies of Some Typical Processes

Sr. No.	*Process*	*Name of the Process*	*DH (kcal)*	*Temp. (°C)*
1.	(l) = H_2O (g)	Enthalpy of evaporation	9.7	100
2.	$H_2O(s) = H_2O(l)$	Enthalpy of fusion	1.44	0
3.	$I_2(s) = I_2(g)$	Enthalpy of sublimation	14.88	25
4.	$H_2(g) + \frac{1}{2}O2(g)$ $H_2O(l)$	Enthalpy of reaction	–68.32	25

When the steam condenses to liquid water, heat is evolved. Hence condensation of vapour results in decrease in enthalpy. When one mole of steam condenses at 25°C, the decrease in enthalpy is 10.51 kcal which may be expressed as

$$H_2O\,(g) \rightarrow H_2O\,(l) = -\,10.51 \text{ kcal.}$$

The enthalpy of vaporisation may thus be defined as *the change in enthalpy when a liquid evaporates into vapour state or when a vapour condenses into liquid state*. The enthalpy of vaporisation in the former case is positive while it has a negative value in the latter case.

(ii) *Enthalpy of fusion.* A solid changes into liquid on absorbing heat. On the other hand, a liquid freezes to a solid on giving out heat. The enthalpy of the system increases in the first case whereas it decreases in the latter case. The change in enthalpy in such type of phase change is termed as enthalpy of fusion. 1.44 kcal is the enthalpy of fusion for one mole of water at 0°C and this may be expressed in the form of thermo chemical equations as:

$$H_2O\,(s) \rightarrow H_2O\,(l)\,;\, \Delta H = +\,1.44 \text{ kcal}$$

$$H_2O\,(l) \rightarrow H_2O\,(s)\,;\, \Delta H = -\,1.44 \text{ kcal}$$

Physical Significance of Enthalpy

Enthalpy is the sum of internal energy of pressure, volume and work terms. It signifies the amount of energy stored in system that is available for conversion into heat. The increase in enthalpy is equal to heat absorbed in a process at constant pressure only when pressure volume work is done. Ordinarily heat effects are measured at constant pressure therefore these heat effects indicate the change in enthalpy content and not the changes in internal energy. It may be noted that like internal energy, the enthalpy too can not be calculated or obtained experimentally. We can however, measure ΔH (change in enthalpy). Enthalpy is an extensive property. The enthalpy of an ideal gas is a function of temperature only.

Exercise 1. Calculate the work done when 32.75g of zinc dissolves in excess of dil HCl (i) in an open beaker (ii) in a closed beaker at 300K. Atomic weight of zinc is 65.5.

Solution.

The reaction between zinc metal and HCl can be represented as

$$Zn + 2HCl \rightarrow ZnCl_2 + H_2 \uparrow$$

From the equation we find that 1 mole of zinc reacts with dil HCl to liberate 1 mole of hydrogen. The liberated gas comes out and pushes the surrounding atmosphere and thus does work of expansion. It is given by

$$W = P \times \Delta V$$

where P is the atmospheric pressure, and ΔV is the change in volume. This change in volume may be approximately taken as the volume of hydrogen produced.

$$\therefore \quad W = P\Delta V$$

$$= P \times VH_2$$

or $\quad W = n_{H_2} . R.T \quad [\because \ PV = nRT]$

where n_{H_2} is the number of moles of hydrogen gas. R and T are gas constant and absolute temperature respectively.

According to reaction, one mole of Zn produces 1 mole of hydrogen. 32.75g of zinc is equal is 0.5 mole and so it will produce 0.5 mole of hydrogen.

Hence $\quad nH_2 = 0.5$ mole

$$R = 8.314 \text{ JK}^{-1} \text{ mole}^{-1}$$

$$T = 300K$$

Thus $\quad W = 0\ 5 \times 8.314 \times 300$

$$= 1.2471 \text{ KJ.}$$

Exercise 2.

If one mole of ice at 0°C and 4.6 mm, Hg pressure is converted to water vapour at the same temperature and pressure by increasing the space above ice sufficiently, calculate the enthalpy and intrinsic energy change. Given that the latent heat of fusion of ice is 80 calories per gram and latent heat of liquid water at 0°C is 596 calories per gram.

Solution. Heat absorbed by 1 mole (i.e., 18 gm) of ice in being converted to 1 mole of water at 0°C and 4.6 mm pressure

$$= 80 \times 18 = 1440 \text{ calories}$$

Heat absorbed by 1 mole of water in being converted to 1 mole of water vapour

$$= 596 \times 18 = 10728 \text{ calories.}$$

$\therefore$ Total heat absorbed will be

$$q = 1440 + 10728 = 12168 \text{ calories}$$

Work done is given by

$$w = PV = RT$$

$$= 2 \times 273 \text{ calories} \quad (\because R = 2 \text{ cal, and } T = 0°C = 273°A)$$

$$= 546 \text{ calories}$$

$\therefore$ $\Delta E = q - w$ (by first law)

$$= 12168 - 546 = 11622 \text{ calories}$$

At pressure is kept constant,

$\therefore$ DH = q_p = 12168 calories as calculated above.

Note : The work done '*w*' can also be calculated as follows:

Volume of 1 mole water vapour at 0°C and 4.6 mm. Hg pressure

$$\text{(by Gas eqn.)} = 22.4 \times \frac{760}{4.6}$$

$$= 3717 \text{ litres.}$$

Ignoring the volume of ice in comparison with the volume of water vapour.

$$\Delta V = 3717 \text{ litres}$$

$$w = P \times \Delta V = \frac{4.6}{760} \times 3717 = 22.4 \text{ litres atmospheres}$$

$$= \frac{22.4 \times 10^3 \times 76 \times 13.6 \times 980}{4.2 \times 10^7} = 543 \text{ calories.}$$

Exercise 3.

If ΔH for the reaction

$$CO(g) + \tfrac{1}{2} O_2(g) = CO_2(g)$$

at 298K is 282.85KJ mol^{-1}. Calculate ΔE of the reaction.

Solution

$$\Delta n = n_{CO_2} - n_{CO} - \tfrac{1}{2} n_{O_2}$$

$$= 1 - 1 - \frac{1}{2}$$

$$= -\frac{1}{2}$$

∴ $$\Delta E = \Delta H - \Delta nRT$$

$$= 282.85 - (-\tfrac{1}{2})\,(8.314)\,(298)$$

$$= 283.85\text{KJ mol}^{-1}$$

Relation Between Heat of Reaction at Constant Volume (q_v) and at constant Pressure (q_p).

We know

$$\Delta H = \Delta E + P\Delta V$$

Moreover $q_v = \Delta E$

and $q_p = \Delta H$

In case of an ideal gas,

$$PV = nRT$$

where V is the volume occupied by n moles of the gas.

If n_1 are the number of moles of the gaseous reactants and n_2 are the number of moles of gaseous products.

∴ Difference in the number of moles of products and reactants

$$= n_2 - n_1 = \Delta_n.$$

Hence the change in volume, $\Delta V = \dfrac{V}{n} \times \Delta n$

Work done (w) by the system at constant pressure P is given by the expression,

$$w = P\Delta V = P\frac{V}{n}\Delta n$$

$$= RT \times \Delta n \qquad [\because PV = nRT]$$

$$\therefore P\Delta V = \Delta n \times RT$$

Susbtituting the value of PΔV in equation. $q_p = \Delta E_p + P\Delta V$

Hence $q_p = q_0 + \Delta nRT$

$$\therefore \quad \Delta H = \Delta E + \Delta nRT$$

ΔV, i.e., change in volume in reactions involving only solids or liquids is negligible, $q_p = q_v$, i.e., heat of reaction at constant pressure at constant volume is almost equal.

Heat Capacity

The quantity of heat required to carry out a change of state depends on how the process is carried out only for a reversible process where the path is fully specified we can relate the heat to a property of the system. On this bases *heat capacity* is given by

$$C_x = \left(\frac{\delta\theta}{dT}\right)_x$$

where x indicates that the process is reversible and the path is fully specified. Only two heat capacities are in common use. These are C_v, heat capacity at constant volume and C_p, heat capacity at constant pressure. In both these cases, the system is presumed closed and of constant composition.

$$dC_v = \left(\frac{\delta Q}{dT}\right)_v \text{ or } \left(\frac{\partial E}{\partial T}\right)_v \qquad (1)$$

and

$$C_p = \left(\frac{\partial H}{\partial T}\right)_p \qquad (2)$$

From the above equations we, have

$$dE = C_V dT \text{ (constant volume)}$$

$$dH = C_P\, dT \text{ (constant pressure)}$$

These equations relate properties only and are independent of the path. Thus for these equations no restriction with respect to reversibility is necessary. It is only essential that the change takes place between equilibrium states.

This new function which relates the heat changes to temperature changes at constant pressure or at constant volume is denoted by *C* and is known as *heat capacity*.

It may be defined as follows:

(a) *The heat capacity of a system*, between any two temperatures is the quantity of heat required to raise the temperature of the system from the lower to the higher temperature divided by the temperature difference.

If the temperature of the system is raised from T_1 to T_2 its heat capacity be expressed as

$$C(T_2, T_1) = \frac{q}{T_2 - T_1}$$

where *q* is the quantity of heat absorbed and C is the heat capacity of the system

(i) *Heat Capacity at Constant Volume.* According to 1st law, we have

$$q = \Delta E + W$$

where $W = P\Delta V$, ΔV is the change in volume

But at constant volume

$$\Delta V = O$$

$$\therefore \qquad W = O$$

Hence $q = \Delta E$

Therefore, the heat capacity of the system at constant volume

$$C_v(T_2, T_1) = \left(\frac{q}{T_2 - T_1}\right)_v$$

$$= \left(\frac{\Delta E}{T_2 - T_1}\right)_v$$

$$= \left(\frac{\partial E}{\partial T}\right)_v$$

Heat capacity at constant volume of a given system is defined as the rate of change of internal energy with temperature.

(ii) Heat Capacity at Constant Pressure. Under these conditions, work is done by the heat energy supplied. Let the change in volume be ΔV then work done (w) is given by

$$w = P\Delta V$$

$$\therefore \quad q = \Delta E + w \qquad \text{(1st law)}$$

$$= \Delta E + P\Delta V$$

Heat capacity at constant pressure

$$C_P(T_2, T_1) = \left(\frac{q}{T_2 - T_1}\right)_P$$

$$= \left(\frac{\Delta E + P\Delta V}{T_2 - T_1}\right)_P$$

We also know

$$\Delta H = \Delta E + P\Delta V \qquad \text{(definition)}$$

$$\therefore \quad C_P(T_2, T_1) = \left(\frac{\Delta H}{T_2 - T_1}\right)_P$$

or $$C_P = \left(\frac{\partial H}{\partial T}\right)_P$$

In general, $C_P = \frac{dH}{dT}$

Hence heat capacity at constant pressure of a system is defined as the rate of change of enthalpy with temperature. From the above discussion we note that the heat capacity of a system at constant volume (C_V) is numerically equal to increase in internal energy of the system per degree rise of temperature at constant volume and the heat capacity of a system at constant pressure (C_P) is equal to the increase in enthalpy of the system per degree rise of temperature at constant pressure.

It is evident that C_P is more than C_υ.

For one mole of an ideal gas, heat capacities at constant pressure and at constant volume are indicated by C_P and C_V respectively and are known as *molar heat capacities.*

Heat capacity of gases. It is conveniently defined by the following empirical relations:

$$C_P = a + bT + cT^2 + dT^3 + \ldots$$

where a, b, c, d, . . . are constants and T is the absolute temperature.

Intergration of Heat Capacity Equation

The equation $dH = C_P\, dT$ can be integrated and on integration, we have

$$\int dH = H_O + \int C_P\, dT$$

$$H = H_O + \int (a + bT + cT^2 + dT^3 + \ldots)\, dT$$

where H_o is a constant of integration

or $$H = H_O + aT + \frac{bT^2}{2} + \frac{cT^3}{3} + \ldots$$

Molar Heat Capacity

It is the heat capacity of 1 mol, of a substance and *specific heat* is the heat capacity per gram of substance.

Exercise 1.

The temperature of 2 kg of water dropped from 95°C to 40°C when 1.25 kg of a metal piece at 25°C was added to water. Calculate the heat capacity of metal if heat capacity of water is 4.184 $Jg^{-1} K^{-1}$.

Solution

Let the specific heat of metal be 'C'. Then

Heat gained by metal = $m \times c \times \theta$

$$= (1.25 \times 1000) \times c \times 15$$

Heat lost by water = $(2 \times 1000)(4.184)(55)$

Heat gained = Heat lost

Thus

$$1.25 \times 1000 \times 15 \times c = 2 \times 1000 \times 4.184 \times 55$$

or

$$c = \frac{2 \times 1000 \times 4.184 \times 55}{1.25 \times 1000 \times 15}$$

$$= 24.55 \text{ J k}^{-1} \text{ g}^{-1}.$$

Molecular Interpretation of Internal Energy and Enthalpy

It is possible to calculate internal energy and enthalpy on the basis of molecular approach. However, these calculations are only for thermal part of internal energy and enthalpy. Moreover we, can apply the molecular approach only to ideal gases for appreciable accuracy. It may not be possible to apply this approach to liquid and solid systems. Inspite of all these limitations, calculations of these properties based on molecular model leads to better understanding of these thermodynamic terms.

In molecular approach the energy of a molecule is divided into two terms:

(i) The energy of the system for all molecules in their lowest allowed energy level, denoted by E_0.

(ii) The thermal energy or excitation energy denoted by $(E–E_0)$.

The sum of these two energies obviously is E.

Since enthalpy is given by

$$H = E + PV$$

so $\quad H - E_0 = (E–E_0) + PV$

But $\quad PV = RT \quad$ (for one mole)

Hence $\quad H–E0 = (E - E_0) + RT$

Thus thermal contribution to enthalpy $(H–E_o)$ can be calculated if the thermal energy $(E - E_o)$ is evaluated.

Calculation of Internal Energy and Enthalpy Functions.

We illustrate these calculations taking the example of NO_2. For this non-linear molecule there are three translational degrees of freedom, three rotational degrees of freedom and three vibrational degrees. The three vibrational modes have energy levels with spacing of 1.49, 2.63 and 3.21 × 10^{-20} J/molecule. The energies corresponding to various degrees and modes at 298 K are given below:

(*i*) Energy due to translational mode

$$= \frac{3}{2} R.T.$$

$$= 3716 \text{ J/Mole}$$

(*ii*) Energy due to rotational mode $= \frac{3}{2} RT$

$$= 3716 \text{ J/mole.}$$

(*iii*) Energy due to vibrational mode

$$E_{vib} = \frac{RTx}{e^x - 1}$$

where $x = \frac{\varepsilon}{\kappa T}$

where ε stands for energy level spacing and k is the Boltzmann constant (1.3805×10^{-23} J.k^{-1}).

Thus contributions due to vibrational modes are calculated as under:

ε	x	*Energy (J/mole)*
1.49×10^{-20} J/molecule	3.63	251
2.63×10^{-20} J/molecule	6.40	29
3.21×10^{-20} J/molecule	7.80	8
	Total	288

Thus the value of thermal energy component of internal energy

$$E = E_o + 3716 + 3716 + 288$$

$$= Eo + 7720$$

or $$E - E_o = 7720 \text{ J/Mole.}$$

Similarly, the thermal energy component of enthalpy is given by

$$H–H_o = (E–E_o) + RT$$

$$= 7720 + 2477$$

$$= 10197 \text{ J.}$$

Molecular Interpretationof C_v and C_p

The heat capacities C_v and C_p are temperature derivatives of internal energy and enthalpy, i.e.,

$$C_v = \frac{dE}{dt}$$

and $$C_p = \frac{dH}{dt}$$

The thermal energy $(E - E_o)$ consists of three parts

(i) $\frac{3}{2}$ RT because of translational motion.

(ii) $\frac{3}{2}$ RT because of rotational motion.

(iii) $\frac{RTx}{(e^x - 1)}$ for each vibrational mode, where $x = \frac{\varepsilon}{\kappa . T}$

It is thus possible to calculate the contribution of each mode to the heat capacity.

Mode	*Contribution to energy*	*Contribution to heat capacity*
Translational	$\frac{3}{2}$RT	$\frac{d}{dT}\left(\frac{3}{2}RT\right) = \frac{3}{2}$R
Rotational	$\frac{3}{2}$RT	$\frac{d}{dT}\left(\frac{3}{2}RT\right) = \frac{3}{2}$R
	RT (Non-Linear)	$\frac{d}{dT}$(RT) = R
Vibrational	$\frac{RTx}{e^x - 1}$	$\frac{d}{dt}\left(\frac{RTx}{e^x - 1}\right) = \frac{Rx^2e^x}{(e^x - 1)^2}$

In case of NO_2 at 298 K the contributions of various modes are as under

Mode	*Heat capacity*	*Contribution J/°/mole*
Translational	$\frac{3}{2}$R	6.27
Rotational	$\frac{3}{2}$R	6.27
Vibrational		
(i) $\varepsilon = 1.49 \times 10^{-20}$ x = 3.63	$\frac{Rx^2e^x}{(e^x - 1)^2}$	3.10

(ii) $\varepsilon = 2.63 \times 10–20$ $x = 6.40$	$\frac{Rx^2e^x}{(e^x-1)^2}$	0.58
(iii) $\varepsilon = 3.21 \times 10^{-20}$ $x = 7.80$	$\frac{Rx^2e^x}{(e^x-1)^2}$	0.21
		Total = 16.43

To get C_p we use $C_p - C_v = R$ or $C_p = C_v + R$.

Thus $$C_p = 16.43 + 8.31$$

$$= 24.74 \text{ J/degree/mole}$$

Work

Every change in the condition of matter is to be ascribed to the operation of one or other forms of energy. Whenever a change in the state of a system takes place, there always occurs a change in energy.

The energy change accompanying a physical or chemical process is in the form of (i) work (ii) heat (iii) light etc.

The energy is a composite of two factors, an *intensity factor* and a *capacity factor* and is the product of these two factors. The *intensity factor* is *the measure of the resistance* offered to a change. The change can take place only when this resistance is withdrawn or overcome. Such an intensity factor tending to produce a change is known as *force.* A force, overcoming a resistance to an extent which we may designate as the *capacity factor*, performs *work*, expends energy. The work done depends on the force required to produce motion against the resistance and on the distance through which the object acted upon is moved.

If an object, say a table, is pushed by the hand, it may be displaced from its original position. We consider it as a mechanical work (or simply, work) done by muscular effort (energy). In fact, whenever the application of a force causes a displacement of system, mechanical work is done. The magnitude of work is measured by the product of force (*intensity factor*) and the displacement (*capacity factor*) caused. Thus,

$$\text{Work} = F \times x$$

where F is the force and x is the displacement caused.

The performance of the work need not necessarily be always achieved by muscular effort. The work may be produced by expending energy in diverse ways. The following are some ways by which work is done:

(i) *Pressure-volume* (PΔV) *work.* Work is done when there a difference of pressure between the system and surroundings and is accompanied by change in volume of the system, e.g., work done by the system (the case of a gas formed in a chemical reaction) in pushing back the surrounding atmosphere. In this case if the pressure of the gas with which it pushes the surrounding atmosphere is P (*intensity factor*—force) and ΔV is the volume through which the atmosphere is pushed (*capacity factor*—extent), the work done by the issuing gas = $P\Delta V$.

(ii) *Electrical work.* Electricity passed through a circuit may be made to produce the rotation of the armature of a motor used for lifting a weight or winding a spring. Conversely, the work done in turning the blades of a turbine would generate electricity. This is an instance of the conversion of electrical energy into work.

When an electric current flows through a circuit, work done is equal to the product of the potential difference applied (E volts) which causes the flow of current (*intensity factor*) and the quantity of electricity that flows in a given time [Q coulombs = current strength in amperes (I) × times in seconds (t)].

So electrical work = E.I.t

= E.Q. volts Coulombs or Joules.

(iii) *Gravitational work.* When the position of a mass (body) is changed with respect to the gravitational field, work done is said to be gravitational work.

In the raising of an object of mass m against gravity, g, through a height h, the force exerted in opposition to gravity is mg (intensity factor) and the capacity factor is the height *h*. The work done is *mgh*.

Pressure-Volume Work (Mechanical work). In most processes of interest to a chemist in thermodynamics is pressure-volume or mechanical work. This type of work is associated with increase or decrease of the volume of the system. It means that work is done either on expansion or compression of the system.

Consider an ideal gas contained in a cylinder fitted with a frictionless piston (Fig. 2.3).

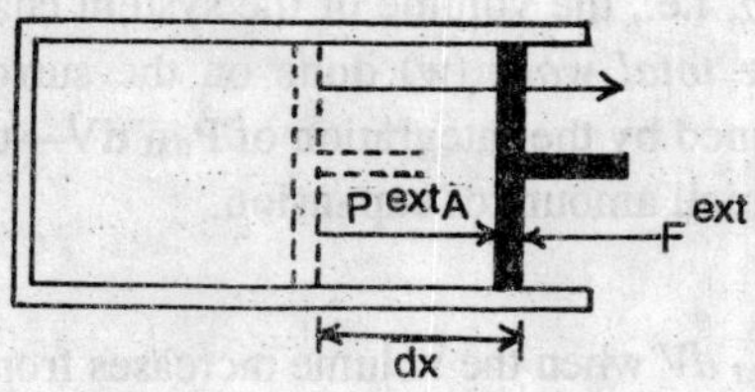

Fig. 2.3 Pressure-volume work or mechanical work in the expansion of an ideal gas.

If P_{ext} is the pressure (of surrounding atmosphere) acting on the piston of cross-sectional area *A*, the force F_{ext} acting on the piston is given by

$$F_{ext} = P_{ext}\,A$$

The force is balanced by an equal force exerted by the gas on the piston.

Suppose the gas in the cylinder expands by an infintesimally small amount and as a result, the piston is pushed out through an infintesimally small distance dx (displacement dx by force F from the point of its application in the direction of the force).

The small amount of work done (*dw*) by the system, i.e., a gas on the surroundings is given by

$$dw = F_{ext} \times dx$$

$$= P_{ext} \times A \times dx$$

$$= P_{ext} \times dV$$

The quantity $A \times dx$ represents infinitesimally small increase in volume in the expansion process and is represented by dV.

It may be noted that here P_{ext} (the force per unit area F_{ext}/A) is always the external pressure or the pressure of the surroundings, against which the gas expands.

If this expansion is brought about infinitesimally slowly in a reversible manner in a number of stages from an initial volume V_1 to the final volume V_2, i.e., the volume of the system changes by a finite quantity $V_2 - V_1$ the *total work* (*w*) done on the surroundings by the system can be obtained by the integration of P_{ext} dV—the work done in an infinitesimally small amount of expansion.

$$w = \int_{V_1}^{V_2} P_{ext}\, dV$$ when the volume increases from v_1 to v_2.

where P_{ext} is variable.

In case the piston is pushed inward by an infinitesimally small distance due to an infinitesimally small amount of contraction, the infinitesimally small amount of work done (*dw*) by the surroundings on the gas, i.e., system by

$$dw = P_{ext}\, dV$$

where *d*V represents infinitesimally small decrease in volume in the contraction process.

If the contraction is carried out infinitesimally slowly in a reversible manner in a number of stages, the work done (*w*) on the system by the surroundings is obtained by the integration of $P_{ext}\, dV$.

$$\therefore \qquad w = \int_{V_1}^{V_2} P_{ext}\, dV$$

when the volume of the system decreases from V_1 to V_2.

Pressure-volume or mechanical work done at constant pressure.

If the pressure P_{ext} remains constant during the process,

the total work done $(w) = P_{ext} \int_{V_1}^{V_2} dV$

$$= P_{ext} . (V_2 - V_1) = P_{ext} . \Delta V$$

where V_1 is the volume of the system in the initial state and V_2 is the volume of the system in the final state.

ΔV, i.e., $(V_2 - V_1)$ is the change in volume of the system. If ΔV is positive, it means the gas expands and (w) the work done will be positive and the work is done by the system (gas) on the surroundings.

If ΔV is negative, it means the gas undergoes contraction and (w) the work done will be negative and the work is done by the surroundings on the system (gas).

It may thus be noted that

(i) When the gas *expands, the work is done by the system on the surroundings* and has a *positive value* (+*ve*).

(ii) When the gas *contracts, the work* (w) *is done by the surroundings on the system* and has a *negative value* (–ve).

Note that it is the *external pressure* on the piston which differs from the pressure of the gas by an infinitesimally small amount that has been used in calculating the work.

Work done in free expansion of a gas against vacuum and maximum work.

No matter what the gas pressure is, a volume change does no work unless the system is linked to the surroundings by an external force represented by P_{ext}. If this external force is zero, there is no mechanical link between the system and the surroundings and no mechanical work can be done by or on the system.

In the above example if the pressure applied on the piston, P_{ext}, is very small as compared with the pressure of gas contained in the cylinder, the gas will expand rapidly (not in small stages), i.e., irreversibly and the work done by the system will be very small as the opposing force is very small. If the gas is allowed to expand against vacuum when P_{ext} the opposing pressure is zero the work done will be zero.

i.e., $P_{ext} \times \Delta V = 0$

Thus for free expansion of gas against vacuum, no work is done since

$$P_{ext} = 0$$

It may be of interest to note here that *the magnitude of the opposing (external) pressure determines the amount of work done by the system in expansion.* The smaller the difference between the opposing (external pressure) and the pressure of the gas (system) contained in the cylinder, the greater will be the work done by it in expansion. *Thus if the difference between the magnitudes of two pressures is infinitesimally small which is a condition for ideal reversible process, work done is maximum.*

The comparison when P_{ext} is only slightly more than that of the gas is called *reversible compression.* The work of reversible expansion and reversible compression are equal and opposite in sign when the initial and the final states of the gas are exchanged. So if the gas is compressed, reversible applied pressure P_{ext} should be infinitesimally greater than the pressure of the gas and the work done should be minimum.

Pressure-Volume or mechanical work done at constant volume (*Isochoric process*). If volume is costant then $V_1 = V_2$ and work done,

$$w = \int_{V_1}^{V_2} PdV = 0 \quad \therefore dV = 0$$

Pressure-Volume or mechanical work done in reversible expansion of an ideal gas at constant temperture *(isothermal process).*

We know, work done

$$\omega = \int_{V_1}^{V_2} PdV$$

and for an ideal gas PV = nRT

or $$P = \frac{nRT}{V}$$

Substituting this value of P in the above equation

$$\omega = \int_{V_1}^{V_2} \frac{nRT}{V} dV$$

$$= nRT = \int_{V_2}^{V_2} \frac{dV}{V}$$

$$= nRT \ 1n \frac{V_2}{V_1}$$

since $V \ \alpha \frac{1}{P}$

$$w = nRT \ln \frac{P_1}{P_2} \quad \text{or} \quad 2.303 \ nRT \log \frac{P_1}{P_2}$$

Work is no Property of the System or Surroundings

It may be understood that work is no property of the system or the surroundings. It is only one of the modes of exchange of energy between the system and the surroundings.

The work done in a process depends on how the change from V_1 and V_2 is accomplished. We can see this more clearly by referring to Fig. 2.4. These are two particularly simple paths by which a system may change its state from P_1, V_1 to P_2, V_2.

In part (a), we first change the volume from V_1 to V_2 at constant pressure P_1. Then we change the pressure from P_1 to P_2, keeping the volume constant. In part (b) we simply reverse the order of changes.

In Fig. 2.4 the work done in following each of the paths is represented as the area under each of the curves followed from the initial state to the final state. It is clear from this drawing that the work done depends on the path followed, even though the initial and final states are the same. Consequently, we must conclude that *work is not a state function*, for its value depends on the path taken between states.

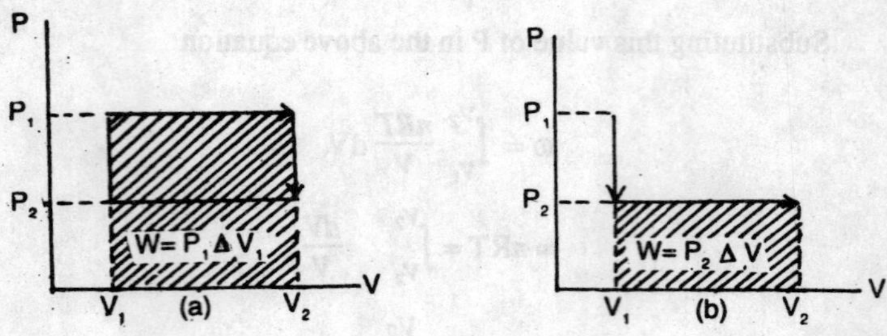

Fig. 2.4. Work done going from the initial to the final state depends on the path followed.

Work is only the means by which energy can be transferred to and from the simple hypothetical systems of mechanics.

Since *w* is not state function and hence *dw* is not an *exact differential.* This can be proved mathematically.

Proof. To prove, let us assume that *w* is a state function and *dw* is exact differential. If by this assumption we arrive at wrong conclusions this assumption will be proved wrong and it will prove that *w* is not a state function and *dw* is not an exact differenuial.

The work done is given by

$$dw = PdV \qquad (1)$$

As V is a function of T and P, it means that

$$V = f(T, P)$$

or $$dV = \left(\frac{\partial V}{\partial T}\right)_P dT + \left(\frac{\partial V}{\partial P}\right)_T dP \qquad (2)$$

Substituting equation (2) in (1), we get

$$dw = P\left(\frac{\partial V}{\partial T}\right)_P dT + P\left(\frac{\partial V}{\partial P}\right)_T dP \qquad (3)$$

If T is constant, dT = 0 and therefore the above equation becomes as

$$\left(\frac{\partial w}{\partial P}\right)_T = P\left(\frac{\partial V}{\partial P}\right)_T \qquad (4)$$

Differentiating the above equation with respect to temperature, we get

$$\left(\frac{\partial^2 w}{\partial T \partial P}\right) = P\frac{\partial^2 V}{\partial T \partial P} \qquad (5)$$

At constant pressure P, $dP = 0$; equation (3) reduce to

$$\left(\frac{\partial w}{\partial T}\right)_P = P\left(\frac{\partial V}{\partial T}\right)_P \qquad (6)$$

Differentiating the above equation with respect to P at constant T, we get

$$\left(\frac{\partial^2 w}{\partial T \partial P}\right) = \left(\frac{\partial V}{\partial T}\right)_P + P\frac{\partial^2 V}{\partial P \partial T} \qquad (7)$$

If dw is exact differential, it has to satisfy the following relation

$$\frac{\partial^2 w}{\partial P \partial T} = \frac{\partial^2 w}{\partial T \partial P}$$

From equations (5) and (7), we have

$$\left(\frac{\partial V}{\partial T}\right)_P + P\frac{\partial^2 V}{\partial P \partial T} = P\frac{\partial^2 w}{\partial T \partial P}$$

But $$\frac{\partial^2 V}{\partial P \partial T} = \frac{\partial^2 V}{\partial T \partial P}$$

$$\therefore \qquad \left(\frac{\partial V}{\partial T}\right)_P = 0 \qquad (8)$$

The above [equation (8)] is not true under ordinary conditions and hence *dw* is not an exact differential. In other words, it proves that *w* is not a state function.

Heat q as a Path Dependent Function. Similar to *w*, it can be proved that the heat q is not a state function but depends on path.

Suppose we assume that *q* is a state function and *dq* is a perfect or exact differential. According to the first law of thermodynamics, we have

$$dq = dE + PdV \qquad (9)$$

But $E = f(T, V)$

$$\therefore \qquad dE = \left(\frac{\partial E}{\partial T}\right)_V dT + \left(\frac{\partial E}{\partial V}\right)_T dV \qquad (10)$$

On substituting equation (10) in (9), we get

$$dq = \left(\frac{\partial E}{\partial T}\right)_V dT + \left(\frac{\partial E}{\partial V}\right)_V dV + PdV \qquad (11)$$

At constant volume dV = 0 and , therefore, equation (11) becomes as

$$\left(\frac{\partial q}{\partial T}\right)_V = \left(\frac{\partial E}{\partial T}\right)_V \qquad (12)$$

On differentiating the above equation with respect to V at constant T, we get

$$\frac{\partial^2 q}{\partial V \partial T} = \frac{\partial^2 E}{\partial V \partial T} \qquad (13)$$

At constant temperature, dT = 0 and therefore, equation (11) becomes as

$$\left(\frac{\partial q}{\partial T}\right)_T = \left(\frac{\partial E}{\partial V}\right)_T + P \tag{14}$$

Differentiation of the above equation with respect to T at constant V yields

$$\frac{\partial^2 q}{\partial T \partial V} = \frac{\partial^2 q}{\partial T \partial V} + \left(\frac{\partial P}{\partial T}\right)_V \tag{15}$$

If *q* is a state function, it means that *dq* is an exact differential. For this the following relation must be valid

$$\frac{\partial^2 q}{\partial V \partial T} = \frac{\partial^2 q}{\partial T \partial V}$$

From equations (13) and (15), we have

$$\frac{\partial^2 E}{\partial V \partial T} = \frac{\partial^2 E}{\partial T \partial V} + \left(\frac{\partial P}{\partial T}\right)_V \tag{16}$$

But E is a state function and

$$\frac{\partial^2 E}{\partial V \partial T} = \frac{\partial^2 E}{\partial T \partial V} \tag{17}$$

From equations (16) and (17), we get

$$\left(\frac{\partial P}{\partial T}\right)_V = 0 \tag{18}$$

The above condition is not possible because of wrong assumption about *q*. Thus *q* is a path dependent function. However *q* may become a state function under certain conditions.

Irreversible Work of Expansion

Let us consider a gas enclosed in a cylinder. The initial and final states of the system are shown in Fig. 2.5. Suppose the external pressure is released suddenly from a initial pressure P_1 to a final pressure P_2 and then gas expands against this new final pressure P_2.

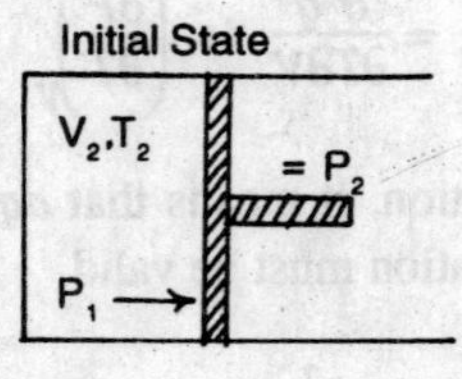

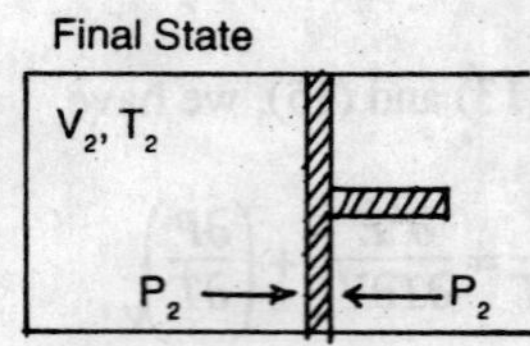

Fig. 2.5. Irreversible expansion of an ideal gas against a constant pressure.

$$\text{Work done, } w_{irr} = \int_{V_1}^{V_2} P_2 dV = P_2 \int_{V_1}^{V_2} dV$$

$$= P_2 (V_2 - V_1) \qquad (1)$$

But $\qquad P_1V_1 = n\text{RT}$

and $\qquad P_2V_2 = n\text{RT}$

$$\therefore \qquad w_{irr} = P_2 \left[\frac{nRT}{P_2} - \frac{nRT}{P_1}\right]$$

$$= n\text{RT}\left[\frac{1}{P_2} - \frac{1}{P_1}\right] P_2$$

$$= nRT\left[1 - \frac{P_2}{P_1}\right] \qquad (2)$$

Comparison of w_{rev} and w_{irrev}

The work done in an *isothermal reversible expansion* is always greater than done during irreversible expansion of an ideal gas, we know

$$w_{rev} = nRT \ln \frac{P_1}{P_2} \qquad (1)$$

and $$w_{irr} = nRT\left[1 - \frac{P_2}{P_1}\right] \qquad (2)$$

Equation (1) can be written as

$$w_{rev} = nRT \ln\left[1 - 1 + \frac{P_1}{P_2}\right]$$

$$= nRT \ln\left[1 - \left(1 - \frac{P_1}{P_2}\right)\right] \qquad (3)$$

We know $\ln(1-x) = -x - \frac{1}{2}x^2 \cong -x$

$$\therefore \qquad \ln\left[1 - \left(1 - \frac{P_1}{P_2}\right) = -\left(1 - \frac{P_1}{P_2}\right)\right] \qquad (4)$$

On substituting equation (4) in (3), we get

$$w_{rev.} = -nRT\left(1 - \frac{P_1}{P_2}\right) \qquad (5)$$

$$= nRT\left(\frac{P_1}{P_2} - 1\right) \qquad (6)$$

On subtracting equation (2) from (6), we get

$$w_{rev} - w_{irr} = -nRT\left[\frac{P_1}{P_2} - 1\right] - nRT\left[1 - \frac{P_2}{P_1}\right]$$

$$= nRT\left[\frac{P_1}{P_2} - 2 + \frac{P_2}{P_1}\right]$$

$$= nRT\left(\frac{P_1^2 - 2P_1P_2 + P_2^2}{P_1P_2}\right)$$

$$= \frac{nRT}{P_1P_2}\ (P_1 - P_2)^2 \qquad (7)$$

As $(P_1 - P_2)^2$ is always a positive quantity, it means that L.H.S. of equation (7) should also be a positive quantity which means that $w_{rev} - w_{irr} > 0$ or $w_{rev} > w_{irr}$ for isothermal expansion of an ideal gas.

Work done in Adiabatic Expansion of an Ideas Gas (Reversible Expansion)

In adiabatic changes the system is not allowed to exchange energy with its surroundings, i.e.,

$$\delta q = 0$$

Insertion of this criterion into equation $\delta q = d\mathbf{E} + \delta w$ leads to

$$dE = -\delta w \qquad (1)$$

Equation (1) implies that *all the work performed will be done at the expense of the internal energy of the system.*

Let one mole of an ideal gas occupy a volume *V* at pressure *P* and absolute temperature *T*. Suppose the gas undergoes an adiabatic expansion by performing external work. As work is done by the gas, the internal energy of the gas decreases by an amount equivalent to the external work done, and consequently the temperature falls.

Suppose in the expansion of a gas, the volume increases from V to V+dV and temperature drops from T to T–*d*T. The work done by the gas at constant pressure is given by P *d*V, i.e.,

$$w = P.dV \qquad (2)$$

Since this work is done by at the expense of the internal energy of the gas, the internal energy must also decrease from E (initial value) to E–dE (final value). Therefore, the decrease in the internal energy is given by E – (E – dE) or dE.

Setting up the equality demanded by equation (1), we obtain.

$$P.\,dV = -\,dE = -\,C_v.dT \qquad (3)$$

An ideal gas must obey the equation of state

$$PV = nRT$$

For one mole of an ideal gas, PV = RT [∵ n = 1]

or $$P = \frac{RT}{V}$$

By substituting this value of P in equation (3), we get

$$\frac{RT}{V}\,dV = -\,C_u\,.\,dT$$

or $$\frac{dV}{V} = -\frac{C_v}{R}\cdot\frac{dT}{T}$$

Integrating this equation between the limits V_1 at T_1 and V_2 at T2, we obtain the expression

$$\int_{V_1}^{V_2}\frac{dV}{V} = -\frac{C_v}{R}\int_{T_1}^{T^2}\frac{dT}{T}$$

In this integration process C_v and R are assumed to be constant.

$$1n\frac{V_2}{V_1} = -\frac{C_v}{R}\,1n\frac{T_2}{T_1} = \frac{C_v}{R}\;1n\left(\frac{T_1}{T_2}\right) \qquad (4)$$

$$\left[\because\; 1n\frac{T_2}{T_1} = -\,1n\frac{T_1}{T_2}\right]$$

$$1n\frac{V_2}{V_1} = 1n\left(\frac{T_1}{T_2}\right)^{Cv/R} \qquad (5)$$

or $$V_1 T_1 Cv/R = V_9 T_2^{Cv/R} = K$$

where K is a constant.

Other forms of equation (5) can be obtained by putting $C_p - C_v = R$ in equation (4),

$$1n \frac{V_2}{V_1} = \frac{C_v}{C_p - C_v} 1n \left(\frac{T_1}{T_2}\right)$$

or $$\left(\frac{C_v - C_v}{C_v}\right) 1n \frac{V_2}{V_1} = 1n \left(\frac{T_1}{T_2}\right)$$

or $$(\gamma - 1)\, 1n \frac{V_2}{V_1} = 1n \frac{T_1}{T_2}$$

[because γ is the sp. heat ratio i.e. C_p/C_v]

or $$1n \left(\frac{V_2}{V_1}\right)^{\gamma-1} = 1n \frac{T_1}{T_2}$$

or $$\left(\frac{V_2}{V_1}\right)^{\gamma-1} = \left(\frac{T_1}{T_2}\right)$$

or $$V_2^{\gamma-1}\, T_2 = V_1^{\gamma-1}\, T_1 \qquad (5)$$

Another form of equation (5) can also be obtained by substituting $T_1 = \frac{P_1 V_1}{R}$ and $T_2 = \frac{P_2 V_2}{R}$ in equation (5)

Therefore $$(V_2^{\gamma-1}) \frac{P_2 V_2}{R} = (V_1^{\gamma-1}) \frac{P_1 V_1}{R}$$

or $$P_2 V_2^{\gamma} = P_1 P_1^{\gamma} \qquad (6)$$

or $$PVg = K \qquad (7)$$

or $$P = KV^{-\gamma} \text{ where K is a constant} \qquad (8)$$

Substituting for P from equatin (2) in equation (8) we obtain the expression,

$$w_{rev} = \int_{V_1}^{V_2} KV^{-\gamma}\, dV$$

$$w_{rev} = \int_{V_1}^{V_2} V^{-\gamma}\, dV = \frac{K}{1-\gamma}\left[V^{1-\gamma}\right]_{V_1}^{V_2}$$

$$= \frac{K}{1-\gamma}\left[V_2^{1-\gamma} - V_1^{1-\gamma}\right] = \frac{K}{1-\gamma}\left[\frac{K}{V_2^{\gamma-1}} - \frac{K}{V_1{\gamma-1}}\right]$$

But $P_1V_1^{\gamma} = P_2V_2^{\gamma} = K$ [See equation (6)]

Therefore, $w_{rev} = \frac{1}{\gamma-1}\left[\frac{P_1V_1\gamma}{V_1^{\gamma-1}} - \frac{P_2V_2\gamma}{V_2^{\gamma-1}}\right] = \frac{1}{\gamma-1}\,[P_1V_1 - P_2V_2]$

$$= \frac{1}{\gamma-1}\,[RT_1 - RT_2]$$

[$\because P_1V_1 = RT_1$ and $P_2V_2 = RT_2$]

$$= \frac{R(T_1 - T_2)}{\gamma-1} \qquad (9)$$

This equation gives the work done in adiabatic expansion of an ideal gas

Whenever $T_1 > T_2$, $w_{rev} > 0$ i.e. work done is on the gas.

ADIABATIC IRREVERSIBLE EXPANSION

The pressure is constant in such an expansion and the work done is given by

$$w_{irr} = \int_{V_1}^{V_2} P_2 dv$$

$$= P_2\,(V_2 - V_1)$$

Comparison of Isothermal and Adiabatie Expansion of an Ideal Gas. (i) For a reversible adiabatic process, the relationship between pressure and volume is given by

$$PV^{\gamma} = \text{constant.}$$

On the other hand, for a reversible isothermal process, the relationship is simply given by

$$PV = \text{constant.}$$

(ii) Now, since C_p is always greater than C_v, therefore, the ratio $\frac{C_p}{C_v} = \gamma$ is greater than unity. Hence the increase of volume for a given decrease of pressure will be less in an adiabatic than in an isothermal expansion. Thus, the plot of pressure against volume i.e., pressure-volume curve will be steeper for an adiabatic than for an isothermal process starting at the same point (Fig. 2.6).

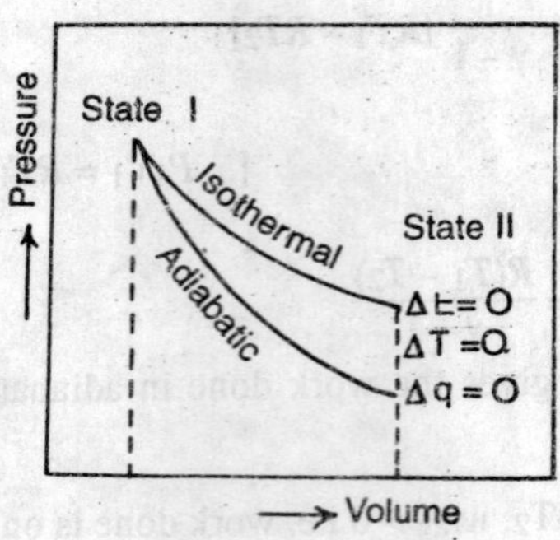

Fig. 2.6. Isothermal adiabatic curves.

(iii) The smaller volume change can be explained by the fact that in the adiabatic expansion, the internal energy decreases and consequently the temperature falls. The cooling effect leads to the shrinkage in volume. In an isothermal expansion the temperature remains constant.

(iv) Since the work of expansion is equal to the area under the curve, it is evident from the figure that the work done by ideal gas is greater in isothermal expansion than that in adiabatic expansion.

Example 1. Two moles of hydrogen are compressed adiabatically from N.T.P. conditions to occupy a volume of 4.48 litres. Calculate the final pressure and temperature, (γ for hydrogen = 1.41).

Solution.

Here, V_1 = 2 moles at N.T.P.; P_1 = 1 atm; T_1 = 273°A; γ = 1.41

= 22.4 × 2 litres $\quad P_2 = ? \quad T_2 = ?$

= 44.8 litres

V_2 = 4.48 litres.

Using the formula,

$$P_1V_1\gamma = P_2V_2^{\gamma}$$

i.e., $\dfrac{P_1}{P_2} = \left(\dfrac{V_2}{V_1}\right)^{\gamma}$, we get

$$\frac{1}{P_2} = \left(\frac{44.8}{4.48}\right)^{1.41} = \left(\frac{1}{10}\right)^{1.41}$$

or $\quad P_2 = (10)^{1.41}$

$\therefore \quad \log P_2 = 1.41 \log 10 = 1.41$

$\therefore \quad P_2 = 25.7$ atms.

The new temperature (T_2) can be calculated either by using the formula $T_1V_1^{\gamma-1} = T_2V_2^{\gamma-1}$

or as follows:

$n = 2$ moles

$P_2 = 25.7$ atms

$V_2 = 4.48$ litres

and $\quad R = 0.082$

Now, $P_2V_2 = nRT_2$

or

$$T_2 = \frac{P_2 \times V_2}{nR}$$

$\therefore$

$$T_2 = \frac{25.7 \times 4.48}{0.082 \times 2} = 702.1\text{A}^\circ = (702.1-273)^\circ\text{C}$$

$$= 429.1^\circ\text{C}$$

i.e.,

$$T_2 = 4291^\circ\text{C}$$

Example 2. A quality of air at 25°C was allowed to expand reversibly and adiabatically from 200 atm to 20 atm. Assuming the ideal behaviour and taking C_v = 5 cals./degree/mole, calculate the final temperature of the air.

Solution.

The relation between pressure and temperature is:

$$\frac{T_2}{T_1} = \left(\frac{P_1}{P_2}\right)\frac{1-\gamma}{\gamma}$$

$$C_v = 5 \qquad \therefore C_p = C_v + R = 5 + 2 = 7.$$

$\therefore$

$$\gamma = \frac{7}{5} \text{ and } 1 - \gamma = 1 - \frac{7}{5} = -\frac{2}{5}$$

and

$$\frac{1-\gamma}{\gamma} = \frac{-2/5}{7/5} = \frac{-2}{5} \times \frac{5}{7} = -\frac{2}{7}$$

$\therefore$

$$\frac{T_2}{T_1} = \left(\frac{P_1}{P_2}\right)^{\frac{-2}{7}}$$

or

$$\frac{T_2}{T_1} = \left(\frac{P_2}{P_1}\right)^{\frac{-2}{7}}$$

Now

$$T_1 = 273 + 25 = 298^\circ\text{A}$$

$$P_1 = 200 \text{ atm.}$$

$P_2 = 20$ atm.

$T_2 = ?$

Putting the values, we have

$$T_2 = T_1 \times \left(\frac{P_2}{P_1}\right)^{\frac{2}{7}}$$

$$= 298 \times \left(\frac{20}{200}\right)^{\frac{2}{7}} = 154^{\circ}\text{A}$$

$$= 119^{\circ}\text{C}.$$

Example 3. Two litres of helium at 0°C are compressed to one litre adiabatically and reversibly. Calculate the change in temperature of the gas, γ for helium = 1.66.

Solution.

For an adiabatic process

$$\frac{T_2}{T_1} = \left(\frac{V_1}{V_2}\right)^{\gamma - 1}$$

Now $\gamma = 1.66$, $T_1 = (273 + 0) = 273^{\circ}$K, $T_2 = ?$

$V_1 = 2$ litres, $V_2 = 1$ litre

Substituting these values

$$\frac{T_2}{273} = \left(\frac{2}{1}\right)^{1.66 - 1}$$

or $T_2 = 273 \times (2)1.66 = 431.4^{\circ}\text{K} - 273 = 158.4^{\circ}\text{C}$

Hence the change in temperature is 158.4°C.

Example 4. An enclosed mass of gas occupies a certain volume at NTP. Calculate the pressure and temperature when it is adiabatically (*i*) expanded to double the volume and; (*ii*) compressed to half the volume ($\gamma = 1.4$).

Solution.

(i) Using $P_1V_1^\gamma = P_2 V_2^\gamma$

Pressure $P_2 = P_1 \left(\frac{V_1}{V_2}\right)^\gamma$

$= \frac{0.76}{2^{1.4}}$

$= .28$ metre of mercury

Again $V_1^{\gamma-1} . T_1 = V_2^{\gamma-1} . T_2$

$\therefore \quad T_2 = T_1 \left(\frac{V_1}{V_2}\right)^{\gamma-1}$

$= \frac{273}{2^{0.4}}$

$= 206.9°K$

(ii) $P_2 = P_1 \left(\frac{V_1}{V_2}\right)^\gamma$

$= .76 \times 2^{1.4}$

$= 2.006$ metre of mercury

$T_2 = T_1 \left(\frac{V_1}{V_2}\right)^{\gamma-1}$

$= 273 \times 2^{0.4}$

$= 363.2°K.$

Example 5. An enclosed mass of oxygen is adiabatically cooled from 50°C to 0°C. Calculate the ratio by which the volume and pressure change

$$\left[\text{Hint Use} \left(\frac{V_2}{V_1}\right)^{\gamma-1} = \frac{T_1}{T_2} \text{ and } \left(\frac{P_2}{P_2}\right)^{\gamma-1} = \left(\frac{T_2}{T_1}\right)^{\gamma} \right]$$

$$\left[\frac{V_2}{V_1} = 1.523,\ \frac{P_2}{P_1} = 0.555\right]$$

Example 6. An enclosed mass of gas at NTP is expanded isothermally to double the volume. Then it is compressed to earlier volume adiabatically. If $\gamma = 1.4$ find the final temperature and pressure.

Solution.

(i) It is an isothermal expansion

and Final Volume = 2V

$$\text{Final Pressure} = \frac{P}{2}$$

Final Temperature = T. (remains constant)

(ii) It is an adiabatic compression

so $PV^{\gamma} = \text{constant}$

$$\therefore \quad \frac{P}{2}(2\,V)^{\gamma} = P_1 V^{\gamma}$$

$$\therefore \quad P_1 = \frac{P}{2} \times 2\gamma$$

$$= P \times 2^{0.4}$$

$$= 1.32\ P$$

$$= 1.32\ \text{atms}$$

Using, $V_1^{\gamma-1} \cdot T_1 = V_2^{\gamma-1} \cdot T_2$

$$T_2 = T_1 \left(\frac{V_1}{V_2}\right)^{\gamma-1}$$

$$= 273 \times 2^{.04}$$

$$= 360.2^{\circ}\text{K}.$$

Example 7. An enclosed mass of gas at NTP is expanded isobasically to double the volume. It is then compressed isothermally to the earlier volume. Calculate the final pressure and volume.

[*Hint.* For isobasic process $\frac{V}{T}$ = constant]

[Ans. $T = 546°K$, $P = 2$ atm.]

Example 8. An enclosed mass of gas exerts a pressure of 10^5 N/m^2 at 27°C. It is heated isobasically to 47°C. It is then compressed isothermally to the earlier volume. It is cooled adiabatically to the earlier temperature and brought to the intial pressure isochorically. Find the temperature and volume. If the initial volume is 10 litres and γ = 1.4.

Solution.

$$\text{Initial } P = 10^5 \text{ N/m}^2 \text{ ; } T = 300°K$$

Let the volume be V.

(i) After isobaric change

$$\frac{V}{T} = \frac{V_1}{T_1}$$

$$\therefore \quad V_1 = V \times \frac{320}{300}$$

$$= \frac{16V}{15}$$

$T_1 = 320°K$ and $P_1 = 10^5$ N/m^2.

(ii) After isothermal change

$$P_1V_1 = P_2V_2$$

$$\therefore \quad P_2 = 10^5 \times \frac{16V}{15V}$$

$$= 10^5 \times \frac{16}{15} N/m^2$$

$$T_2 = 320°K.$$

(iii) After adiabatic cooling $T_3 = 300°K$

Using, $T_2V_2^{\gamma-1} = T_3V_3^{\gamma-1}$

$$V_3 = V_2 \times \left(\frac{T_2}{T_3}\right)^{\frac{1}{\gamma - 1}}$$

$$= V \times \left(\frac{320}{300}\right)^{2.5}$$

$$= 1.175\ V$$

Using, $P_3V_2\gamma = P_3V_3\ \gamma$

$$P_3 = P_2 \left(\frac{V_2}{V_3}\right)^{\gamma}$$

$$= 10^5 \times \frac{16}{15} \times \left(\frac{V}{1.175}V\right)^{1.4}$$

$$= .85 \times 10^5\ N/m^2.$$

(iv) After the isochoric change,

$P_4 = 10^5\ N/m^2$, $V_4 = 1.175\ V$

We have $P_3 = .85 \times 10^5\ N/m^2$

$V_3 = 1.175\ V$

$T_3 = 300°K$

Using, $$\frac{P_3}{T_3} = \frac{P_4}{T_4}$$

$$T_4 = T_4 = \frac{300 \times 10^5}{.85 \times 10^5}$$

$$= 353°K$$

Hence Final Temperature = 353°K

Final Pressure $= .85 \times 10^5\ N/m^2$

Final Volume $= 1.175 \times 10$

$= 11.75$ litres.

Example 9. Calculate the work done when 1 mole of a perfect gas expands isothermally and reversibly from a volume of 1 litre to a volume of 10 litres at 25°C. (*K.U. 1979*)

Solution.

We know

$$w = 2.303\ nRT \text{ long } \frac{V_2}{V_1}$$

Thus $$w = 2.303 \times 1 \times 8.314 \times 298 \log \frac{10}{1}$$

$$= 5705 \text{ J.}$$

Example 10. Calculate the work done by 10 litres of an ideal gas at 0°C and 10 atmosphere in an isothermal reversible expansion to a final pressure of 1 atmosphere (R = 8.314 Joules) (*G.N.D.U. 1981*)

Solution.

We know

$$w = 2.303\ n\text{RT} \log \frac{P_1}{P_2}$$

To calculate *n*.

W know 22.4 litres of gas at NTP is equal to 1 mole. Thus changing the volume of gas to NTP, we get

$$\text{Volume at N.T.P.} = \frac{10 \times 10 \times 273}{273}$$

$$= 100 \text{ litres}$$

Since 22.4 litres = 1 mole

$\therefore$ $$100 \text{ litres} = \frac{100}{22.4} = 4.46 \text{ moles.}$$

Thus $$w = 2.303 \times 4.46 \times 8.314 \times 273 \log \frac{10}{1}$$

$$= 23.3 \text{ KJ.}$$

Example 11. Three moles of a perfect gas with $C_v = 5.0$ cal/mole/degree are to be compressed adiabatically and reversibly from a volume at 75 litres at 1 atmosphrere to a pressure of 100 atmosphere. Predict : (a) the final volume of the gas; (b) the final temperature of the gas. *(K.U. 1979)*

Solution.

Here

$n = 3$ $\quad P_2 = 100$ atm

$V_1 = 75$ litres $\quad V_2 = ?$

$P_1 = 1$ atm $\quad T_2 = ?$

γ can be calculated as under

$$\gamma = \frac{C_p}{C_v}$$

$$= \frac{C_v + R}{C_v}$$

$$= \frac{5+2}{5} \quad \text{[use R = 2 calories]}$$

$$= \frac{7}{5} = 1.4$$

(*a*) Final volume of gas

$$\left(\frac{V_1}{V_2}\right)^{\gamma} = \frac{P_2}{P_1}$$

Thus $$\left(\frac{75}{V_2}\right)^{1.4} = \frac{100}{1}$$

or $V_2 = 2.796$ litres.

(*b*) Final Temperature

Using $T_1 V_1^{\gamma-1} = T_2 V_2^{\gamma-1}$, we can get the value of final temperature.

It can also be obtained, if we use

$$P_2V_2 = nRT_2 \quad \text{or} \quad T_2 = \frac{P_2V_2}{nR}$$

Thus $$T_2 = \frac{100 \times 2.796}{3 \times 0.0821}$$ [use $R = 0.0821$]

$$= 1135.2 \text{ K or } 862.2^\circ\text{C}.$$

Applications of the First Law of Thermodynamics

1. *Relation between C_p and C_v*. A relation between C_p and C_v can be deduced with the help of first law of thermodynamics.

Consider one gram mole of an ideal (or perfect) gas enclosed in a cylindrical vessel fitted with a weightless and frictionless piston. Raise the temperature of the gas from T to (T + 1), i.e., by 1°C, then, Heat supplied at constant volume = molar heat capacity at constant volume

$$C_v = \frac{\Delta E}{[(T+1)-T]} = \Delta E \qquad (1)$$

Heat supplied at constant pressure = Molar heat capacity at constant pressure

$$C_p = \frac{\Delta H}{[(T+1)-T]} = \Delta H \qquad (2)$$

From equations (1) and (2),

$$C_p - C_v = \Delta H - \Delta E$$

$$= \Delta E + P\Delta V - \Delta E \qquad (\because \ \Delta H = \Delta E + P\Delta V)$$

$$= P \times \Delta V.$$

If during the expansion work, the volume of system changes from V_1 to V_2 then $\Delta V = V_2 - V_1$.

$$\therefore \quad C_p - C_v = P(V_2 - V_1) = PV_2 - PV_1 \qquad (3)$$

For 1 mole of ideal gas:

$$PV = RT$$

$$PV_1 = RT$$

$$PV_2 = R(T + 1)$$

as the temperature rises from T to T + 1

$$\therefore \quad C_p - C_v = PV_2 - PV_1$$

$$= R(T + 1) - RT$$

$$= RT + R - RT = R$$

$$= 2 \text{ calories} \qquad (4)$$

Hence, the difference between the molar heat capacity of an ideal gas at constant pressure and at constant volume is equal to the gas constant (R) or 2 calories.

1. Relations Between C_p and C_v. The first law of thermodynamics may be used to deduce various relations between C_p and C_v on the basis of differential calculus.

Relation I. Internal energy, a state function, is regarded as a function of two independent variables V and T, i.e.,

$$E = f(V, T)$$

Increase in dE can be expressed as

$$dE = \left(\frac{\partial E}{\partial T}\right)_V dT + \left(\frac{\partial E}{\partial V}\right)_T dV$$

Dividing by dT, keeping pressure constant, we get

$$\left(\frac{\partial E}{\partial T}\right)_P = \left(\frac{\partial E}{\partial T}\right)_V + \left(\frac{\partial E}{\partial V}\right)_T \left(\frac{\partial E}{\partial T}\right)_P \qquad (1)$$

But we know that

$$C_p = \left(\frac{\partial H}{\partial T}\right)_P \qquad \text{[By def.]}$$

$$= \left[\frac{\partial}{\partial T}(E + PV)\right]_P$$

or
$$C_P = \left(\frac{\partial E}{\partial T}\right)_P + P\left(\frac{\partial V}{\partial T}\right)_P$$

Substituting the value of $\left(\frac{\partial E}{\partial T}\right)_P$ from equation (1) in the above equation, we get

$$C_P = \left(\frac{\partial E}{\partial T}\right)_V + \left(\frac{\partial E}{\partial V}\right)_T \left(\frac{\partial V}{\partial T}\right)_P + P\left(\frac{\partial V}{\partial T}\right)_P$$

$$C_P = \left(\frac{\partial E}{\partial T}\right)_V + \left(\frac{\partial V}{\partial T}\right)_P \left[\left(\frac{\delta E}{\delta V}\right)_T + P\right]$$

or
$$C_p = C_v + \left(\frac{\partial V}{\partial T}\right)_P \left[\left(\frac{\partial E}{\partial V}\right)_T + P\right] \left[C_v = \left(\frac{\partial E}{\partial T}\right)_V\right] \quad (2)$$

But for ideal gases, we have

$$\left(\frac{\partial E}{\partial V}\right)_T = 0 \text{ and } P\left(\frac{\partial V}{\partial T}\right)_P = R$$

Therefore, equation (2), becomes

$$C_p = C_v + R$$

or
$$C_p - C_v = R$$

Relation II. A second relation may also be obtained by simplifying the term $\left(\frac{\partial E}{\partial V}\right)_T$ in equation (2)

since

$$\left(\frac{\partial E}{\partial V}\right)_T = \left[\frac{\partial}{\partial \varsigma}(H - PV)\right]_T \qquad \text{(because } H = E + PV\text{)}$$

$$= \left(\frac{\partial H}{\partial V}\right)_T - P - V\left(\frac{\partial P}{\partial V}\right)_T$$

By combining this equation with equation (2), we obtain

$$C_p = C_v + \left(\frac{\partial V}{\partial T}\right)_P \left[\left(\frac{\partial H}{\partial V}\right)_T - P - V\left(\frac{\partial P}{\partial V}\right)_T + P\right]$$

$$C_V + \left(\frac{\partial V}{\partial T}\right)_P \left[\left(\frac{\partial H}{\partial V}\right)_T - V\left(\frac{\partial P}{\partial V}\right)_T\right]$$

But $\left(\frac{\partial H}{\partial V}\right)_T \left(\frac{\partial H}{\partial P}\right)_T = \left(\frac{\partial P}{\partial V}\right)_T$

Therefore $C_p = C_v + \left(\frac{\partial V}{\partial T}\right)_P \left[\left(\frac{\partial H}{\partial P}\right)_T \left(\frac{\partial P}{\partial V}\right)_T - V\left(\frac{\partial P}{\partial V}\right)_T\right]$

$$C_v = \left(\frac{\partial V}{\partial T}\right)_P \left(\frac{\partial P}{\partial V}\right)_T \left[\left(\frac{\partial V}{\partial T}\right)_P - V\right]$$

But $\left(\frac{\partial V}{\partial T}\right)_P \left(\frac{\partial P}{\partial V}\right)_T = -\left(\frac{\partial P}{\partial T}\right)_V$ (from partial differentiation)

so
$$C_p = C_v - \left(\frac{\partial P}{\partial T}\right)_V \left[\left(\frac{\delta H}{\partial P}\right)_T - V\right] \qquad (3)$$

But for ideal gases $\left(\frac{\partial H}{\partial P}\right)_T = 0$ and $V\left(\frac{\partial P}{\partial T}\right)_V = R$

$$C_p = C_v + R$$

or
$$C_p - C_v = R$$

Relation III. If the derivative $\left(\frac{\partial H}{\partial P}\right)_T$ is replaced by another derivative in equation (3), a new relation between C_p and C_v will be obtained.

$$H = f(T, P)$$

$$dH = \left(\frac{\partial H}{\partial T}\right)_P dT + \left(\frac{\partial H}{\partial P}\right)_T dP \qquad (5)$$

If H is kept constant, it means that $\Delta H = 0$ and therefore equation (5) becomes

$$\left(\frac{\partial H}{\partial T}\right)_P \left(\frac{\partial T}{\partial P}\right)_H + \left(\frac{\partial H}{\partial P}\right)_V = 0$$

or
$$\left(\frac{\partial H}{\partial P}\right)_T = -\left(\frac{\partial H}{\partial T}\right)_P \left(\frac{\partial T}{\partial P}\right)_H \qquad (6)$$

On substituting equation (6) in (3), we get

$$C_P - C_V = \left[V + \left(\frac{\partial H}{\partial T}\right)_P \left(\frac{\partial T}{\partial P}\right)_H\right] \left(\frac{\partial P}{\partial T}\right)_V \qquad (7)$$

The relation (7) can also be deduced by using second law of thermodynamics.

2. Important Thermodynamics Relations

(*a*) To prove $C_P - C_V = R$. We have earlier proved that

$$C_P - C_V = \left[\left(\frac{\partial E}{\partial V}\right)_T + P\right] \left(\frac{\partial V}{\partial T}\right)_P \qquad \text{[Sec eq. (2)]}$$

For an ideal gas the change of internal energy with the change in volume at constant temperature is zero, i.e.,

$$\left(\frac{\partial E}{\partial V}\right)_T = 0 \qquad (8)$$

We know $PV = RT$ (for one mole of ideal gas)

or
$$PdV + VdP = RdT$$

At constant pressure, the above equation reduces to

$$P\left(\frac{\partial V}{\partial T}\right)_P = R \qquad (9)$$

On substituting equations (8) and (9) in (2), we get

$$C_p - C_v = R$$

(b) To prove $\left(\frac{\partial E}{\partial V}\right)_P = \frac{C_V P}{R}$. We know,

$$E = f(T, V)$$

$$dE = \left(\frac{\partial E}{\partial T}\right)_V dT + \left(\frac{\partial E}{\partial V}\right)_T dV$$

$$= C_V dT + \left(\frac{\partial E}{\partial V}\right)_T dV \quad (10)$$

But $\left(\frac{\partial E}{\partial V}\right)_T = 0$ for ideal gas. On substituting this equation in (10), we get

$$dE = C_v dT$$

or
$$\left(\frac{\partial E}{\partial V}\right)_V = C_V \left(\frac{\partial T}{\partial V}\right)_P \quad (11)$$

But $PV = RT$ or $PdV + VdP = RdT$

or
$$\left(\frac{\partial T}{\partial V}\right)_P = \frac{P}{R} \quad (12)$$

On substituting equation (12) in (11), we get

$$\left(\frac{\partial E}{\partial V}\right)_P = C_V \frac{P}{R} \quad (13)$$

(c) To prove $\left(\frac{\partial C_V}{V}\right)_T = 0$. We know

$$C_v = \left(\frac{\partial E}{\partial T}\right)_V \quad (14)$$

On differentiating the above equation with respect to volume at constant T, we get

$$\frac{\partial}{\partial V}\left(Cv\right)_T = \frac{\partial}{\partial V}\left[\left(\frac{\partial E}{\partial T}\right)_V\right] = \frac{\partial^2 E}{\partial V \partial T} \tag{15}$$

As dE is an exact differential, it means that

$$\frac{\partial^2 E}{\partial V \partial T} = \frac{\partial^2 E}{\partial V \partial T} = \frac{\partial}{\partial T}\left[\left(\frac{\partial E}{\partial V}\right)_V\right]_V \tag{16}$$

On substituting equation (16) in (15), we get

$$\frac{\partial}{\partial v}\left(Cv\right)_T = \frac{\partial}{\partial T}\left[\left(\frac{\partial E}{\partial V}\right)_T\right]_V \tag{17}$$

For an ideal gas $(\partial E/\partial V)_T$ is zero; therefore equation (17) becomes as

$$\frac{\partial}{\partial V}\left(Cv\right)_T = \frac{\partial}{\partial T}(\text{O}) = 0$$

or

$$\frac{\partial}{\partial V}\left(Cv\right)_T = 0 \tag{18}$$

(d) To prove $\left(\frac{\partial E}{\partial V}\right)_P = C_P \left(\frac{\partial T}{\partial V}\right)_P - \text{P}$. We know,

$$H = E + PV$$

or

$$E = H - PV \tag{19}$$

On differentiating the above equation (19) with respect to V at constant P, we get

$$\left(\frac{\partial E}{\partial V}\right)_P = \left(\frac{\partial H}{\partial V}\right)_P - \text{P}$$

$$\left(\frac{\partial E}{\partial V}\right)_P = \left(\frac{\partial H}{\partial T}\right)_P \left(\frac{\partial T}{\partial V}\right)_P - \text{P} \tag{20}$$

But $\left(\frac{\partial H}{\partial T}\right)_P = C_P$ (21)

On substituting equation (21) in (20), we get

$$\left(\frac{\partial E}{\partial V}\right)_P = C_P\left(\frac{\partial T}{\partial V}\right)_P - P \tag{22}$$

(e) To prove $\left(\frac{\partial H}{\partial V}\right)_T = 0$. We know

$$H = E + PV$$

On differentiating w.r.t V at constant T we get

$$\left(\frac{\partial H}{\partial V}\right)_T = \left(\frac{\partial E}{\partial V}\right)_T + P + V\left(\frac{\partial P}{\partial V}\right)_T \tag{23}$$

For an ideal gas, $\left(\frac{\partial E}{\partial V}\right)_T = 0$ (24)

Also $PV = RT$ (for one mole of ideal gas)

or $PdV + VdP = RdT$

At constant temperature, dT = 0 and therefore the above equation reduces to

$$P + V\left(\frac{\partial P}{\partial V}\right)_T = 0 \tag{25}$$

On substituting equations (24) and (25) in (23), we get

$$\left(\frac{\partial H}{\partial V}\right)_T = 0.$$

(f) To prove $\left(\frac{\partial E}{\partial T}\right)_P = C_P - P\left(\frac{\partial V}{\partial T}\right)_P$. We know

$$H = E + PV$$

or

$$E = H - PV$$

On differentiating the above equation with respect to T at constant pressure P, we get

$$\left(\frac{\partial E}{\partial T}\right)_P = \left(\frac{\partial H}{\partial T}\right)_P - P\left(\frac{\partial V}{\partial T}\right)_P$$

By definition, $C_P = \left(\frac{\partial H}{\partial T}\right)_P$

$$\therefore \quad \left(\frac{\partial E}{\partial T}\right)_P = C_P - P\left(\frac{\partial V}{\partial T}\right)_P$$

(g) To prove $\left(\frac{\partial E}{\partial P}\right)_V = C_v\left(\frac{\partial T}{\partial P}\right)_V$. From the R.H.S. of the equation, we have

$$C_v\left(\frac{\partial T}{\partial P}\right)_V = \left(\frac{\partial E}{\partial T}\right)_V \left(\frac{\partial T}{\partial P}\right)_V$$

$$= \left(\frac{\partial E}{\partial P}\right)_V \left(\frac{\partial T}{\partial T}\right)_V$$

$$= \left(\frac{\partial E}{\partial P}\right)_V$$

or
$$C_P\left(\frac{\partial T}{\partial P}\right)_V = \left(\frac{\partial E}{\partial P}\right)_V$$

3. Variation of Heat of Reaction with Temperature

Kirchoff's Equation. The variation of the heat of reaction with temperature can be deduced easily with the help of the first law of thermodynamics.

Consider a simple hypothetical reaction in which the reactants in the state A at the temperature T_1 are to be converted into products in state B at the temperature T_2. It can be done by two paths.

Let us suppose that the reaction is carried at constant pressure and so there is a change in heat content or enthalpy of the system.

In the *first path*, the reactants are first heated from temperature T_1 to T_2 and then allowed to react at temperature T_2 to form products. If C_p (R) is the heat capacity of the reactants at constant pressure, the heat required to raise the reactants from temperature T_1 to T_2 at constant pressure, $q = C_p(R) \times (T_2 - T_1)$.

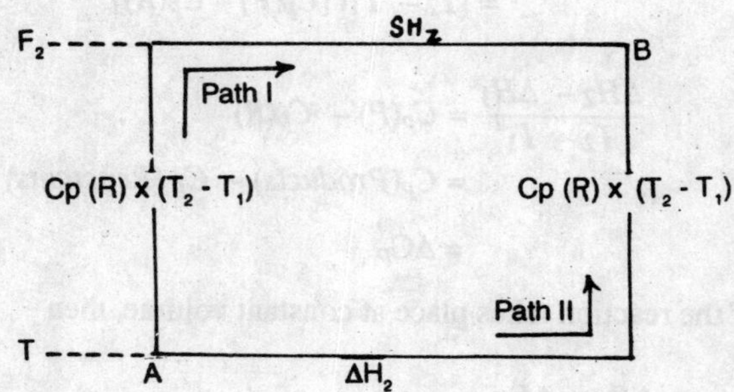

Fig. 2.7.

Let ΔH_2 be the heat of the reaction at constant pressure and temperature T_2.

The net increase in the heat content or enthalpy of the system,

$$\Delta H = C_P(R) \times (T_2 - T_1) + \Delta H_2 \qquad \text{(i)}$$

In the *second path*, the reactants are first allowed to react at temperature T_1 to form products and then the products are heated from temperature T_1 to T_2.

Let ΔH_1 be the heat of reaction at constant pressure and temperature T_1.

If $C_p(P)$ is the heat capacity of the products at constant pressure, the heat requireed to raise the products from temperature T_1 to T_2 at constant pressure $q' = C_p(P) \times (T_2 - T_1)$.

The net increase in the heat content or enthalpy of the system $\Delta H' = C_p(P) \times (T_2 - T_1) + \Delta H_1$.

According to the first law of thermodynamics, the two heat changes must be the same, since the initial and the final states of the system are the same in both cases.

$$\therefore \quad \Delta H = \Delta H'$$

or $\quad C_p(R) \times (T_2 - T_1) + \Delta H_2 = C_P(P) \times (T_2 - T_1) + \Delta H_1$

$$\Delta H_2 - \Delta H_1 = [C_p(P) \times (T_2 - T_1)] - [C_p(R) \times (T_2 - T_1)]$$

$$= [T_2 - T_1][C_p(P) - C_p(R)]$$

$$\frac{\Delta H_2 - \Delta H_1}{T_2 - T_1} = C_p(P) - C_p(R)$$

$$= C_p(Products) - C_p(Reactants)$$

$$= \Delta C_p$$

If the reaction takes place at constant volume, then

$$\frac{\Delta E_2 - \Delta E_1}{T_2 - T_1} = \text{Cv (products)} - C_v\text{ (Reactants)} = \Delta C_v$$

Where ΔE_1 and ΔE_2 are the changes in the internal energies of the system at temperatures T_1 and T_2 respectively and ΔC_v is the difference in the heat capacities of the products and reactants at constant volume.

This relationship was first deduced by Kirchhof in 1858 and hence it is known as **Kirchhoff's equation** or **Kirchhoff's law.**

Although the molar heat capacity varies with temperature, for small differences of temperature, it can be assumed to be constant.

This equation can be easily used to calculate the heat of reaction at a given temperature, when it is known at any other temperature. The calculations can be made only if we know the molar heat capacities of the reactants and the products.

Kirchhoff's Equation on the basis of Differential Calculus

It has been observed that the heat of reaction varies with temperature.

$$\underset{\text{(Reactants)}}{A} \longrightarrow \underset{\text{(Products)}}{B}$$

Suppose H_A is the total heat content of the reactants and H_B that of the product at the same temperature and pressure.

The enthalpy change accompanying the reaction is given by:

$$\Delta H = H_B - H_A$$

Differentiating this equation with respect to temperature T, pressure remaining constant, we get.

$$\left(\frac{\partial(\Delta H)}{\partial T}\right)_p = \left(\frac{\partial H_B}{\partial T}\right)_p - \left(\frac{\partial H_A}{\partial T}\right)_p \qquad (1)$$

Heat capacity of a system at constant pressure $C_p = \left(\frac{\partial H}{\partial T}\right)_p$

or $$\left(\frac{\partial H_B}{\partial T}\right)_p = C_{p}(B) \quad \text{and} \quad \left(\frac{\partial H_A}{\partial T}\right)_p = C_{p}(A)$$

Substituting these values in equation (1), we get

$$\left(\frac{\partial(\Delta H)}{\partial T}\right)_p = C_{p(B)} - C_{p(A)}$$

But $C_{p(B)} - C_{p(A)} = \Delta C_p$, the increase in the heat capacity of the system.

$$\therefore \quad \left(\frac{\partial(\Delta H)}{\partial T}\right)_p = \Delta C_p \qquad (2)$$

This equation is known as *Kirchhoff's equation* which tells that "*The rate of variation of the heat of reaction with temperature, at constant pressure is equal to the increase in the heat capacity accompanying the reaction.*"

Since the term ΔH is used for the heats of reaction at constant pressure, the restrictive subscript p can be dropped from equation (2)

$$\frac{d\Delta H}{dT} = \Delta C_p \qquad \text{or} \qquad d\Delta H = \Delta C_p \times dT \qquad (3)$$

This equation is only for small temperature difference, dT. For large temperature difference, the expression can be obtained by integrating equation (3). Let the value of ΔH at $T_1 = \Delta H_1$.

and value of ΔH at T_2 (which is greater than T_1) = ΔH_2.

On integrating equation (3) between these limits.

$$\int_{\Delta H_1}^{\Delta H_2} d\Delta H = \int_{\Delta T_1}^{\Delta T_2} d\Delta C_p dT \text{ or } \Delta H_2 - \Delta H_1 = \int_{\Delta T_1}^{\Delta T_2} \Delta C_p \times dT$$

or
$$\Delta H_2 = \Delta H_1 + \int_{\Delta T_1}^{\Delta T_2} \Delta C_p \Delta\ dT \quad (4)$$

Now if ΔC_p *is zero*, i.e., the sums of the heat capacities of the products and the reactants are equal, $\Delta H_2 = \Delta H_1$,

If ΔC_p *is constant*, i.e., the difference between the sums of the heat capacities of the products and reactants do not change with the temperature the equation (4) may be written as

$$\Delta H_2 = \Delta H_1 + \Delta C_P \int_{T_1}^{T_2} dT$$
$$= \Delta H_1 + \Delta C_p (T_2 - T_1)$$

or

$$\Delta H_2 - \Delta H_1 = \Delta C_p (T_2 - T_1) \quad \text{or} \quad \frac{\Delta H_2 - \Delta H_1}{T_2 - T_1} = \Delta C_p \quad (5)$$

In this case ΔH either decreases or increases regularly with temperature. For most of the reactions equation (5) is valid if (T_2–T_1) is not very large. Although the molar heat capacity varies with temperature, for small differences of temperature, it can be assumed to be constant.

If ΔC_p *changes with temperature, the equation (4) has to be integrated by expressing ΔC_p as a function of temperature.*

In a similar manner the energy of reaction varies according to the equation,

$$\Delta E_2 = \Delta E_1 = \int_{T_1}^{T_2} \Delta C_v\, dT$$

Where ΔE_2 and ΔE_1 are the energies at temperature T_2 and T_1 respectively and ΔC_v is the difference between the sum of the heat capacities at constant volume for the products and reactants.

It is evident that equation (2) can be integrated at constant pressure

$$\int d(\Delta H) = \int \Delta C_p \, dT$$

$$\Delta H = \int \Delta C_p \, dT + K$$

where K is a constant of integration. When T = 0, ΔH_1 = K. Hence equation (4) takes the form.

$$\Delta H_2 = \Delta H_1 + \int \Delta C_p \, dT \qquad (4A)$$

where ΔH_1 is the difference in heat contents of products and reactants at the absolute zero.

But $\quad (C_P)_A = \alpha A + \beta_A T + \gamma_A T^2 + ...$

$\quad (C_P)_B = \alpha_B + \beta_B T + \gamma_B T^2 + ...$

$\therefore \quad C_P = (C_P)_B - (C_P)_A$

$$= (\alpha_B + \beta_B T + \gamma_B T^2 \ldots) - (\alpha_A + \beta_A T + \gamma_A T^2 + \ldots)$$

$$= (\alpha_B - \alpha_A) + T(\beta_B - \beta_A) + T^2 (\gamma_B - \gamma_A) + \ldots$$

i.e., $\quad \Delta C_p = \Delta\alpha + T\Delta\beta + T^2\Delta\gamma + \ldots \qquad (5)$

Inserting this value of ΔC_p in equatin (4A) and carrying out the integration,

$$\Delta H_2 = \Delta H_1 + T\Delta\alpha + \frac{T^2}{2}\Delta\beta + \frac{T^3}{3}\Delta\gamma + \ldots \qquad (6)$$

The values of $\Delta\alpha$, $\Delta\beta$, $\Delta\gamma$, . . . can be calculated from heat capacity data. Equation (6) is another form of Kirchoff's equation. This equation can be used to calculate ΔH_1 provided the value of ΔH_2 and constants $\Delta\alpha$, $\Delta\beta$, $\Delta\gamma$ . . .are known.

If we consider the reaction,

$$n_1 A + n_2 B = n_3 C + n_4 D$$

where n_1 and n_2 are the number of molecules of reactants A and B and

n_3 and n_4 are the number of molecules of products C and D,

Now $(C_P)_A = n_1\ (\alpha_1 + \beta_1 T + \gamma_1 T^2 + \delta_1 T^3 + \ldots\ldots)$

$(C_P)_B = n_2\ (\alpha_2 + \beta_2 T + \gamma_2 T^2 + \delta_2 T^3 + \ldots\ldots)$

$(C_P)_C = n_3\ (\alpha_3 + \beta_3 T + \gamma_3 T^2 + \delta_3 T^3 + \ldots\ldots)$

$(C_P)_D = n_4\ (\alpha_4 + \beta_4 T + \gamma_4 T^2 + \delta_4 T^3 + \ldots\ldots)$

The change in heat capacity (ΔC_P) is given by

$$\Delta C_P = [\{\alpha_3 + \alpha_4) - (\alpha_1 + \alpha_2)\}] + \{(\beta_3 + \beta_4) - (\beta_1 + \beta_2)\}T + [\{\gamma_3 + \gamma_4) - (\gamma_1 + \gamma_2)\}T2 + \ldots] (n_3 + n_4 - n_1 - n_2)$$

or $\Delta C_p = (\alpha + \beta T + \gamma T^2 + \ldots)\ (n_3 + n_4 - n_1 - n_2)$

Comparing it with Kirchhoff's equation, we get

$$\left[\partial \frac{\Delta H}{\partial T}\right]_P = \Delta C_P = \alpha + \beta T + \gamma T^2 + \ldots \qquad (6A)$$

On integrating this equation, we get

$$\Delta H_2 = \Delta H_1 + \alpha T + \beta \frac{T^2}{2} + \gamma \frac{T^3}{3} + \ldots \qquad (6B)$$

where ΔH_1 is the integration constant and

$\alpha = \alpha_3 + \alpha_4 - \alpha_1 - \alpha_2$

$\beta = \beta_3 + \beta_4 - \beta_1 - \beta_2$

$\gamma = \gamma_3 + \gamma_4 - \gamma_1 - \gamma_2$

$\delta = \delta_3 + \delta_4 - \delta_1 - \delta_2$

Equation (6B) is the general expression for the variation of ΔH with temperature.

Joule Thomson Effect

Introduction. The behaviour of real gases was thoroughly studied by Joule in collaboration with Thomson by performing a series of experiments (1852 to 1862) that led to the discovery of the phenomenon known as the "Joule Thomson effect." This may be stated as follows:

"When a gas under higher constant perssure is allowed to pass through a porous plug (or "throttle") to a region of lower pressure, its temperature is lowered."

Joule-Thomson Experiment. Joule and Thomson employed a long tube being thermally insulated from outside. They fitted it with a porous plug to allow the gas to be kept at different pressures P_1 and P_2 on the two sides of the tube. When the flow became steady, its temperature at the two extreme ends of the tube was recorded by using very sensitive thermometers and pressure difference was measured with a gauged.

Experimental Results. Various gases were subjected to the Joule-Thomson experiment and following results were reported:

(i) All the gases show a change in temperature on passing through the porous plug.

(ii) The change in temperature is found to be proportional to the difference in pressure on the two sides of the porous plug.

(iii) At sufficiently low temperature, all gases show a drop in temperature in passing through the porous plug. But at ordinary temperature, most gases suffer a fall in temperature while gases like hydrogen and helium record a rise in temperature.

(iv) Every gas has inversion temperature at a given initial pressure. Above this temperature, the gas shows heating effect. Below this temperature, the gas shows a fall in temperature.

Experimental set up. The experimental set up of Joule and Thomson is illustrated schematically in Fig. 2.7. A tube thoroughly insulated (so as to be nearly under adiabatic conditions) was fitted with a porous plug, as indicated. The pressure on the left and right side of a porous plug were kept at P_1 and P_2 respectively ($P_1 > P_2$). By applying pressure on the piston A on the left slowly enough (so as not to change the pressure P_1), a volume of gas V_1 was forced slowly through the porous plug and then allowed to expand to the pressure P_2 and volume V_2 by moving the piston B on the right outward. While the expansion was taking place, accurate temperature readings were taken on the gas in the two chambers. *A* fall in temperature was observed in all cases except for hydrogen and helium.

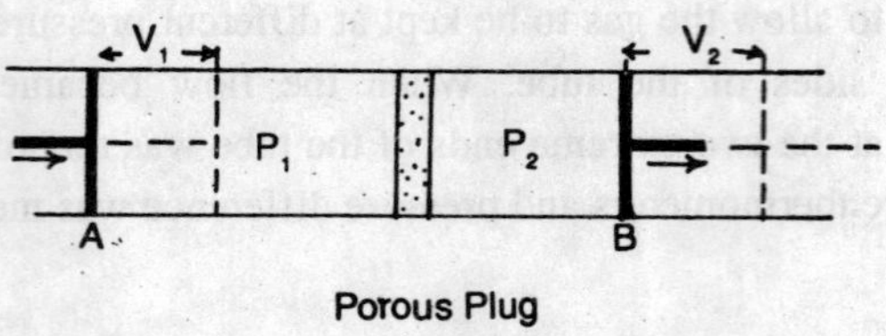

Fig. 2.8. Expansion of gas through Porous Plug (Joule-Thomson effect).

Explanation*. The work done *on* the system at the left piston is P_1V_1 and the work done *by* the system at the right piston is P_2V_2. Hence the net work done *by* the system is:

$$w = P_2V_2 - P_1V_1$$

Since the process was conducted adiabatically.

$$q = 0$$

$$\therefore \qquad \Delta E = -w$$

$$\text{or} \qquad E_2 - E_1 = -(P_2V_2 - P_1V_1)$$

---

*A simple explanation of Jole-Thomson effect based upon the fact that the molecules of the gas are moving and so possess a definite amount of kinetic energy. When the gas expands, it does some work in overcoming the Van der Walls forces of attraction existing between the molecules. Thus a part of the energy is used up in doing this work and, therefore, the energy of the system falls and as a result the temperature also falls.

or $$E_2 + P_2V_2 = E_1 + P_1V_1$$

i.e., $$H_2 = H_1$$

$$dH = 0$$

Thus *during the free expansion of real gas, the heat content or enthalpy remains constant.* Under these conditions, the extent of the temperature change is found to depend both on the initial temperature and the pressure of the gas. Joule and Thomson defined a term, μ, known as *Joule-Thomson co-efficient* as:

$$\mu = \left[\frac{\partial T}{\partial P}\right]_H$$

Thus μ *may be defined as the temperature change in degrees experienced by a gas a free expansion by a pressure drop of one atmosphere under conditions of constant enthalpy.* For cooling, μ is positive, while for an observed heating, μ is negative. The pressure at which the gas neither heats nor cools on expansion, i.e., where $\mu = 0$, is called an *Inversion Point*. This point varies with the temperature. It is observed that at a higher pressure, there is a temperature range within which μ is positive, i.e., cooling takes place. Below or above this range μ is negative, i.e., heating takes place, (see table for values of μ for N_2).

Thus at the constant pressure, there are two inversion temperatures, an upper one and a lower one. Since most of the gases at room temperature are below their upper inversion temperature, hence they undergo cooling on adiabatic expansion. Hydrogen and helium on the other hand, are still above their upper inversion temperatures and hence they show heating. It may be expected that these gases would also exhibit cooling if their temperatures were lowered below their upper inversion point at a given pressure. This is actually the case.

Relationship of Joule-Thomson Co-efficient with other Thermodynamical Quantities. As seen earlier, heat content of a system depends only upon the state of the system. Hence *d*H is a complete differential. Thus, taking pressure and temperature as the variables, we have:

Table 2.2. Joule-Thomson Co-efficient For N_2

P,	μ at					
atms.	–150°C	–100°C M	0°C	100°C	200°C	300°C
1	1.266	0.6490	0.2656	0.1292	0.0558	0.0140
20	1.125	0.5998	0.2694	0.1173	0.0472	0 0096
33.5	0.1704	0.5494	0.2377	0.1100	0.0430	0.0050
60	0.0601	0.4506	0.2088	0.0975	0.0372	0.0013
100	0.0202	0.2764	0.1679	0.0768	0.0262	0.0079
140	–0.0056	0.1373	0.1316	0.0582	0.0168	1.1122
180	–0.0211	0.0765	0.1015	0.0462	0.0094	1.0160
203	–0.0084	0.0084	0.0891	0.0419	0.0070	0.0171

$$dH = \left(\frac{\partial H}{\partial P}\right)_T dP + \left(\frac{\partial H}{\partial T}\right)_P dT$$

But $\left(\frac{\partial H}{\partial T}\right)_P$ represents C_p

Substituting the value of $\left(\frac{\partial H}{\partial T}\right)_P$ in the above equation

$$dH = \left(\frac{\partial H}{\partial P}\right)_T dP + C_p\, dT$$

Since during the process the heat content remains constant, therefore dH = 0.

$$\therefore \quad 0 = \left(\frac{\partial H}{\partial P}\right)_T dP + C_p\, dT$$

$$\therefore \quad C_p dT = -\left(\frac{\partial H}{\partial P}\right)_T dP$$

or

$$\left(\frac{\partial T}{\partial P}\right)_H = -\frac{\left(\frac{\partial H}{\partial P}\right)_T}{C_p}$$

or

$$\left(\frac{dT}{dP}\right)_H = -\frac{1}{C_p}\left(\frac{\partial H}{\partial P}\right)_T \tag{1}$$

But
$$\mu = \left(\frac{\partial T}{\partial P}\right)_H$$

$\therefore$
$$\mu = \frac{1}{C_p}\left(\frac{\partial H}{\partial P}\right)_T$$

Assuming μ to be constant over a small range of pressure, the equation may be written as

$$\frac{\Delta T}{\Delta P} = -\frac{1}{C_p}\left[\frac{\partial H}{\partial P}\right]_T$$

$$\Delta T = -\frac{1}{C_p}\left[\frac{\partial H}{\partial P}\right]_T \times \Delta P \tag{2}$$

Joule-Kelvin Effect for a Van der Waal's gas.

The Van der Waal's equation is

$$\left[P + \frac{a}{V^2}\right][V-b] = RT$$

In van der Waal's theory it is assumed that the internal pressure arising from intermolecular forces between the molecules of a a real gas is given by $\frac{a}{V^2}$ i.e.,

Internal pressure $= \frac{a}{V^2}$

where a is constant. Therefore, the internal work done against these forces from V_1 to V_2 will be given by

$$w_2 \int_{V_1}^{V_2} \frac{a}{V^2} dV = a \left[\frac{1}{V_1} - \frac{1}{V_2}\right] \quad (3)$$

We have already proved that the external work done by the gas in the Joule-Thomson experiment is P_2V_2–P_1V_1. i.e.,

$$w_1 = P_2V_2 - P_2V_1$$

so the total work done by the gas is given by

$$w = w_1 + w_2$$

or $$w = \left[P_2V_2 - P_1V_1\right] + a\left[\frac{1}{V_1} - \frac{1}{V_2}\right] \quad (4)$$

Since the system is thermally insulated, the external work must be done at the expense of its internal energy which is independent of temperature only i.e., dE = w.

Therefore, the decrease in internal energy (dE) is given by

$$dE = CV\,(T_1 - T_2)$$

where T_1 and T_2 are the temperatures on the high and low pressure sides respectively and C_v be the molar heat capacity of the gas at constant volume. When a state is approached, it therefore follows that

$$dE = w$$

$$Cv\,(T_1 - T_2) = (P_2V_2 - P_2V_1) + a\left[\frac{1}{V_1} - \frac{1}{V_2}\right]$$

$$Cv\,(T_1 - T_2) = (P_2V_2 - P_1V_2) + a\left(\frac{P_1}{RT_1} + \frac{P_2}{RT_2}\right) \quad (5)$$

$(\because P_1V_1 = RT_1$ and $P_2V_2 = RT_2)$

On multiplication and neglecting a very small fraction $\frac{ab}{V^2}$, Van der Waal's equation takes the form

$$PV = RT - \frac{a}{V} + Pb$$

$$= RT - \frac{aP}{RT} + Pb \qquad [PV = RT]$$

Thus, $$P_1V_1 = RT_1 - \frac{aP_1}{RT_1} + P_1b$$

and $$P_2V_2 = RT_2 - \frac{aP_1}{RT_2} + P_2b$$

or $$P_2V_2 - P_1V_1 = R(T_2 - T_1) + \frac{aP_1}{RT_1} - \frac{aP_2}{RT_2} + (P_2 - P_1)b$$

$$= R(T_2 - T_1) + a\left(\frac{P_1}{RT_1} - \frac{P_2}{RT_2}\right) + (P_2 - P_1)b$$

By putting this value in equation (5), we get

$$C_v(T_1 - T_2) = R(T_2 - T_1) + 2a\left(\frac{P_1}{RT_1} - \frac{P_2}{RT_2}\right) + b(P_2 - P_1) \quad (6)$$

As the fall in temperature is very small, we can put the term $\frac{2a}{RT_1}$ equal to $\frac{2a}{RT_2}$. Therefore, equation (6) takes the form

$$C_p(T_1 - T_2) = R(T_2 - T_1) + \frac{2a}{RT_1}(P_1 - P_2) + b(P_2 - P_1)$$

$$= R(T_2 - T_1) + (P_1 - P_2)\left(\frac{2a}{RT_1} - b\right)$$

or $$C_p(T_1 - T_2) = (P_1 - P_2)\left(\frac{2a}{RT_1} - b\right) \quad (7)$$

$$[\because C_p - C_v = R]$$

or $$\frac{T_1 - T_2}{P_1 - P_2} = \frac{1}{C_p}\left(\frac{2a}{RT_1} - b\right)$$

or $$\frac{\Delta T}{\Delta P} = \frac{1}{C_p}\left(\frac{2a}{RT_1} - b\right)$$

or

$$\mu = \frac{1}{C_p}\left(\frac{2a}{RT_1} - b\right) \tag{8}$$

Where μ the Joule-Thomson effect per unit pressure difference. From equation (8) it is evident that

(i) at low temperatures, $\frac{2a}{RT_1}$ > b ; μ > 0 i.e., the gas will be cooled in Joule-Thomson expansion.

(ii) at high temperature, $\frac{2a}{RT_2}$ < b ; μ < 0 i.e., the heating of the gas will take place.

(iii) at moderate temperature, $\frac{2a}{RT_2}$ = b; μ = 0 i.e., there is no change in temperature. This temperature is known as *temperature of inversion* and is usually denoted by T_i i.e.,

$$\frac{2a}{RTi} = b \quad \text{or} \quad Ti = \frac{2a}{Rb} \tag{9}$$

Therefore, if vander Waal's constant are known, T_i can be calculated.

REVERSIBLE ISOTHERMAL EXPANSION OF REAL GAS OBEYING VAN DER WAAL'S EQUATION

Work of Expansion. We have already proved that

$$dw = PdV \tag{1}$$

or

$$w = \int_{V_1}^{V_2} PdV \tag{2}$$

For *n* moles, the Van der Waal's equation is

$$P = \frac{nRT}{V - nb} - \frac{an^2}{V^2} \tag{3}$$

On substituting equation (3) in (2), we get

$$w = \int_{V_1}^{V_2} \frac{nRT}{V - nb} dV - \int_{V_1}^{V_2} \frac{an^2}{V^2} dV$$

or

$$w = nRT \left(\frac{V_1 - nb}{V_1 - nb} \right) + an^2 \left[\frac{1}{V^1} - \frac{1}{V_1} \right] \qquad (4)$$

Internal Energy. In Van der Waal's equation, the factor an^2/V^2 represents the internal pressure and $(\partial E/\partial V)_T$ represents the corresponding internal pressure, i.e.,

$$\left(\frac{\partial E}{\partial V} \right)_V = \frac{an^2}{V^2}$$

or

$$dE = \frac{an^2}{V^2} dV \qquad \text{(at constant T)} \qquad (5)$$

On integrating the above equation between suitable limits, we get

$$\int_1^2 dE = \int_{V_1}^{V_2} \frac{an^2}{V^2} dV$$

or

$$E_1 - E_1 = -an^2 \left[\frac{1}{V_2} - \frac{1}{V_2} \right] \qquad (6)$$

or

$$\Delta E = -an^2 \left[\frac{1}{V_2} - \frac{1}{V_1} \right] \qquad (7)$$

Enthalpy Change : We know

$$H = E + PV$$

For the initial state, the enthalpy is

$$H_1 = E_1 + P_1 V_1$$

and for the final state, the enthalpy is

$$H_2 = E_2 + P_2 V_2$$

Enthalpy change = $H_2 - H_1$

or
$$\Delta H = (E_2 + P_2V_2) - (E_1 + P_1V_1)$$

$$\Delta H = (E_2 - E_1) + (P_2V_2 - P_1V_1)$$

$$= \Delta E + (P_2V_2 - P_1V_1) \quad (8)$$

The van der Waal's equation for *n* moles is

$$P = \frac{nRT}{V-b} - \frac{an^2}{V^2}$$

On multiplying the above equation by V, we get

$$PV = \frac{nRTV}{V-nb} - \frac{an^2}{V}$$

For the initial state,

$$P_1V_1 = \frac{nRTV_1}{V_1 - nb} - \frac{an^2}{V_1} \quad (9)$$

for the final state,

$$P_2V_2 = \frac{nRTV_2}{V_2 - nb} - \frac{an^2}{V_2} \quad (10)$$

On substituting equations (9) and (10) in (8), we get

$$\Delta H = \Delta E + nRT\left[\frac{V_2}{V_2 - nb} - \frac{V_1}{V_1 - nb}\right] + an^2\left[\frac{1}{V_1} - \frac{1}{V_2}\right] \quad (11)$$

Here

$$\frac{V_2}{V_2 - nb} - \frac{V_1}{V_1 - nb} = \frac{V_2(V_1 - nb) - V_1(V_2 - nb)}{(V_2 - nb)(V_1 - nb)}$$

$$= \frac{nb(V_1 - V_2)}{(V_2 - nb)(V_1 - nb)} = \frac{nb}{V_2 - nb} - \frac{nb}{V_1 - nb} \quad (12)$$

On substituting equation (12) in (11), we get

$$\Delta H = \Delta E + nRT\left[\frac{nb}{V_2 - nb} - \frac{nb}{V_1 - nb}\right] + an^2\left(\frac{1}{V_1} - \frac{1}{V_2}\right)$$

We have already proved that

$$\Delta E = -an^2\left(\frac{1}{V_1} - \frac{1}{V_2}\right) \text{[see eq. (7)]}$$

$$\therefore \qquad \text{DE} = -an^2\left(\frac{1}{V_1} - \frac{1}{V_2}\right) + n\text{RT}\left[\frac{nb}{V_2 - nb} - \frac{nb}{V_1 - ab}\right]$$

$$+ an^2\left[\frac{1}{V_1} - \frac{1}{V_2}\right]$$

$$= n^2bRT\left[\frac{1}{V_2 - nb} - \frac{1}{V_1 - nb}\right] + 2an^2\left[\frac{1}{V_1} - \frac{1}{V_2}\right] \qquad (13)$$

Heat Change q : According to the first law of thermodynamics.

$$q = \Delta E = w$$

We have already proved that

$$\Delta \text{E} = -an^2\left[\frac{1}{V_1} - \frac{1}{V_2}\right] \text{ [see eq. (7)]}$$

$$w = n\text{RT} \ln\left(\frac{V_2 - nb}{V_1 - nb}\right) + an^2\left(\frac{1}{V_2} - \frac{1}{V_1}\right) \text{[see eq. (4)]}$$

$$\therefore \; q = -an^2\left[\frac{1}{V_2} - \frac{1}{V_1}\right] + \text{nRT} \ln\left(\frac{V_2 - nb}{V_1 - nb}\right) + an^2\left(\frac{1}{V_2} - \frac{1}{V_1}\right)$$

$$= n\text{RT} \ln\left[(\text{V}_2 - nb)/(\text{V}_1 - nb)\right] \qquad (14)$$

REVERSIBLE ADIABATIC EXPANSION OF REAL GAS OBEYING VAN DER WAAL'S EQUATION

When reversible adiabatic expansion of real gas takes place, there is no exchange of heat between the system and the surroundings, i.e.,

$$q = 0$$

First law of thermodynamics gives,

$$\Delta E = -w \quad (1)$$

If we know ΔE, w can be calculated.

As E is a state function of T and V, it means that

$$E = f(\text{T, V})$$

$$dE = \left(\frac{\partial E}{\partial T}\right)_V dT + \left(\frac{\partial E}{\partial T}\right)_V dV \quad (3)$$

For a van der Waal's equation,

$$\left(P + \frac{n^2 a}{V^2}\right)(\text{V} - nb) = \text{nRT}$$

$$\left(\frac{\partial E}{\partial T}\right)_V = nCv \text{ and } \left(\frac{\partial E}{\partial V}\right)_T = \frac{an^2}{V^2} \quad (4)$$

On substituting equation (4) in (3), we get

$$dE = n\,\text{C}_\text{v}d\text{T} + \frac{an^2}{V^2}dV \quad (5)$$

If it is assumed that C_v is independent of temperature T, the equation (5) can be integrated between the suitable limits to yield

$$\int_1^2 dE = n\,C_v \int_{T_1}^{T_2} dT + an^2 \int_{V_1}^{V_2} \frac{dV}{V^2}$$

or

$$\text{E}_2 - \text{E}_1 = \Delta\text{E} = n\text{C}_v(\text{T}_2 - \text{T}_1) - an^2\left[\frac{1}{V_2} - \frac{1}{V_1}\right]$$

or

$$\Delta E = n\, C_v\, (T_2 - T_1) - an^2\left[\frac{1}{V_2} - \frac{1}{V_1}\right] \quad (6)$$

When the value of ΔE calculated from equation (5) is substituted in equation (1) the value of w can be calculated.

Relation Between T and V: For a differential change in an adiabatic procss, the first law may be put as

$$dE = -PdV \quad (7)$$

$$dE = n\, C_v dT + \left(\frac{\partial E}{\partial V}\right)_V dV \quad (8)$$

For a van der Waal's equation, we know

$$\left(\frac{\partial E}{\partial V}\right)_T = \frac{a}{V_2} \text{ (for } n = 1 \text{ mole)} \quad (9)$$

and

$$P = \frac{RT}{V - b} - \frac{a}{V^2} \text{ (for } n = 1 \text{ mole)} \quad (10)$$

$$dE = CvdT + \frac{a}{V^2} dV = -\left[\frac{RT}{V-b} - \frac{a}{V^2}\right] dV$$

or

$$C_v dT = -\left(\frac{RT}{V-b}\right) dV$$

or

$$\frac{C_v}{R}\frac{dT}{T} = -\frac{dV}{V-b}$$

On integrating the above equation between the suitable limits, we get

$$\frac{Cv}{R}\int_{T_1}^{T_2} d\ln T = -\int_{V_1}^{V_2}\frac{dV}{V-b}$$

or

$$\frac{Cv}{R}\ln\frac{T_2}{T_1} = \ln\frac{(V_1 - b)}{(V_2 - b)}$$

or

$$(T_2)^{Cv/R}(V_2 - b) = (T_1)^{Cv/R}(V_1 - b) \quad (11)$$

or $$(T)^{(Cv/R)}\,(V-b) = \text{constant} \tag{12}$$

Equation (10) can be simplified to get

$$T_2 = T_1 \left(\frac{V_1 - b}{V_2 - B}\right)^{R/C_v} \tag{13}$$

SOLVED EXAMPLES

Example 1. Express the following quantities in the alternative forms specified:

(a) 2000 kW in MW (b) 002 m^3 in dm^3

(c) 10 bar in kPa (d) 1000 pF in μF

(e) 10 MJ in J (f) 1 gm/cm^3 in kg/m^3.

Ans.

(a) $2000 \text{ kW} = \dfrac{2000 \times 10^3}{10^6} = 2 \text{ MW}$

(b) $.002 \text{ m}^3 = .002 \times \left(\dfrac{10}{1}\right)^2 = 2 \text{ dm}^3$

(c) $10 \text{ bar} = 10 \times \dfrac{10^5}{10^3} = 1000 \text{ kPa}$

(d) $1000 \text{ pF} = 1000 \times \dfrac{10^{-12}}{10^{-6}} = .001 \ \mu\text{F}$

(e) $10 \text{ MJ} = 10 \times 10^6 = 1 \times 10^7 \text{J}$

(f) $1 \text{ g/cm}^3 = \dfrac{1 \times 10^{-3}}{(1/100)^3} = 1000 \text{ kg/m}^3.$

Example 2. What will be the volume occupied by 3 moles of an ideal gas at 150 kPa and 300 K. (R = 8.314 J/mol K)

Solution. $PV = nRT$ (gas equation)

$$V = \frac{nRT}{P}$$

$$= \frac{3 \times 8.314 \times 300}{150 \times 1000}$$

$$= 0.04988 \text{ m}^3.$$

[Note that the answer has been obtained in m^3 by converting pressure directly to Pa (or N/m^2), since R and T were already in base units.]

Example 3. Calculate the minimum work required to form the surface associated with a water mist having a particle radius of 10 μm? 1 dm^3 of water at 50°C is to be turned into mist at 50°C. Surface tension of water at 50°C is 63 mN/m.

Solution. $$A = \frac{3V}{r}$$

$$= \frac{3 \times 10^{-3}}{10 \times 10^{-6}}$$

$$= 300 \text{ m}^2$$

$$W = -\int_0^A \gamma dA = -\gamma A$$

$$= -63 \times 10^{-3} \times 300$$

$$= -18.9 \text{ J}.$$

Example 4. A car is moving down a hill with a velocity of 30 m/s. The driver sees a red light at the bottom of the hill when he is at a vertical height 30 m above the hill. If the weight of car and driver is 1350 kg how much energy as heat must be dissipated by the brakes [wind and other frictional effects are negligible].

Solution. $$\Delta E = Q - W$$

ΔE is made up of ΔE_k (K.E.) and ΔE_p (P.E.)

$$\Delta E_k = \frac{m\Delta u^2}{2} \quad \text{and} \quad \Delta E_p = mg\,\Delta h$$

Since in this case $W = 0$ [∵ no moving surface force]

$$\therefore \quad Q = \frac{m\Delta u^2}{2} + mg\,\Delta h$$

$$= \frac{1350 \times (0 - 30^2)}{2} + 1350 \times 9.8066\ (0\text{–}30)$$

$$= -\frac{1215000}{2} - 397167$$

$$= -1004667\ \text{J} \quad \text{or} \quad -1004.667\ \text{kJ.}$$

Example 5. One mole of a diatonic gas is isobarically heated from 27°C to 87°C. Calculate the increase of internal energy and the external work done.

Solution. $\Delta E = C_v\,(T_2 - T_1)$

$$= \frac{5}{2} R\,(T_2 - T_1)$$

$$= \frac{5}{2} \times 8.37 \times 60$$

$$= 1.256 \times 10^3\ \text{J}$$

$$W = P\,(V_2 - V_1)$$

$$= R\,(T_2 - T_1)$$

$$= 8.37 \times 60$$

$$= 502.2\ \text{J.}$$

Example 6. 14 gms of nitrogen are heated isochorically when the pressure increases from 2×10^5 N/m^2 to 2.8×10^5 N/m^2. Calculate the change in internal energy. Initial temperature is 27°C, $R = 8.4$ J/mole°K.

Solution. $\frac{P_1}{T_1} = \frac{P_2}{T_2}$ [isochoric process]

$$\therefore \quad T_2 = 300 \times \frac{2.8}{2} \times \frac{10^5}{10^5}$$

$$= 420\text{°K.}$$

Change in internal Energy $= C_v \frac{m}{M}(T_2 - T_1)$

$$= \frac{5}{2} R \times \frac{14}{28} \times 120$$

$$= 1260\ \text{J.}$$

Example 7. 2 gms of hydrogen at 127°C are adiabatically expanded so that final pressure is half the initial pressure. Calculate the final temperature and the work done. $g = 1.4$, $R = 8.4$ J/mole°K.

Solution. We know, for an adiabatic process

$$\frac{T_1\gamma}{P_1^{\gamma-1}} = \frac{T_2^{\gamma}}{P_2^{\gamma-1}}$$

$$\therefore \quad T_2 = T_1\left(\frac{P_2}{P_1}\right)^{\frac{\gamma-1}{\gamma}}$$

$$= 400 \times \left(\frac{1}{2}\right)^{\frac{.4}{1.4}}$$

$$= 328\ ^\circ\text{K}$$

$$W = n.\frac{R}{\gamma - 1}(T_1 - T_2)$$

$$= 1 \times \frac{8.4}{.4}(400 - 328)$$

$$= 1512 \text{ J.}$$

Example 8. At 500°C and between pressure 0–60 atm. μ j.t. for nitrogen can be represented by

$$\mu \text{ j.t.} = 0.0140 - 2.50 \times 10^{-4}\,\text{P}$$

Find the expected temperature drop when the gas is expanded from 60 to 20 atmosphere pressure.

Solution. By definition,

$$\mu_{J.T.} = \left(\frac{\partial T}{\partial P}\right)_H$$

$$\therefore \quad dT = \mu_{J.T.}\,dP$$

$$= (.0140 - 2.5 \times 10^{-4}\,P)dP$$

Integrating with in limits

$$\Delta T = .0140 \quad /P \Big|_{60}^{20} \quad -2.5 \times 10^{-4} \Big/ \frac{P_2}{2} \Big|_{60}^{2C}$$

$$= .0140 \times (-40) - \frac{2.5 \times 10^{-4}}{2}(20^2 - 60^2)$$
$$= -.560 + .4$$
$$= -.16^\circ C.$$

Example 9. Two moles of a Vander Walls gas at 27°C expands isothermally and reversibly. The initial volume is 20 *l* and the final volume becomes 40 *l*. Calculate *Q, W, ΔE* and *ΔH*. Given that a = 5.49 l^2—atm. $mole^{-2}$, b = .064 l $mole^{-1}$.

Solution. In case of Van der Walls gas

$$W = nRT \ln\left(\frac{V_2 - b}{V_1 - b}\right) + a\left(\frac{1}{V_2} - \frac{1}{V_1}\right)$$

$$= 2 \times .082 \times 300 \times 2.303 \log\left(\frac{40 - .064}{20 - .064}\right)$$

$$+ 5.49\left(\frac{1}{40} - \frac{1}{20}\right)$$

= 226.5 litre-atm.

$$\Delta E = -a\left(\frac{1}{V_2} - \frac{1}{V_1}\right)$$

$$= -5.49\left(\frac{1}{40} - \frac{1}{20}\right)$$

= – .135 litrs atm.

$$Q = nRT \ln \frac{V_2 - b}{V_1 - b}$$

$$= 2 \times .082 \times 300 \times 2.303 \times \log\left(\frac{40 - .064}{20 - .064}\right)$$

= 226.63 litre-atm

$$\Delta H = nRT\, b\left[\frac{1}{V_2 - b} - \frac{1}{V_1 - b}\right] - 2a\left[\frac{1}{V_2} - \frac{1}{V_1}\right]$$

$$= 2 \times .082 \times 300 \times .064\left[\frac{1}{40 - .064} - \frac{1}{20 - .064}\right]$$

$$-2 \times 5.49\left[\frac{1}{40}-\frac{1}{20}\right]$$

$= -.92$ litre-atm.

Example 10. An ideal gas whose equation of state is $PV = nRT'$ and whose internal energy is a function of T' only is used as the working substance in a carnot cycle. Show that the ideal gas temperature T' is equal to the thermodynamic temperature T.

Solution.

According to first law of thermodynamics

$$q = \Delta E + w$$

For an isothermal process (since T' is constant) $\Delta E = 0$,

$$\therefore \quad Q_1 = nRT_1' \ln \frac{V_b}{V_c}$$

Similarly

$$Q_2 = nRT_2' \ln \frac{V_c}{V_d}$$

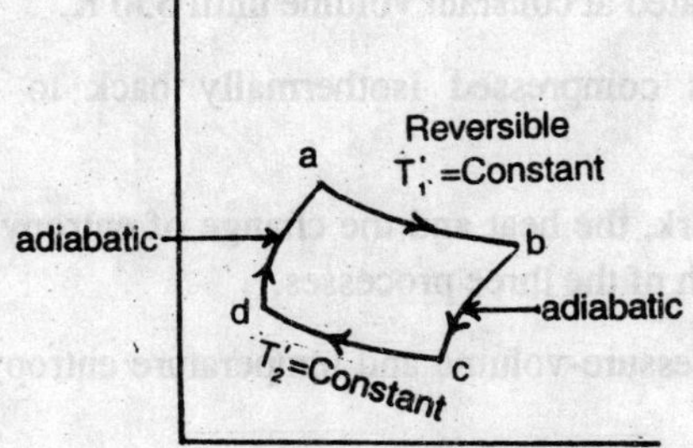

Fig. 2.9

But for any reversible adiabatic process

$$TV^{\gamma-1} = \text{constant}$$

$$\therefore \quad \frac{T_1'}{T_2'} = \left(\frac{V_c}{V_b}\right)^{\gamma-1} = \left(\frac{V_d}{V_a}\right)^{\gamma-1}$$

Hence

$$\frac{V_b}{V_a} = \frac{V_c}{V_d}$$

or $$\frac{Q_1}{Q_2} = \frac{nRT_1' \ln\left(\frac{V_b}{V_a}\right)}{nRT_2' \ln\left(\frac{V_c}{V_a}\right)} = \frac{T_1'}{T_2'}$$

Thus $\frac{T_1}{T_2} = \frac{T_1'}{T_2'}$ $\quad \therefore \frac{Q_1}{Q_2} = \frac{T_1}{T_2}$

So from above it follows that

$$T = T'$$

Hence the ideal gas temperature T' is equal to the thermodynamic temperature *T*.

Example 11. An ideal gas with a constant value of C_p = 29.6 Joules/g mol/°K is made to undergo a cycle consisting of the following reversible processes in a closed system:

(i) The gas expands adiabatically from 5×10^6 N/m^2 and 550K to 1×10^6 N/m^2.

(ii) The gas is heated at constant volume until 550 K.

(iii) The gas is compressed isothermally back to its initial condition.

Calculate the work, the heat and the change of entropy per gram mole of the gas for each of the three processes.

Solution. The pressure-volume and temperature entropy diagram are shown below.

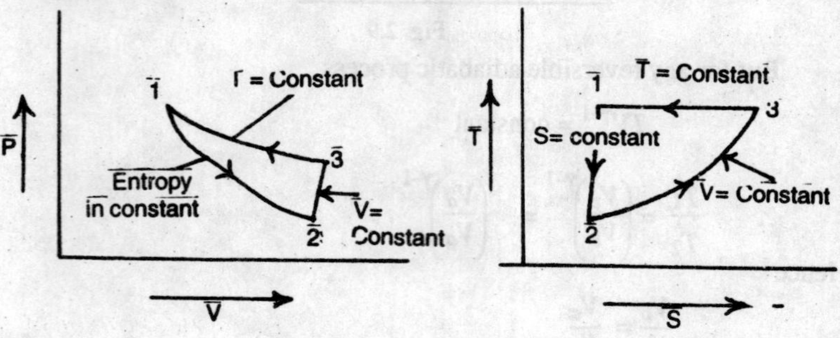

Fig. 2.10.

Given

$$p_1 = 5 \times 10^6 \text{ N/m}^2 \qquad T_1 = 550 \text{ K}$$

$$p_2 = 1 \times 10^6 \text{ N/m}^2 \qquad T_3 = 550 \text{ K}$$

We know that for an ideal gas

$$C_p - C_v = R$$

$$C_v = C_p - R = 29.6 - 8.31$$

$$= 21.3 \text{ J/(g mole) (°K)}$$

and $\gamma = \dfrac{C_p}{C_v} = \dfrac{29.6}{21.3} = 1.39$

Again, since

$$\frac{T_2}{T_1} = \left(\frac{p_2}{p_1}\right)^{\frac{\gamma-1}{\gamma}}$$

$$\therefore \qquad T_2 = T_1 \left(\frac{p_2}{p_1}\right)^{\frac{\gamma-1}{\gamma}} = 550 \left(\frac{1 \times 10^6}{5 \times 10^6}\right)^{\frac{1.39-1}{1.39}} = 350 \text{ K.}$$

From ideal gas equation, we obtain

$$\upsilon_1 = \frac{RT_1}{p_1} = \frac{8.31 \times 550}{5 \times 10^6}$$

$$= 914 \times 10^{-6} \text{ m}^2\text{/g mole}$$

and

$$\upsilon_2 = \frac{RT_2}{p_2} = \frac{8.31 \times 350}{1 \times 10^6} = 2910 \times 10^{-6} \text{ m}^3\text{/g mole}$$

Now since $\upsilon_3 = \upsilon_2$,

so $$p_3 = \frac{RT_3}{\upsilon_3} = \frac{8.31 \times 550}{2910 \times 10^{-6}} = 1.57 \times 10^6 \text{ N/m}^2$$

Thus the value of p_3 may also be obtained by

$$p_3 = p_2 \frac{T_3}{T_2} = (1 \times 10^6) \times \frac{550}{350}$$

$$= 1.57 \times 10^6 \text{ N/m}^2$$

Since the process from 1 to 2 is reversible and adiabatic, it follows that

$$q_{12} = 0 \quad \text{and} \quad \Delta S_{12} = S_2 - S_1 = 0$$

From first law of thermodynamics, we have

$$w_{12} = q_{12} - \Delta E_{12} = -\Delta E_{12} = E_1 - E_2 = C_v (T_2 - T_2)$$

$$= 21.3\,(550 - 350) = 4260 \text{ J/g mole}$$

Since the process from 1 to 2 follows the relation pv^γ = constant,

So, $$w_{12} = \int_1^2 pd_v = \frac{p_2 v_2 - p_1 v_1}{1-\gamma} = \frac{R\,(T_2 - T_1)}{1-\gamma}$$

$$= \frac{8.31\,(350 - 550)}{1 - 1.39} = 4260 \text{ J/g mole}$$

Since the process from 2 to 3 is at constant volume, it follows.

$$w_{23} = \int_2^3 pdv = 0$$

and $$q_{23} = \Delta E_{23} + w_{23} = E_3 - E_2 = C_v (T_3 - T_2)$$

$$= 21.3\,(550 - 350) = 4260 \text{ J/g mole.}$$

The entropy change ΔS_{23} is given by

$$\Delta S_{23} = S_3 - S_2 = C_v\, l_n \frac{T_3}{T_2} R\, l_v \frac{V_3}{V_2} = C_v\, l_n \frac{T_3}{T_2}$$

$$= 21.3 \ln \frac{550}{350} = 9.62 \text{ J/(g mol) (°K)}$$

For isothermal process 3 to 1, we have

$$\Delta S_{31} = S_1 - S_3 = C_p \ln \frac{T_1}{T_3} - R \ln \frac{P_1}{P_3} = -R \ln \frac{P_1}{P_3}$$

$$= -8.31 \ln \frac{5 \times 10^6}{1.57 \times 10^6} = -9.62 \text{ J/(g mole) (°K)}$$

which is equal to $-\Delta S_{23}$.

The heat rejection is given by $q_{31} = T_1 (\Delta S_{31}) = 550\ (-9.62) = -5290$ J/g mole

Since $\Delta E_{31} = 0$ (isothermal process)

$\therefore \quad w_{31} = q_{31} = -5290$ J/g mole (From first law of thermodynamics)

Since the process from 3 to 1 follows the relation $pv = RT =$ constant, the value of w_{31} can be obtained by using the formula

$$w_{31} = \int_3^1 p dv = p_1 v_1 \ln \frac{v_1}{v_3} = RT_1 \ln \frac{p_3}{p_1}$$

$$= 8.31 \times 550 \ln \frac{1.57 \times 10^6}{5 \times 10^6} = -5290 \text{ J/g mol}$$

The minus sign for q_{31} and w_{31} indicates that heat is removed from the gas and work is done as the gas.

Both the net work and the net heat of the cycle are having the same value of (4260 – 5290) = – 1030 J/g mole.

3

Thermochemistry

***The branch of physical chemistry in which we study the heat changes accompanying a chemical process is known as* thermo-chemistry.**

A chemical change is generally associated with energy change. The energy is given out or absorbed in the form of heat, electrical or photo energy.

Thermochemical equation. The chemical equations in which the heat changes involved in the process are indicated alongwith the chemical change is known as a *thermochemical equation* e.g.,

$$C_{(s)} + O_{2(g)} \rightarrow CO_{2(g)} + 39.4 \text{ kJ}$$

or

$$C_{(s)} + O_{2(g)} \rightarrow CO_{2(g)}; \quad \Delta H = -39.4 \text{ kJ}$$

The above equation tells us that when carbon burns in oxygen to form carbon dioxide, 39.4 kJ of energy are set free for each mole of carbon-dioxide formed.

Some other examples of thermochemical equations

(i) $H_{2(g)} + I_{2(g)} \rightarrow 2HI_{(g)}; \quad \Delta H = +53.6 \text{ kJ}$

(ii) $CH_{4(g)} + 2O_{2(g)} \rightarrow CO_{2(g)} + 2H_2O_{(l)}; \quad \Delta H = -890.3 \text{ kJ}.$

Conventions used in Writing Thermochemical Equations

(*a*) **Physical States of Reactants and Products.** In writing the thermochemical equations knowledge of physical states of reactants and products is essential since it influences the magnitude of heat changes involved. These are indicated in a thermochemical equation by putting symbols (g), (l), (s) after the chemical formula for the gaseous, liquid and solid states respectively.

(*b*) **Allotropic Forms of Substances.** Different quantities of heat changes are found to accompany the same chemical reaction with different allotropic forms of a substance. To give the right information it is therefore essential to name the allotropic form in a thermochemical equation, e.g. In case of sulphur. We use S_R (rhombic), S_M (Monoclinic) etc.

(c) **Reactions in Solutions.** To write the thermochemical equations for the reactions taking place in solutions, the solvent must be indicated e.g., (aq) after the formula of a substance indicates that water used as solvent is present in large excess. The excess is so large that any further dilution does not bring about any heat change.

(d) **Heat Changes Accompanying the Process.** There are two conventions to indicate the heat changes accompanying the process:

(i) *Old conventions.* The heat accompanying a process was written as a part of the equation on the product side. If heat is evolved it is represented by a (+) sign and if heat is absorbed it is indicated by (–) sign e.g.,

$$N_{2(g)} + 3\,H_{2(g)} \rightarrow 2NH_{3(g)} + 93.7 \text{ kJ}$$

$$N_{2(g)} + O_{2(g)} \rightarrow 2NO_{(g)} - 180.7 \text{ kJ}.$$

(ii) *Modern convention.* Based on thermodynamics, this convention is generally used now-a-days. Heat evolved or absorbed is indicated by ΔH, the difference in the total heat content (enthalpy) of the products and the reactants. It is written separately, (+) sign of ΔH indicates that heat is absorbed and (–) sign indicates that heat is evolved.

e.g., $$C_{(s)} + O_{2(g)} \rightarrow CO_{2(g)}; \qquad \Delta H = -39.4 \text{ kJ}$$

$$CH_{4(g)} + 2O_{2(g)} \rightarrow CO_{2(g)} + 2H_2O_{(l)}; \qquad \Delta H = -890.3 \text{ kJ}$$

(e) *Temeperature and Pressure of Substances.* Heat changes accompanying a chemical process are also dependent on the temperature and pressure of the reaction. These are also mentioned in the thermochemical equation. For reactions taking place under standard conditions i.e., 25°C and 1 atm, pressure, the heat change is written as ΔH°.

Importance Terms Used

(i) ***Heat of Reaction. The reactions which are accompanied by the evolution of heat*** i.e., in which heat is given to the surroundings are called *exothermic reactions.* In these reactions ΔH is negative i.e., heat content (enthalpy) of the reactants is greater than that of the products. So in exothermic reaction the enthalpy of the system is lowered, e.g.,

$$C_{(s)} + O_{2(g)} \rightarrow CO_{2(g)} + 393.5 \text{ kJ}$$

$$\Delta H = H_{CO_2} - (H_C + H_{O_2})$$

so the above reaction may be written as

$$C_{(s)} + O_{2(g)} \rightarrow CO_{2(g)}; \qquad \Delta H = -393.5 \text{ kJ}$$

The reactions which are accompanied by absorption of heat are called *endothermic reactions.* In endothermic reactions the enthalpy of the reactants is less than that of products, i.e., ΔH is positive, so the heat content of the system is raised in such reactions e.g.,

$$H_2O_{(l)} + 290.3 \text{ kJ} \rightarrow H_{2(g)} + O_{2(g)}$$

$$2HgO_{(s)} + 181.6 \text{ kJ} \rightarrow 2Hg_{(l)} + O_{2(g)}$$

or

$$2HgO(s) \rightarrow 2Hg_{(l)} + O_{2(g)}; \qquad \Delta H = 181.6 \text{ kJ}$$

These can be depicted graphically as in Fig. 3.1.

Let us consider a chemical system having two reactants *A* and *B* in the initial state. Suppose it undergoes a chemical reaction at a constant temperature and pressure, forming the products *C* and *D*.

$$A + B \rightarrow C + D$$

Suppose H_1 is the enthalpy of the system in the initial state (i.e.,

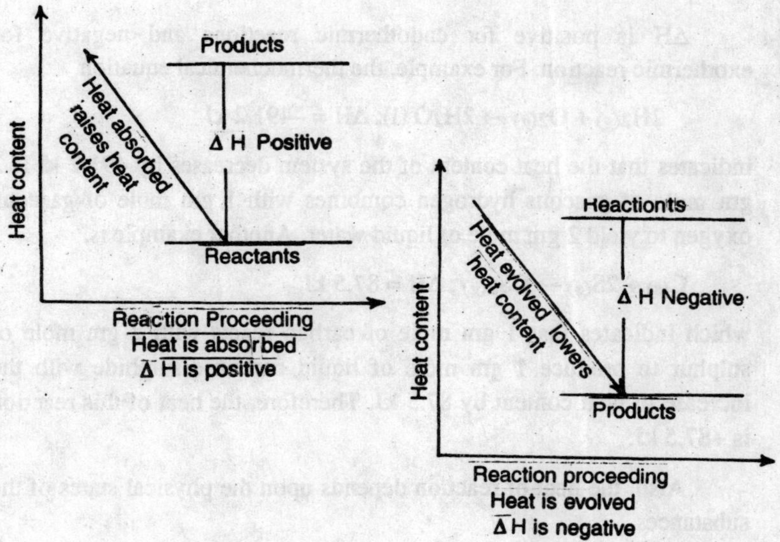

Fig. 3.1. Heat content changes in exothermic and endothermic processes.

enthalpy of reactants *A* and *B*) and H_2 is the enthalpy in the final state at the same temperature and pressure (i.e. enthalpy of products *C* and *D*). Then,

$$\Delta H = H_2 - H_1$$

Thus, at constant pressure and temperature the heat of reaction is a measure of the enthalpy change, ΔH. In simple words,

"The heat of reaction at a given temperature and pressure is the difference between the heat contents of the products and the reactants when the number of gram molecules of substances indicated by the chemical reaction have undergone complete reaction."

Mathematically, it may be put as

ΔH = Heat contents of the products—Heat contents of the reactants

= Heat contents of 1 mole of C and 1 mole of D — Heat

contents of mole of A and 1 mole of B.

ΔH is positive for endothermic reactions and negative for exothermic reaction. For example, the thermochemical equation

$$2H_{2(g)} + O_{2(g)} \rightarrow 2H_2O\ (l);\ \Delta H = -491.2\ kJ$$

indicates that the heat content of the system decreases by 491.2 kJ if 2 gm mole of gaseous hydrogen combines with 1 gm mole of gaseous oxygen to yield 2 gm mole of liquid water. Another example is,

$$C_{(s)} + 2S_{(s)} \rightarrow CS_{2(g)}\ ;\ \Delta H = 87.5\ kJ.$$

which indicates that 1 gm mole of carbon reacts with 2 gm mole of sulphur to produce 1 gm mole of liquid carbon disulphide with the increase of heat content by 87.5 kJ. Therefore, the heat of this reaction is +87.5 kJ.

Also, the heat of reaction depends upon the physical states of the substances.

Heat of reaction at constant volume (ΔE). When a reaction is being carried out at a constant volume, the heat change is called the heat of reaction at constant volume. Example is the combustion of a substance in a bomb calorimeter.

Heat of reaction at constant pressure ΔH: When pressure is kept constant throughout the reaction, the heat change is then termed as heat of reaction at constant pressure. The enthalpy change in the heat exchanged with the surroundings in the course of reaction when the reaction occurs at constant pressure.

Relationship between ΔH and ΔE. In the case of reactions involving solids and liquids, the difference between ΔH and ΔE is almost negligible. On the other hand, ΔH may not be equal to ΔE in reactions which involve gases. Let us consider the following reactions:

$$2NH_{3(g)} \rightarrow N_{2(g)} + 3H_{2(g)}$$

2moles　1 mole　3 moles

$$2H_{2(g)} + O_{2(g)} \rightarrow 2H_2O_{(g)}$$

2 moles　2 moles　1 mole

When these reactions occur at constant volume, no work will be done by or on the system as there is neither expansion nor contraction. However, if the reaction are carried out at constant pressure, then the work will be performed by the system in the first case as there is an increase in volume

$$(2NH_3 \rightarrow N_2 + 3H_2)$$

and upon the system in the second case

$$(2H_2 + O_2 \rightarrow 2H_2O)$$

as there is a decrease in volume. Consequently, the heat of reaction at constant volume (ΔE) will be more than heat of reaction at constant pressure (ΔH) when contraction takes place during the course of reaction where as ΔE will be less than ΔH if expansion takes place during the course of reaction. Let us consider the following gaseous reaction.

$$aA + bB + ... = cC + dD + ...$$

Suppose n_1 = No. of moles of gaseous reactants, i.e., a + b + ...

n_2 = No. of moles of gaseous products, i.e., c + d + ...

Then $\Delta n(g) = (c + d + ...) - (a + b + ...)$

$= (n_2 - n_1)$,

V_1 = volume of the gaseous reactants,

V_2 = volume of the gaseous products,

P = the external pressure of the gas, and

T = temperature of the system on absolute scale.

For an ideal gas,

$$PV = nRT$$

$$PV_1 = n_1RT$$

and $$PV_2 = n_2RT$$

Now, work done by the gas in expansion or contraction from volume. V_1 to V_2 in the reaction

$$aA + bB = cC + dD + ...$$

is given by

$$w = \text{pressure} \times \text{change in volume}$$

$$= P(V_2 - V_1) = PV_2 - PV_1$$

$$= n_2RT - n_1RT = RT(n_2 - n_1)$$

$$= \Delta n(g)\, RT$$

Thus, heat of reaction at constant pressure is given by, the relation, ΔH = heat of reaction at constant volume + work done by or on the gas during change of volume V_1 to V_2

$$\Delta H = \Delta E + \Delta n(g)\, RT$$

or $$\Delta H = \Delta E + 2\Delta n(g) T \quad [\because R = 2 \text{ cal/mole/degree}) \qquad (3)$$

Case I. When $n_2 > n_1$, Δn (g) is positive. Therefore,

$$\Delta H > \Delta E$$

Case II. When $n2 < n1$, $\Delta n(g)$ is negative.

Therefore, $\Delta H < \Delta E$

Case III. When there is no change in volume, $n_2 = n_1$. Therefore,

$$\Delta n(g) = 0,$$

or $$\Delta H = \Delta E$$

Factors on Which? Heat of a Reaction Depends

Various factors that effect the heat of a reaction are

(i) Physical State of Reactants and Products

Different values of heat of reaction are obtained depending upon the state of reactants and products. Even different allotropic modifications yield different quantities of heat e.g.

$$S_{(R)} + O_2(g) \rightarrow SO_2(g)\ ;\ \Delta H = -\ 302\ KJ$$

$$S_{(M)} + O_2(g) \rightarrow SO_2(g);\ \Delta H = -304.7\ KJ$$

So we observe that there is a difference in heats of reactions of the two allotropic forms of sulphur.

(ii) Temperature

The temperature at which a reaction is carried out also effects the heat of a reaction. The variation of heat of a reaction with temperature can be calculated with the help of Kirchhoff's equation.

(iii) Heat of Solution of Reactants and Products

Different values for heat of a reaction have been obtained when reactions are carried out with anhydrous reactants or in solution, e.g.,

$$H_2S_{(g)} + I_{2(g)} \rightarrow 2HI_{(g)} + S;\ \Delta H = -\ 71.9\ kJ$$

$$H_2S_{(g)} + I_2 \quad \rightarrow 2HI + S\ ;\ \Delta H_2 = -91.5\ kJ$$

(Saturated solution) (solution)

Various Types of Heats of Reactions

(i) *Heat of formation* (ΔH_f) At any given temperature it is defined as the heat change or enthalpy change when *one mole* of a given compound is formed starting from the elements in their stable state of aggregation. At 25°C the heat of formation of

$H_2O_{(l)}$, $CO_{2(g)}$ and $C_2H_{4(g)}$ are

–68.3; –94.3 and + 53.1 kcal

respectively. It means the following reactions:

$$H_{2(g)} + \frac{1}{2}O_2(g) \rightarrow H_2O_{(l)}\ \Delta H_{298} = -68.3\ kcal$$

$$= -285.8\ kJ$$

$$C_{(s)} + O_{2(g)} \rightarrow CO_{2(g)}\ \Delta H_{298} = -94.3\ kcal$$

$$= -393.5 \text{ kJ}$$

$$2C_{(s)} + 2H_{2(g)} \rightarrow C_2H_{4(g)}\ \Delta H_{298} = +53.1 \text{ kcal}$$

$$C + 2H_{2(g)} \rightarrow CH_{4(g)}\ \Delta H = -74.8 \text{ kJ}$$

(graphite)

In case the heat of formation of a compound is negative it is known as *exothermic compound* and if the heat of formation of a compound is positive it is known as an *endothermic compound.*

The stability of a compound is dependent upon its heat content. The heat content of a substance is equal to its heat of formation. The elements of their stable state of aggregation at 1 atmosphere pressure and at 298°A is assigned a zero value. A negative heat of formation means that the heat content of the compound is less than the total heat content of its element thus it will be more stable than its elements. Similarly the compounds with positive heat of formation will be associated with more energy than that of its elements and so will be less stable as compared to its elements.

From the forgoing discussion it can be concluded that more negative the heat of formation of a compound more will be its stability e.g. The heats of formation of HCl, HBr, HI are –20.06, –8.66 and +6.24 K cal respectively means that HCl is most stable of these three and HI is least stable.

All these values of the heats of formation are based on the convention of taking the enthalpy of elements in their standard state as zero and these are not absolute values of enthalpy.

(2) *Heat of Combustion.* It is defined as the enthalpy change when one mole of a compound is burnt in oxygen to give products in their stable forms e.g., the standard heat of combustion of carbon is –94.05 kcal.

$$C_{(graphite)} + O_{2(g)} \rightarrow CO_{2(g)}\ \Delta H_{298}^{\circ} = -94.05 \text{ K cal}$$

$$= -393.5 \text{ kJ}$$

$$C_2H_5OH_{(l)} + 3O_{2(g)} \rightarrow 2kO_{2(g)} + 3H_2O_{(l)}\ \Delta H = -1365.6 \text{ kJ}$$

(3) *Heat of Hydration.* For an anhydrous or partly hydrated salt it is the enthalpy change that accomapnies the combination of requisite amount of water to form a new hydrated stable salt. The heat of hydration of anhydrous copper sulphate is given by

$$\underset{\text{anhydrous}}{CuSO_{4(s)}} + 5H_2O \rightarrow \underset{\text{hydrated}}{CuSO_4.5H_2O} \; \Delta H = -18.85 \text{ k cal}$$

Enthalpy of solvation is a special case of hydration. It is defined as the enthalpy change when one more of gaseous ions became one mole of solvated (hydrated) ions e.g.,

$$Na^+ (g) \rightarrow Na^+ (aq) \qquad \Delta H = ?$$

(4) *Heat of Neutralization.* It is defined as the heat change accompanying the neutralization of one gm-equivalent of an acid (or base) in dilute aqueous solution by one gm-equivalent of a dilute aqueous solution of a base (or acid). In case of NaOH and HCl the heat of neutralization is found to be 13.7 k cal at 25°C.

In case of neutralization of a strong acid and a strong base the heat of neutralization is always found to be equal to 13.7 k cal. It can easily be explained that in all such cases, irrespective of the acid or alkali used, it is always the combination of H^+ and OH^- ions to form H_2O. Thus in such cases heat of neutralization is simply equal to the heat of formation of water.

$$A^- + H^+ + B^+ + OH^- \rightarrow B^+ + A^- + H_2O; \Delta H_2 = 13.7 \text{ kcal}$$

or $$H^+ + OH^- \rightarrow H_2O; \Delta H = -13.7 \text{ kcal}$$

$$= -57 \text{ kJ}$$

In case any one of the two i.e. acid or base is weak then heat of neutralization is less than –13.7 k cal. This is due to the fact that some heat is utilized for the dissociation of the weak acid or weak base. e.g.

$$HCl_{(aq)} + NH_4OH_{(aq)} \rightarrow NH_4Cl_{(aq)} + H_2O; \Delta H = -12.4 \text{ kcal}$$

$$= -51.5 \text{ kJ}$$

The change of [–13.7 to –12.3] heat of neutralization is due to the fact that some heat is used for dissociation of NH_4OH into NH_{4+} and

OH^- ions. The difference of 1.4 cal or (5.5 kJ) is the heat of dissociation of NH_4OH.

(5) *Heat of Ionization.* We have noticed in the heat of neutralization that whenever a weak acid or base reacts with a strong base or acid the value of heat of neutralization is less than 13.7. This is due to the fact that some heat is used for ionising the weak acid or base. This amount of heat energy involved is known as heat of ionization e.g. In case NH_4OH it has been calculated as equal to 1.4 kcal. or 5.5 kJ.

(6) *Heat of Precipitation.* The enthalpy change accompanying the formation of one mole of a precipitate is known as heat of precipitation.

$$BaCl_{2(aq)} + Na_2SO_{4(aq)} \rightarrow BaSO_{4(s)} + 2NaCl_{(aq)}$$

$$\Delta H = -5.8 \text{ kcal}$$

In this the heat of precipitation of $BaSO_4$ is –5.8 kcal.

(7) *Heat of Transition.* It is defined as the change in heat content when one mole of a substance changes from one allotropic modification to another allotropic modification e.g.,

$$S_R \rightarrow S_M; \qquad \Delta H = 80 \text{ cal.}$$

$$C_{(graphic)} \rightarrow C_{(diamond)}; \Delta H = 454 \text{ cal.}$$

(8) *Heat of Solution.* It is the amount of heat change accompanying the dissolution of one mole of a solute in such a large excess of solvent that any further dilution of the solution does not bring about any heat changes.

(9) *Heat of Dilution.* It has been observed that the heat of solution of a solute depends on its concentration or it varies with its concentration which means that further dilution of solution by a solvent is accompanied by heat changes.

Integral heat of dilution is the change of enthalpy when a solution containing one mole of solute it diluted with the addition of solvent thus changing its concentration. This integral heat of dilution is equal to the difference in integral heats of solutions at these two concentrations.

In all such cases it is customary to state the number of moles of solvent containing are mole of solute e.g.

$$HCl_{(g)} + 50\, H_2O_{(l)} \rightarrow HCl\,(50\, H_2O)\,; \qquad \Delta H = -17.51 \text{ Kcal}$$

In this example 50 moles of water (solvent) contain 1 mole of HCl.

For infinite dilution, furthur dilution is not accompanied by any heat changes, the heat change is known as integral heat of solution and is written as

$$HCl_{(g)} + aq = HCl_{(a)}\,; \qquad \Delta H = -17.96 \text{ k cal}$$

From these two equations we can calculate the integral heat of dilution of HCl(g) from a solution containing 1 mole of HCl per 50 mole of water to that containing 1 mole of HCl at infinite dilution

$$HCl\,(50\, H_2O) + aq = HCl_{(aq)} \qquad {}^{b}\Delta H = -.45 \text{ k cal}$$

Differential Heats of Dilution is defined as the heat change when 1 mole of solvent is added to a large volume of solution at specified concentration.

Differential Heat of Solution is defined as the change in enthalpy per mole of solute when an infinitesimal amount of solute is added to a solution of definite concentration. The integral and differential heats of dilution and solution can be obtained by plotting ΔH of solution at various concentrations against the number of moles of solvent associated with a definite quantity of solute.

In the figure 3.2, it is assumed that the process of dissolution is endothermic.

Integral heat of solution at a particular concentration.

$$= \frac{\Delta H}{m}$$

where ΔH is the enthalpy change and m is the number of moles of solute dissolved at a particular concentration. From this curve it can be easily seen that the value of slope depends on the concentration of solution.

The *differential heat of solution* will depend on number of moles

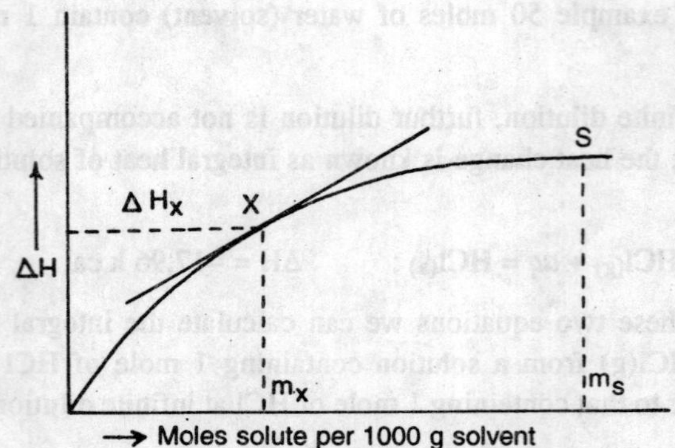

Fig. 3.2. Variation of heat of solution with the number of moles of solute per 1000 g of solvent.

of solvent and solute present in given solution in addition to temperature and pressure. It is for this reason that the concentration of solution is mentioned in defining the differential heat of solution. The slope of the curve and the shape of the curve between ΔH and *m* gives the following information:

(i) At smaller values of *m*, the curve is linear and slope has a constant value so the differential and integral heats of solutions are equal and given by $\frac{\Delta H}{m}$ at infinite dilution or very dilute solutions.

(ii) With increase in values of *m*, the slope decreases. Thus the differential heat of solution decreases and it becomes zero for a saturated solution.

Example 1. Calculate ΔH° for the following reaction

$$TiO_2(s) + 2Cl_2\,(g) \rightarrow TiCl_4\,(l) + O_2(g)$$

$$\Delta H_f\,(TiO_2,\,s) = -912.11 \text{ kJ mol}^{-1},$$

$$\Delta H_f^{\circ}\,(TiCl_4,\,l) = -750.19 \text{ kJ/mol}.$$

Solution. We know

$$\Delta H^\circ = H^\circ_{\text{(Products)}} - H^\circ_{\text{(Reactants)}}$$

So $\Delta H^\circ = (-750.19 + 0) - (-912.11 + 0)$

$= 161.92$ kJ.

[**Note.** The standard state enthalpy of formation of elements at 298 K is considered as equal to zero].

Example 2. A current of 5 A and 10 V was passed for 15 minutes through a liquid A. The amount of vapour collected is 25.6 g. If this much heat is only used to vapourise the liquid, calculate ΔH_{vap}. Given mole. wt. of A = 84.

Solution. We know

$$Q = I \times t \times V$$

$$= 5 \times 15 \times 60 \times 10$$

$$= 45000 \text{ J}$$

$$= 45 \text{ kJ}.$$

Hence $$\Delta H_{vap} = \frac{Q \times \text{mol. wt}}{\text{weight of liquid vapourised}}$$

$$= \frac{45 \times 84}{25.6}$$

$$= 147.65 \text{ kJ mol}^{-1}.$$

Example 3. The ΔH° for the reaction

$$C_2H_5OH\,(l) + 3O_2\,(s) \rightarrow 2CO_2\,(g) + 3H_2O\,(l)$$

is -1366.95 kJ. Find out the value of ΔE°.

Solution. We know

$$\Delta H^\circ = \Delta E^\circ + RT\,\Delta n$$

In this case

$-n = 2-3$ (Consider only gaseous products and reactants)

$= -1$

Thus $\Delta E = -1366.95 - (-1 \times 8.314)(298)(10^{-3})$

$= 1364.47$ kJ.

(10) Standard Enthalpy of atomisation (ΔH°). It is the enthalpy change accompanying the formation of one mole of a gaseous single atoms from elements in its standard state. e.g.

$$\frac{1}{2} H_2(g) \longrightarrow H(g)\ ;\ \Delta H = +218 \text{ kJ mol}^{-1}$$

and *not* $\quad H_2(g) \longrightarrow 2H(g)\ ;\quad \Delta H = ?$

In this case the enthalpy change will be double the enthalpy of atomisation.

In case of compounds, the **standard enthalpy of atomisation** is the enthalpy change when one mole of the compound in its standard state is split completely into single gaseous atoms.

Some times it is confused with 'reversed formation' e.g. the equation given here is *not* for the enthalpy of atomisation of methanl.

$$HCHO\ (g) \rightarrow H_2(g) + C(s) + \frac{1}{2} O_2(g)$$

Enthalpy of atomisation both for elements and compounds are already positive.

(11) Lattice Enthalpy. It is the enthalpy change, when our mole of a compound is formed from its gaseous ion.

Lattice enthalpies are always negative.

(12) Enthalpies of Fusion (ΔH°_{fers}) and vapouristion (ΔH°_{vap}). The enthalpy change when one mole of a substance changes state (i.e. from solid to liquid; liquid to gas etc.) or Heat it melts or boils, at the normal temperature of that state under standard conditions.

When it melts the enthalpy change is called Enthalpy of Fusion or *Enthalpy of Melting.*

When it boils the enthalpy change is called Enthalpy of Vapourisation.

Bond Energies (or Enthalpies)

3.7. *Bond Enthalpies (or Energies).* By heat of formation of a compound we mean the heat evolved when one mole of the substance is formed from its elements. The starting elements which go to form a mole of the compound, are taken to be in their stable forms under the conditions of the experiment. As for example, heat of formation of HI at 25°C at one atmosphere pressure is represented by the following thermochemical equation:

$$\frac{1}{2} H_2(g) + \frac{1}{2} I_2(s) \rightarrow HI(g) \; ; \Delta H = +5.95 \text{ k cal.}$$

While considering the bond formed in the molecule of HI, it is assumed that the requisite H and I free atoms are already available in the atomic gaseous state and not in their stable or molecular forms. Therefore, *the heat of bond formation between the gaseous atoms of the elements combining to form a mole of the compound as represented by* the equation:

$$H(g) + I(g) \rightarrow HI(g)$$

may not be equal to the heat of formation of HI.

When a bond is formed in a reaction in which the reactants are free atoms in the gaseous state and the resulting molecular product is also in the gaseous state, the heat **evolved is** known as the **bond formation energy** or the **bond energy or the bond enthalpy.** All bond formations are exothermic processes and, therefore, the ΔH values for them are negative. As this is a general character of bond energies, tables showing bond energies, ignore the specific use of the negative sign.

Bond energy of a given bond is the average of energies required to dissociate the said bond present in different gaseous compounds into free atoms or radicals in the gaseous state.

Bond dissociation energy. Some groups of molecules have similar dissociation energies which display an obvious trend among successive members of the group. For example, consider the diatomic molecules of the alkali metals. The bond energy of each member of the series is comparatively small and decreases as the atomic number of the alkali atom increases.

Among the hydrogen halides, the bond energies are comparatively large and again decrease as the atomic number of halogen increases.

In contrast to these smooth trends among chemically related molecules, there can be notable difference between molecules of atoms that are neat to each other in the periodic table. The dissociation energy of O_2 is only slightly over half of its neighbour N_2, but is more than three times as the dissociation energy of F_2.

We shall find that there are remarkably simple explanations for some of the relations between bond energies, while others are not understood and offer us a fine chance to use our imagination.

It is possible to define bond dissociation energy for the bonds in polyatomic molecules. The dissociation may involve fragmenting the molecule into an atom and a group of atoms, called a radical as in

$$H_2O \text{ gas} = H(g) + OH(g),\ D(H\text{–}OH) = 119.7 \text{ K cal.}$$

In other cases, the dissociation may be into two radicals

$$OH - OH(g) = 2OH(g),\ D(HO - OH) = 48 \text{ K cal.}$$

When a number of similar bonds exist in one and the same molecule as NH_3 (three N–H bonds) or CH_4 (four C–H bonds), the bond energies of the individual bonds are not necessarily identical. Thermal data gives a collective value of bonds from which only an average value of the individual bond energies can be determined.

Let us take the example of methane. The dissociating energy of C–H bond is

$$CH_4(g) = CH_3(g) + H(g)\ D(C\text{–}H) = 102 \text{ kcal.}$$

The heat of atomisation of methane:

$$CH_4(g) = C(g) + 4H(g)$$

as obtained from thermochemical data is 397.1 kcal. *The bond dissociation energies D is the enthalpy change of the reaction in which the gaseous molecule is separated into gaseous atoms.*

For example,

$H_2(g) = 2H(g)$; $\Delta H = 103.2$ kcal/mole.

From spectroscopic data, the dissociation energy in this case has been calculated to be 103.2 kcal/mole. Bond energies calculated from spectroscopic data are, in general, more reliable than that calculated from thermal data.

Bond dissociation energy may thus be defined as *the enthalpy change associated with the dissociation or cleavage of a bond.*

It is evident that the enthalpy changes in the dissociation and that in the formation of a bond are numerically equal but have opposite sign.

Usually bond dissociation energies are given in unit of kilo calories per mole of bonds broken. Table 3.1 lists the bond dissociation energies for some common diatomic molecules.

The bond energy is some what different and is the mean value of the energies required to separate four successive atoms of hydrogen atoms from methane molecule. In fact,

where 397.1 is the heat of the complete dissociation which involves breaking of four C–H bonds.

So the *bond energy* (*or enthalpy*) for any particular type of bond in a compound may be defined *as the average amount of energy required to dissociate* (i.e., break) *bonds of that type present in one molecule of the compound.* Bond energy is sometimes referred to as the *heat of formation* of the bond.

In case of diatomic molecules, however, the bond energy and the dissociation energy of a given bond must be the same.

In calculation of thermal changes in chemical process, the bond energies are useful and some of the accepted values are given in Table 3.2.

The bond enthalpies or bond energies can be used for calculating the enthalpy of a reaction and enthalpy of formation for a compound whose structural formula is known.

Table 3.1: Dissociation Energies of Diatomic Molecules (kcal/mole).

$CH_4(g) = C(g) + 4H(g)$, D(C–H) = 397.1/4 = 99.3 kcals.

Molecule	*Dissociation Energy*	*Molecule*	*Dissociation Energy*
Li_2	25	LiH	58
Na_2	17	NaH	471
K_2	12	KH	43
Rb_2	11	RbH	39
CS_2	10.4	CSH	42
N_2	37	NF	135
Cl_2	59	HBr	103
Br_2	46.1	HBr	87.4
I_2	36.1	HI	71.4
N_2	226	NO	150
C_2	119	CO	256
H_2	104		

Estimation of Heat of formation from bond energies. To calculate the heat of formation of $Se_2Cl_2(g)$, the followng sequence of reactions be used. The required equation is

$$2Se(g) + 2Cl(g) \rightarrow Se_2\,Cl_2(g) \qquad \Delta H = ?$$

$$\Delta H = [-166 \text{ k cal/mole}]$$

can be obtained by

$$Cl_{2(g)} \rightarrow 2Cl_{(g)} \qquad \Delta H = 58.15 \text{ k cal}$$

$$2Se_{(Hexagonal)} \rightarrow 2Se_{(g)} \qquad \Delta H = 2 \times 48.4 \text{ k cal}$$

or $\quad 2Se_{(Hexagonal)} + Cl_{2(g)} \quad Se_2Cl_{2(g)} \qquad \Delta H = -11 \text{ k cal}$

the value of –166 kcal is the enthalpy change for the formation of

Se_2Cl_2 gaseous molecule from $\dot{S}e_{(g)}$ and $Cl_{(g)}$ in which states these elements do not normally occur. The value of –11k cal represents the heat of formation of Se_2Cl_2 from $Se_{(Hexagonal)}$ and $Cl_{2(g)}$ which are the normal states for these elements.

Table 3.2: Bond (Enthalpies) Energies in kcal.

Bond	*Dissociation Energy*	*Bond*	*Dissociation Energy*
C–C	83.1	H–H	104.2
C=C	140.0	O–O	33.0
C=C	198.0	O=O	118.0
C–H	99.0	Cl–Cl	57.8
C–O	84.0	Br–Br	46.0
C=O	173.0	I–I	36.0
C–Cl	78.5	N≡N	225.0
C–N	69.7	O–H	110.57
N–H	93.4	Cl–H	103.0
C–Br	66.0	O–Cl	49.0

The heat of reaction values can also be calculated from the values of bond energies, e.g.

$$C_2H_4(g) + HCl(g) \rightarrow C_2H_5Cl_{(g)} \qquad \Delta H = ?$$

ΔH is calculated as

$$\Delta H = [4(C\text{–}H) + (C{=}C) + (H\text{–}Cl)]\ \}\ [-(5\ C\text{–}H) - (C\text{–}C) - (C\text{–}Cl)]$$

$$= [(C{=}C) + (H\text{–}Cl)] - [(C\text{–}H) + (C\text{–}C) + (C\text{–}Cl)]$$

$$= (140 + 103.0) - (99 + 83.1 + 78.5)$$

$$= -17.6 \text{ k cal.}$$

Bond Energies and Resonance. There is a close agreement in the values of heat of formation obtained by the use of bond energies and by other methods. However, in case of compounds having alternate single and double bonds, large deviations have been observed. e.g.,

$$C_6H_{6(g)} \rightarrow 6C_{(g)} + 6H_{(g)}$$

There is a large deviation in the experimental value than calculated from bond energy values. This deviation can be attributed to resonance due to delocalisation of electrons and in benzene there is no localisation of single and double bonds. The molecule is resonating from one from to another.

NUMERICAL PROBLEMS

Example 1. Calculate the bond energy of C=C from the following data:

(i) $2C_{(s)} + 2H_{2(g)} \rightarrow C_2H_{4(g)}$ $\Delta H = 12.5$ k cal

(ii) $2C_{(s)} \rightarrow 2C_{(g)}$ $\Delta H = 2 \times 170.9$ k cal

(iii) $2H_{2(g)} \rightarrow 4H_{(g)}$ $\Delta H = 4 \times 52.1$ k cal

Solution. Adding equation (ii) and (iii) and subtracting equation (i) from the sum

$$2C_{(g)} + 4H_{(g)} = C_2H_{4(g)} \qquad \Delta H = 537.7 \text{ kcal}$$

The energy of 4(C–H) $= 4 \times 99$

$= 396$ kcal

C = C bond energy is given by

537.7×396

$= 141.7$ k cal

Example 2. Calculate the enthalpy of the reaction

$$CH_3CHO_{(g)} + H_2(g) \rightarrow CH_3CH_2OH_{(g)}$$

Solution. In terms of structural formula the reaction may be represented as

$$
\begin{array}{c}
\quad H \;\; H \qquad\qquad\qquad\qquad\quad H \;\; H \\
\quad | \;\;\; | \qquad\qquad\qquad\qquad\quad\; | \;\;\; | \\
H—C—C=O + H–H \rightarrow H—C—C—OH \\
\quad | \qquad\qquad\qquad\qquad\qquad\; | \;\;\; | \\
\quad H \qquad\qquad\qquad\qquad\qquad H \;\; H
\end{array}
$$

Thus one C=O and one H–H bond are broken and one C–O, one O–H and one C–H bonds are formed.

$\therefore \quad \Delta H = \Delta H(C{=}O) + \Delta H(H{-}H) - \Delta H\ (C{-}C) - \Delta H(O{-}H) - \Delta H(C{-}H)$

Substituting the values from table

$\Delta H = 172 + 104.2 - 84.0 - 110.6 - 98.8$

$= 17.2$ kcal

Example 3. Calculate the enthalpy of formation of isoprene

$$
\begin{array}{c}
H_2C = CH—C = CH_2 \\
\qquad\quad | \\
\qquad\quad CH_3
\end{array}
$$

from the bond energy data given in Table 3.2. Enthalpy of sublimation of graphite is 171.70 kcal/mole.

Solution. $C_5N_8(g) = 5C(g) + 8H(g)$

Since in this change two C–C, two C = C and eight C–H bonds are formed

$\Delta H_2 = 2\Delta H\ (C{-}C) + 2\Delta H\ (C{=}C) + 8\Delta H\ (C{-}H)$

$= 2 \times 83.1 + 2 \times 140 + 8 \times 99.0$

$= 166.2 + 280 + 792 = 1238.2$ kcal.

Enthalpy of sublimation of graphite is 171.70 kcal./mole.

Now 5 C graphite = 5C(g) $\qquad \Delta H_2 = 5 \times 171.70$

$= 858.5$

$4H_2(g) = 8H \qquad \Delta H_2 = 4 \times 104.2$

$= 416.8$

For the reaction

$$5C_{(graphite)} + 4H_2(g) = C_2H_{8(g)}$$

$\therefore$ ΔH_f of isoprene $= 858.5 + 416.8 - 1238.2$

$= 37.1$ kcal.

Example 4. 1 mole of $MgSO_4$ is dissolved in water and 21.8 kcal of heat is evolved. One mole of $MgSO_4.7H_2O$ when dissolved to give a solution of same composition it absorbed 3.3 kcal of heat. Find the heat of hydration.

Solution. It is required to calculate ΔH for the reaction

$$MgSO_{4(s)} + 7H_2O_{(l)} \rightarrow MgSO_47H_2O_{(s)}$$

from the following data given

(i) $MgSO_4 + aq \rightarrow MgSO_4\ (aq)$ $\qquad \Delta H_1 = -21.8$ kcal

(ii) $MgSO_4.7H_2O + aq \rightarrow MgSO_4\ (aq)$ $\qquad \Delta H_2 = +3.3$ kcal

Equation (i) can be supposed to proceed in following two steps:

$MgSO_{4(s)} + 7H_2O_{(l)} \rightarrow MgSO_4.7H_2O_{(s)}$ $\qquad \Delta H_{hyd} = ?$

$MgSO_4.7H_2O(s) + aq \rightarrow MgSO_4(aq)$ $\qquad \Delta H_2 = 3.3$ kcal

In accordance with Hess's law

$$\Delta H_{hydration} + \Delta H_2 = \Delta H_1$$

$\therefore$ $\Delta H_{hydration} = -21.8 - 3.3$

$= -25.1$ kcal.

Example 5. The integral heat of solution of one mole of H_2SO_4 in *n* moles of water is given by the equation

$$\Delta H = -\frac{18070n}{n + 1.798} = -\text{ cal}$$

Calculate ΔH for the following processes (1) 1 mole of H_2SO_4 dissolved in 20 moles of water (2) 1 mole of H_2SO_4 dissolved in large excess of water.

$$\Delta H_1 = -\frac{18070 \times 20}{20 + 1.798} = -16580 \text{ cal}$$

$$\Delta H_2 = -\frac{18070}{1+\frac{1.798}{\alpha}}$$

$$= -18070 \text{ cal.}$$

The Laws of Thermochemistry

The two important laws are (i) The Lavoisier and Laplace's law and (ii) Hess's law:

Hess's law is more important and has a number of applications.

(i) The Lavoisier and Laplace's Law

This law was stated by Lavoisier and Laplace in the yeat 1780. According to this law.

"The total amount of heat required to decompose a compound to its elements is equal to the heat of formation of that compound".

Or

The heat of decomposition of a compound and the heat of formation of that compound are numerically equal but have opposite sign.

This law is applicable to reactions of all types. Thus a thermochemical equation can be reversed, e.g.,

$$C_{(s)} + O_{2(g)} \rightarrow CO_{2(g)} \qquad \Delta H = -99.052 \text{ kcal}$$

can be reversd as

$$CO_{2(g)} \rightarrow C_{(s)} + O_{2(g)} \qquad \Delta H = 94.052 \text{ kcal}$$

(ii) Hess's Law of Heat Summation

This law was given by G.H. Hess in 1840. According to this law "The amount of heat evolved or absorbed in a chemical change is always constant and does not depend on the path adopted". e.g. CO_2 may be obtained by any of the two processes.

Ist Process

$$C_{(s)} + O_{2(g)} \rightarrow CO_{2(g)} \qquad \Delta H = -94 \text{ kcal}$$

IInd Process

$$C_{(s)} + \frac{1}{2} O_{2(g)} \rightarrow CO_{(g)} \qquad \Delta H = -26.4 \text{ kcal}$$

$$CO_{(g)} + \frac{1}{2} O_{2(g)} \rightarrow CO_{2(g)} \qquad \Delta H = -67.6 \text{ kcal}$$

$$C_{(s)} + O_{2(g)} \rightarrow CO_{2(g)} \qquad \Delta H = -94.0 \text{ kcal}$$

Heat evolved is thus same in both the processes.

Hess's Law is a Corollary of the first law of Thermodynamics

Consider the conversion of A (Reactants) into B (Products) by two methods involving Q_1 and Q_2 kJ of energy respectively.

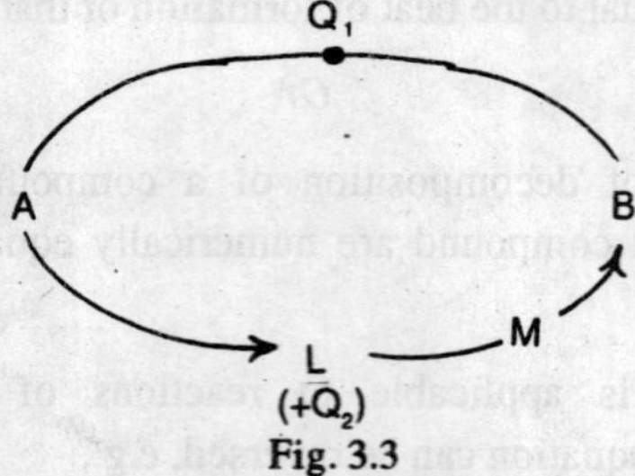

Fig. 3.3

Suppose A is changed to B by method 1 and then converted back into A by method II.

Total enthalpy change = $(Q_2 - Q_1)$ kJ. As it is a cyclic process because the initial and final states are same so there should be no energy change. Thus $Q_2 - Q_1 = 0$

or $$Q_1 = Q_2$$

Hence energy change in a chemical reaction is independent of the path adopted.

Importance of Hess's law. This law has proved to be of immence value for a large number of thermochemical calculations. The

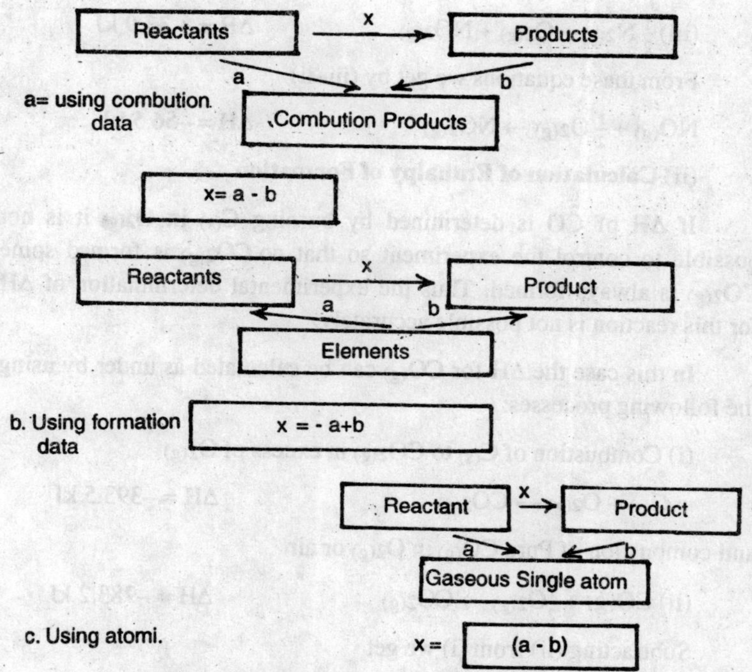

Fig. 3.4 Generalised Hess's law cycle showing three most commonly used types of enthalpy data.

experimental determination of some type of enthalpy changes are not accurate or are difficult. In all such cases this law is found useful.

Some of the important applications of this law are:

(i) *Calculation of enthalpy of combustion.* If the enthalpies of formation are known, the enthalpy of combustion of a particular substance can be calculated or predicted.

Suppose it is required to predict the enthalpy of combustion of $NO_{(g)}$. It can be represented as

(i) $NO_{(g)} + \frac{1}{2}O_{2(g)} \rightarrow NO_{2(g)}$; $\Delta H = ?$

If the enthalpies of formation of $NO_{(g)}$ and $NO_{2(g)}$ are +90.4 and 33.9 kJ mol^{-1} respectively

(ii) $\frac{1}{2}N_{2(g)} + \frac{1}{2}O_{2(g)} \rightarrow NO_{(g)}$ $\Delta H = +90.4$ kJ

(iii) $\frac{1}{2}N_{2(g)} + O_{2(g)} + NO_{2(g)}$ $\Delta H = +33.9$ kJ

From these equations we get by (iii–ii)

$NO_{(g)} + \frac{1}{2}O_{2(g)} \rightarrow NO_{2(g)}$ $\Delta H = -56.5$ kJ

(ii) **Calculation of Enthalpy of Formation**

If ΔH of CO is determined by burning $C_{(s)}$ in $O_{2(g)}$ it is not possible to control the experiment so that no.$CO_{2(g)}$ is formed some $CO_{2(g)}$ is always formed. Thus the experimental determination of ΔH for this reaction is not possible accurately.

In this case the ΔH for $CO_{(g)}$ can be calculated as under by using thé following processes:

(i) Combustion of $C_{(s)}$ to $CO_{2(g)}$ in excess of $O_{2(g)}$

$C_{(s)} + O_{2(g)} \rightarrow CO_{2(g)}$ $\Delta H = -393.5$ kJ

and combustion of Pure $CO_{(g)}$ in $O_{2(g)}$ or air

(ii) $CO(g) + \frac{1}{2}O_{2(g)} \rightarrow CO_{2(g)}$ $\Delta H = -283.2$ kJ

Subtracting (ii) from (i) we get

$C_{(s)} + \frac{1}{2}O_{2(g)} \rightarrow CO_{(g)}$ $\Delta H = -110.3$ kJ

(iii) **Calculation of Enthalpy of Transition**

When one allotropic from changes to another, there is some enthalpy change. These enthalpy changes to another, accompanying the change from one allotropic form to other is known as transition energy. Some such processes are very slow and the values of transition energy are difficult to be determined experimentally. Such values can be calculated from the enthalpies of combustion of these two allotropic modifications. e.g.,

$C_{(graphite)} \rightarrow C_{(diamond)}$ $\Delta H = ?$

(i) Combustion of $C_{(graphite)}$

$C_{(graphite)} + O_{2(g)} \rightarrow CO_{2(g)}$ $\Delta H = -393.5$ kJ

(ii) Combustion of $C_{(diamond)}$

$C_{(diamond)} + O_{2(g)} \rightarrow CO_{2(g)}$ $\Delta H = -395.0$ kJ

Subtracting (ii) from (i)

$C_{(graphite)} \rightarrow C_{(diamond)}$ $\Delta H = 1.5$ kJ

(iv) Calculation of Enthalpies of Reaction

Following example illustrates this calculation. Let us try to find ΔH for the reaction

$C_2H_{4(g)} + H_{2(g)} \rightarrow C_2H_{6(g)}$ $\Delta H = ?$

Given that enthalpies of combustion of ethylene C_2H_4, H_2 and C_2H_6 are

$-1415.4, -287.3, -1568.6$ kJ mol^{-1} respectively

Following thermochemical equations can be framed from this data

(i) $C_2H_{4(g)} + 3O_{2(g)} \rightarrow 2CO_{2(g)} + 2H_2O_{(g)}$ $\Delta H = -1415.4$ kJ

(ii) $H_{2(g)} + \frac{1}{2}O_{2(g)} \rightarrow H_2O_{(l)}$ $\Delta H = -287.3$ kJ

(iii) $C_2H_{6(g)} \frac{7}{2} O_{2(g)} \rightarrow 2CO_{2(g)} + 3H_2O_{(l)}$ $\Delta H = -1568.8$ kJ

Add (i) and (ii) and subtract (iii) to get the required equation

$C_2H_{4(g)} + H_{2(g)} \rightarrow C_2H_{6(g)}$ $\Delta H = -134.1$ kJ

Problems based on Hess's Law

Example 1. Calculate the heat of formation of acetic acid if its heat of combustion is 867 kJ mol^{-1}. The heat of formation of $CO_{2(g)}$ and $H_2O_{(l)}$ are -393.5 kJ mol^{-1} and -285.9 kJ mol^{-1} respectively.

Solution. We want ΔH for the following reaction

$2C_{(s)} + 2H_{2(g)} + O_{2(g)} \rightarrow CH_3COOH_{(l)}$ $\Delta H = ?$

From the given data the following thermochemical equations can be framed:

(i) $C_{(s)} + O_{2(g)} \rightarrow CO_{2(g)}$ $\Delta H = -393.5$ kJ

(ii) $H_{2(g)} + \frac{1}{2} O_{2(g)} \rightarrow H_2O_{(l)}$ $\Delta H = -285.9$ kJ

(iii) $CH_3COOH_{(l)} + 2O_{2(g)} \rightarrow 2CO_{2(g)} + 2H_2O_{(l)}$

$\Delta H = -867$ kJ

The required equation can be obtained by multiplying equation (i) and (ii) by 2 and then adding them and substracting (iii) from this sum.

$2C_{(s)} + 2H_{2(g)} + O_{2(g)} \rightarrow CH_3COOH_{(l)}$ $\Delta H = -491.8$ kJ.

Example 2. Calculate the enthalpy of formation of Al_2Cl_6 from the following data:

(i) $2Al_{(s)} + 6HCl_{(aq)} \rightarrow Al_2Cl_{6(aq)} + 3H_{2(g)}$ $\Delta H = -1008.0$ kJ

(ii) $H_{2(g)} + Cl_{2(g)} \rightarrow 2HCl_{(g)}$ $\Delta H = -184.8$ kJ

(iii) $HCl_{(g)} + aq \rightarrow HCl_{(aq)}$ $\Delta H = -73.5$ kJ

(iv) $Al_2C_{6(s)} + aq \rightarrow Al_2Cl_{6aq}$ $\Delta H = -645.5$ kJ

Solution. We want to find ΔH for the equation

$2Al_{(s)} + 3\ Cl_{2(g)} \rightarrow Al_2Cl_{6(s)}$ $\Delta H = ?$

From the given data this equation can be obtained by adding (i), $3 \times$ (ii), $6 \times$ (iii) and from the sum substract (iv)

$2Al(s) + 3Cl_{2(g)} \rightarrow Al_2Cl_{6(g)}$ $\Delta H = -1357.9$ kJ

Example 3. Calculate ΔH of the reaction

```
     H
     |
  H—C—Cl(g) → C(g) + 2H(g) + 2Cl(g)
     |
     Cl
```

Bond energies of C—H, C—Cl bonds are 415.0 kJ mol^{-1} and 326.0 kJ mol^{-1} respectively.

Solution. Two C—H bonds dissociation = 2×415

652 kJ " " = 830 kJ

$\therefore$ $\Delta H = 830 + 652$ = 1482 kJ mol^{-1}

Example 4. Calculate the heat of hydrogenation of C_2H_2 from the following data:

Bond enthalpies of H—H, C≡C, C—C, C—H are 435.7, 811.2, 347.5, 413.2 kJ mol^{-1}.

Solution. The chemical equation is

$$H{-}C{\equiv}C{-}H = 2H{-}H \rightarrow \begin{matrix} H & H \\ | & | \\ H{-}C & {-}C{-}H \\ | & | \\ H & H \end{matrix}$$

It involves dissociation of —C≡C— and 2 (H—H) bonds and formation of four (C—H) bonds.

Heat of hydrogenation

$$= 811.2 + 871.4 - 347.5 - 1652.8$$

$$= -317.7 \text{ kJ mol}^{-1}.$$

Example 5. The standard heats of formation at 298K for $CCl_4(g)$, $H_2O(g)$ and $HCl(g)$ are –25.5, –57.8. –94.1 and –22.1 K cal/mole respectively. Calculate ΔH_{298} for the reaction.

$$CCl_4(g) + 2H_2O(g) \rightarrow CO_2(g) + 4HCl$$

[I.I.T. 1982].

Solution. $\Delta H^\circ = \Sigma H_f^\circ{}_{\text{(Product)}} - \Sigma H_f^\circ{}_{\text{(Reactions)}}$

$\therefore$ ΔH°_{298} for the reaction

$CCl_4(g) + 2H_2O(g) \rightarrow CO_2(g) + 4HCl(g)$

$$= H_f^\circ \text{ of } CO_2(g) + 4 \times H_f^\circ \text{ of } HCl\ (g)$$

$$-2 \times H_f^\circ \text{ of } H_2O(g) - H_f^\circ \text{ of } CCl_4(g)$$

$$= -94.1 + 4 \times (-22.1) - 2 \times (-57.8) - (-25.5)$$

$$= 41.4 \text{ kcal.}$$

Example 6. The heat of neutralisation of HCl in water is –40.05k cal. Calculate the heat of formation of chloride ions.

Solution. The given reaction is

$$\frac{1}{2} H_2(g) + \frac{1}{2}Cl_2(g) + aq \rightarrow H^+ (aq) + Cl^- (aq) \quad \Delta H^\circ = -40.05 \text{ kcal}$$

$$\frac{1}{2} H_2(g) + aq \rightarrow H^+(aq) \quad \Delta H^\circ = 0$$

(By convention)

From these two equations, subtracting (ii) from (i), we get

$$\frac{1}{2} Cl_2(g) + aq \rightarrow Cl^-(aq), \quad \Delta H^\circ = -40.05 \text{ kcal}$$

Example 7. Calculate the standard heat of formation of sodium ions in aqueous solution. Given

(i) Heat of formation of NaOH (aq) at 25°C is –112.24 kcal

(ii) Heat of formation of OH^- (aq) at 25°C is –54.6 kcal.

Solution. We are given

(i) $Na(s) + \frac{1}{2}O_2(g) + \frac{1}{2}H_2(g) + aq \rightarrow Na^+ (aq) + OH^- (aq),$

$\Delta H^\circ = -112.24$ kcal

(ii) $\frac{1}{2}H_2(g) + \frac{1}{2}O_2(g) + aq \rightarrow OH^- (aq),$ $\quad \Delta H^\circ = -54.6$ kcal

Subtructing equation (ii) from (i), we get

$Na_{(s)} + aq \rightarrow Na^+ (aq)$ $\quad \Delta H^\circ = -57.64$ kcal.

Example 8. The heat of neutralisation of hydrocyanic acid by sodium hydroxide is –2.90 kcal at 25°C, calculate the heat of ionisation of HCN.

Solution. The given reaction is

(i) $HCN + OH^- \rightarrow CN^- + H_2O,$ $\quad \Delta H^\circ = -2.90$ kcal

We know that

$$H^+ + OH^- \rightarrow H_2O \quad \Delta H^\circ = -13.7 \text{ kcal}$$

Let ΔH_i be the heat of ionisation of HCN. Then

(ii) $HCN \rightarrow H^+ + CN^-,$ $\quad \Delta H^\circ = \Delta H_i$

From equations (i), (ii) and (iii), we get

$$-13.7 + \Delta H_i = -2.9$$

or $$\Delta H_i = +10.8 \text{ kcal.}$$

Example 9. Consider the reaction

$$C(\text{graphite}) + O_2(g) \rightarrow CO_2(g)$$

If $\Delta H^\circ = -393.51$ kJ mol^{-1}, C_p's are given as

$$C_p(C) = -5.293 + 58.61 \times 10^{-3}T - 42.55 \times 10^{-7} T^2$$

$$C_p(O_2) = 25.723 + 12.98 \times 10^{-3} T - 38.62 \times 10^{-7}T^2$$

$$C_p(CO_2) = 26.0 + 43.50 \times 10^{-3}T - 148.32 \times 10^{-7}T^2$$

Calculate ΔH at 1000 K.

Solution. We know that if heat capacity depends on temperature, it is given by the equation.

(i) $$\Delta H_T = \Delta H_{298} + \Delta a(T-298) + \frac{1}{2}\Delta b\,(T^2-298^2) + \frac{1}{3}\Delta c\,(T^3-298^3) + \frac{1}{4}\Delta d\,(T^4-298^4) + ...$$

(ii) $$\Delta H_T = \Delta H_{298} + \Delta a'\,(T-298) + \frac{1}{2}\Delta b'\,(T^2-298^2) - \frac{1}{3}\Delta c'\,(T^3-298^3) + ...$$

or $$\Delta H_T = A + \Delta a.T + \frac{1}{2}\Delta b.T^2 + \frac{1}{3}\Delta c.\,T^3 +$$

or $$\Delta H_T = A' + \Delta a'.T + \frac{1}{2}\Delta b'.T^2 - \frac{1}{3}\Delta c'.T^{-1}$$

(iii) where $$A = \Delta H_{298} - (\Delta a)\,298 - \frac{1}{2}(298^2)\,\Delta b - \frac{1}{3}(298^3)\,\Delta c - \frac{1}{4}\Delta d\,(298^4)$$

(iv) and $$A' = \Delta H_{298} - \Delta a'\,(298) - \frac{1}{2}\Delta b'\,(298^2) = -\frac{1}{3}\Delta c'\,(298^{-1})$$

using these equation we get from (iii)

$$A = -393510 - 5.57 \times 298 \times 1 - \frac{1}{2}(-28.09 \times 10^{-3}) \times (298)^2 - \frac{1}{3}(322.54 \times 10^{-7})\,(298)^3$$

$$= -394207.12 \text{ J.}$$

Thus $$\Delta H_{1000} = -394207.12 + (5.57)(1000) + \tfrac{1}{2}(-28.09 \times 10^{-3})(1000)^2 + \tfrac{1}{3}(322.54 \times 10^{-7})(1000)^3$$

$$= -391.38 \text{ kJ.}$$

Example 10. Calculate the heat of reaction

$$CH{\equiv}CH(g) + H_2(g) \rightarrow CH_2 = CH_2(g)$$

If the bond energies of C=C, H—H, C≡C and C—H are 170, 104, 123 and 99 kcal respectively.

Solution. In this reaction one C≡C and one H—H bonds are broken and on C=C and two C—H bonds are formed.

$\therefore$ ΔH = Bond energy of reactants
—Bond energy of products

$$= \Delta H_{C\equiv C} + \Delta H_{H-H} - \Delta H_{C=C} - 2\Delta H \quad C\text{—}H$$

$$= 123 + 104 - 170 - 2 \times 99$$

$$= -141 \text{ kcal.}$$

Example 11. Find out the bond energy of C—H bond from the following data:

$-H_f^\circ\ (CH_4) = -17.89$ kcal/mol

Heat of sublimation of graphite = 171.7 kcal/mol

Heat of dissociation of $H_2(g) = 104.18$ kcal/mol.

Solution. The data given above can be written as

$$CH_4(g) \rightarrow C(\text{graphite}) + 2H_2(g), \quad \Delta H^\circ = +17.89 \text{ kcal}$$

$$C(\text{graphite}) \rightarrow C(g), \quad \Delta H^\circ = +171.7 \text{ kcal}$$

$$2H_2(g) \rightarrow 4H(g), \quad \Delta H^\circ = 2 \times 104.18 \text{ kcal}$$

Adding these we get

$$CH_4(g) \rightarrow C(g) + 4H(g), \quad \Delta H^\circ = 397.95 \text{ kcal}$$

This much energy is the bond energy of four C—H bonds

$$\text{This C—H bond energy} = \frac{397.95}{4}$$

$$= 99.49 \text{ kcal.}$$

Example 12. The enthalpy for the following reactions (ΔH°) at 298 K are given

$$\tfrac{1}{2}H_2(g) + \tfrac{1}{2}O_2(g) \rightarrow OH(g), \qquad \Delta H^\circ = 10.06 \text{ kcal}$$

$$H_2(g) \rightarrow 2H(g), \qquad \Delta H^\circ = 104.18 \text{ kcal}$$

$$O_2(g) \rightarrow 2O(g), \qquad \Delta H^\circ = 118.32 \text{ kcal}$$

Find out O—H bond energy in hydroxyl radical.

Solution. We want the equation

$$OH(g) \rightarrow H(g) + O(g), \qquad \Delta H = ?$$

Reversing the first equation, dividing the equations 2 and 3 by two and then adding them, we get

$$OH(g) + \tfrac{1}{2}H_2(g) + \tfrac{1}{2}O_2(g) \rightarrow \tfrac{1}{2}H2(g) + \tfrac{1}{2}O_2(g)$$

$$+ H(g) + O(g)$$

$$\Delta H = \tfrac{1}{2} \times 104.18 + \tfrac{1}{2} \times 118.32 - 10.06$$

$$= 101.19 \text{ kcal}$$

Bond energy of OH bond = 101.19 kcal.

4

The Second Law of Thermodynamics

The first law of thermodynamics used two state functions E and H but it failed to provide any criterion on feasibility of processes as the state functions E and H were not sufficient to determine, whether a process is feasible or not. We come across a various processes which are exothermic and spontaneous as well as many processes which are endothermic and spontaneous, e.g.,

(i) evaporation of water

(ii) dissolution of KNO_3 or NH_4Cl in water.

(iii) decomposition of HgO, Ag_2O etc.

Thus it can be concluded that no criterion of the feasibility of a chemical or physical process is provided by the two state functions E and H. We also know that the naturally occurring spontaneous processes are irreversible so the criterion of feasibility of a process may be irreversibility. At attempt will be made to understand irreversibility of a process.

Reversible and Irreversible Processes

If a system moves from state A to state B, changes may be observed in P, V, T, E etc. If the system reverses its path from state B to state A the system completely retraces its path in the reverse order then it is called *reversible* and in such a process thermal and mechanical effects must be exactly reversed. If these quantities are not reversed in a reverse process are do not remain the same during reverse process, it is called irreversible.

Examples of Reversible Processes

(i) The isothermal, adiabatic, isochoric or isobasic compression of expansion of an ideal gas.

(ii) Conversion of liquid into vapour or solid into liquid.

(iii) Slow elongation of a spring.

(iv) Slow stretching of a wire.

Examples of Irreversible Processes

(i) Sudden expansion or compression of a gas.

(ii) Exchange of heat between two bodies at different temperatures by conduction, convection or radiation.

(iii) Diffusion of liquids and gases.

The failure of first law of thermodynamics to predict the feasibility of a process on the basis of E and H compelled us to think that some laws besides the first law of thermodynamics should form the basis of predicting the direction of certain reaction to occur or the feasibility of a reaction.

It is the second law of thermodynamics which introduces a new state function *entropy* (S) which provides a criterion for the feasibility of a process. It will be established in the following discussion that entropy of the universe (system + surroundings) increases for an irreversile process and remains constant for a reversible process.

The function entropy was established from the work of carnot, so it would be useful to discuss carnot cycle before we proceed to, second law of thermodynamics and know about new function entropy (S) which can be used to predict the feasibility of a process.

Carnot Cycle. A cycle is a process in which a system turns to its original state after a number of successive changes. In order to calculate the efficiency of any reversible cycle (or reversible heat engine), the best chosen cycle for this purpose in one known as Carnot cycle (or Carnot engine).

The Carnot cycle consists of four operations which are carried out on a system which consists of *one mole* of an ideal gas as the working substance. The gas is enclosed in a cylinder which is fitted with a weightless, frictionless piston so that all the operations in the cycle are carried out reversibly. It is supposed that there are two large heat reservoirs which remain at constant temperatures, viz. T_2 (upper) and T_1 (lower) respectively. It is further assumed that the system can be so arranged that some changes can be carried out at constant temperature (by placing the cylinder on heat reservoir) and these are known as *Isothermal changes.* For other changes, the system can be isolated in such a way that no heat can either enter or leave it (by surrounding the cylinder in a perfectly non-conducting jacket) and these are known as *adiabatic changes.*

Fig. 4.1 show the various pressure-volume changes during the cycle. The various operations are briefly described below in four steps:

Step I. *Isothermal Expansion.* The cylinder containing *one mole* of an ideal gas, occupying a volume V_1 is placed in contact with the heat reservoir at higher temperture T_2. The gas *expands isothermally* and absorbs heat q_2 from the reservoir and reversibly till its volume has increased to V_2. The path of the procss is represented by the isothermal curve AB.

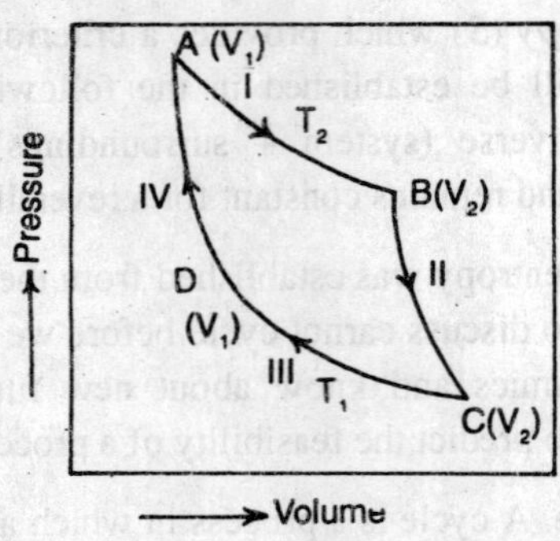

Fig. 4.1 The Carnot Cycle.

Since the gas is ideal, the work done w_1 by the system for 1 mole of the gas on the surroundings will be given by

$$w_1 = \int_{V_1}^{V_2} \text{PdV} = \int_{V_1}^{V_2} \frac{RT_2}{V} dV = RT_2 \ \ln \frac{V_2}{V_1} \qquad (1)$$

In this isothermal reversible expansion

$$\Delta E_1 = 0$$

Since according to First law of thermodynamics,

$$\Delta E_1 = q_2 - w_1$$

heat absorbed by the gas during the expansion (q_2) is equal to the work of expansion done by the gas w_1 during this step, $q_2 = w_1$.

$$\therefore \qquad q_2 = RT_2 \ \ln \frac{V_2}{V_1} \qquad (2)$$

Step II. *Adiabatic Expansion.* The cylinder is now removed from the heat reservoir at T_2 and is surrounded by the non-conducting jacket so that the gas now expands adiabaticaly and reversibly. Work is done in the expansion but since no heat enters or leaves the system the temperature must fall. The reversible expansion is continued till the temperature has fallen to T_1 which is the temperature of the heat reservoir at low temperature and the volume changes from V_2 to V_3. The path is indicated by the adiabatic curve BC in Fig. 4.1.

The work done is given by the equation as under:

$$w_2 = -C_v(T_1 - T_2)$$

$$= C_v(T_2 - T_1) \qquad (3)$$

where C, is the heat capacity of the ideal gas, assumed constant in the given temperature range.

Step III. *Isothermal Compression.* The non-conducting jacket is now removed and the cylinder is placed in contact with the heat reservoir at T_1. The gas is compressed isothermally and reversibly till the volume decreases from V_3 to V_4. The process is represented by the isothermal path CD in Fig. 4.1. In this process the system gives out heat which is taken up by the surroundings.

In this isothermal reversible compression $\Delta E = 0$.

If q_1 is the heat given out by the system to the surroundings at temperature T_1 and w_2 is the work done on the system during this operation, giving proper signs to q and w

$$-q_1 = -w_2$$

(Since according to First law of thermodynamics $\Delta E = q-w$)

$$\int_{V_3}^{V_4} P\,dV = RT_1 \int_{V_3}^{V_4} \frac{dV}{V}$$

$$\text{Now } w_3 = RT_1 \ln \frac{V_4}{V_3} \tag{4}$$

Since $V_3 > V_4$ the value of w_3 will be negative. This is true because work is done on the gas in the compression.

During this operation, an amount of heat q_1, given out by the system to the heat reservoir is exactly equal to w_3.

$$\therefore \quad q_1 = RT_1 \ln \frac{V_4}{V_3} \tag{5}$$

Step IV. *Adiabatic Compression.* The cylinder is removed from the heat reservoir and the non-conducting jacket is replaced. The gas in then compressed adiabatically and reversibly along DA till the initial state is regained. The temperature of the gas rises from T_1 to T_2. In this case work w_4 done on the system is given by

$$w_1 = C_v (T_1 - T_2)$$

$$= -C_v (T_2 - T_1) \tag{6}$$

where C_v has the same value as in equation (3).

As a result of these four steps, the system returns to its original state so that reversible cycle has been completed. The total work done w, is the sum of w_1, w_2, w_3 and w_4.

$$\therefore \quad w = w_1 + w_2 + w_3 + w_4 \tag{7}$$

$$= RT_2 \ln \frac{V_2}{V_1} + C_v (T_2 - T_1) + RT_1 \ln \frac{V_4}{V_3} - C_v (T_2 - T_1)$$

$$= RT_2 \ln \frac{V_2}{V_1} + RT_1 \ln \frac{V_4}{V_3} \tag{8}$$

For an adiabatic change in volume of a gas, the general relationship is:

$$V^{\gamma-1}\,T = \text{constant}$$

For an adiabatic expansion of gas from

V_2 to V_3 (V_2 and V_3 lie on one adiabatic curve)

$$V_2^{\gamma-1}\,T_2 = V_3^{\gamma-1}\,T_1 \qquad \text{(i)}$$

and for an adiabatic compression of gas from V_4 to V_1 (V_4 and V_1 lie on one adiabatic curve)

$$V_1^{\gamma-1}\,T_2 = V_4^{\gamma-1}\,T, \qquad \text{(ii)}$$

Dividing equation (i) by (ii)

$$\left(\frac{V_2}{V_1}\right)^{\gamma-1} = \left(\frac{V_3}{V_4}\right)^{\gamma-1}$$

or

$$\frac{V_2}{V_1} = \frac{V_3}{V_4} \quad \text{or} \quad \frac{V_1}{V_2} = \frac{V_4}{V_3}$$

Putting this value in equation (7), we get

$$w = RT_2 \ln \frac{V_2}{V_1} + RT_1 \ln \frac{V_1}{V_2}$$

$$= RT_2 \ln \frac{V_2}{V_1} - RT_1 \ln \frac{V_2}{V_1}$$

or $$w = R\,(T_2 - T_1) \ln \frac{V_2}{V_1} \qquad (9)$$

The net heat absorbed q by the ideal gas in the whole cycle is given by

$$q = \underset{\text{(heat absorbed)}}{q_2} + \underset{\text{(heat evolved)}}{(q_1)}$$

$$= RT_2 \ln \frac{V_2}{V_1} + RT_1 \ln \frac{V_4}{V_3}$$

Since $$\frac{V_4}{V_3} = \frac{V_1}{V_2}$$

$$q = RT_2 \ln \frac{V_2}{V_1} + RT_1 \ln \frac{V_1}{V_2}$$

$$\therefore \quad q = RT_2 \ln \frac{V_2}{V_1} - RT_1 \ln \frac{V_2}{V_1}$$

$$\therefore \quad q = R(T_2 - T_1) \ln \frac{V_2}{V_1} \quad (10)$$

Thus from eqns. (9) and (10),

$$q = w$$

Therefore, in a cyclic process w = q, a condition of a cyclic process i.e.,

Net work done by the system = Net heat absorbed.

The relationship between w, *net work done by the system and* q_2, *the amount of heat absorbed at the higher temperature is* obtained by dividing equation (9) by equation (2).

Thus

$$\frac{w}{q_2} = \frac{R(T_2 - T_1) \ln \frac{V_2}{V_1}}{RT_2 \ln \frac{V_2}{V_1}}$$

$$\frac{w}{q_2} = \frac{T_2 - T_1}{T_2} \quad (11)$$

Hence $\quad w = q_2 \frac{T_2 - T_1}{T_2}$

Evidently $\quad \frac{T_2 - T_1}{T_2} < 1$

Therefore $\quad w < q_2M \quad (13)$

It is clear from above that *only a part of the heat absorbed by the system at the higher temperature* T_2 *is converted into work.* The remaining heat (q_1) is transferred to the surroundings by the system when it is at a lower temperature T_1.

It may thus be concluded *that the total heat absorbed by a system cannot be completely transformed into work* as we shall see later.

A machine or engine which can convert heat into work, *absorbs* heat from the source at a higher temperature, *converts a parts of* it into useful work and *transfers* the rest of heat to the surroundings (to a sink) at a lower temperature.

Efficiency of a Heat Engine. Efficiency of a heat engine is *the fraction of the heat absorbed at the upper temperature by an engine which it can convert into work.*

$$\text{The efficiency of an engine} = \frac{\text{Work done by the engine}}{\text{Heat absorbed at upper temperature}}$$

$$\eta = \frac{w}{q_2}$$

$$\therefore \qquad \eta = \frac{w}{q_2} = \frac{T_2 - T_1}{T_2} \qquad (14)$$

Since $\frac{T_2 - T_1}{T_2}$ is less than 1, the efficiency of a heat engine is always less than unity.

From equation (14), it is clear that the efficiency of the reversible heat engine depends on the magnitude of difference in two temperatures between which the engine works and is independent of the nature of working substance. That is why when superheated steam is used in a steam engine, the efficiency of the engine is greater than when only steam is used in the engine.

The net heat absorbed q is equal to $q_2 - q_1$.

In a cyclic process net work done by the system

= net heat absorbed

$$\therefore \qquad w = q_2 - q_1 \qquad (15)$$

From equation (12),

$$w = q_2 \frac{T_2 - T_1}{T_2}$$

or

$$q_2 \times \frac{T_2 - T_1}{T_2} = q_2 - q_1$$

$$\text{or} \quad \frac{T_2 - T_1}{T_2} = \frac{q_2 - q_1}{q_2}$$

$$\text{or} \quad \text{efficiency } \eta = \frac{T_2 - T_1}{T_2} = \frac{q_2 - q_1}{q_2} \quad (16)$$

The obvious conclusion of this study is that according to First law of thermodynamics when one form of energy is converted into another, an amount of energy that disappears is equal to the energy that is produced. This law does not tell us any thing about the extent to which such a conversion occurs. The second law, as we shall see, tells us about the extent of conversion. This law conveys us that all other forms of energy are completely convertible into heat whereas heat is not completely converted into any other form of energy without affecting a permanent change in the system. This leads to the statement of second law of thermodynamics which states that it is not possible to convert heat into work without compensation.

Thus the efficiency of a heat engine operating in a Carnot cycle depends only on the two temperatures and primarily on the difference in these temperatures. To get a better efficiency keep the difference in these two temperatures large.

If T_1 and T_2 are same, the process is cyclic no work is done.

To get 100% efficiency either T_1 must be zero or T_2 must be infinity but neither of these two situations are physically realizable so the efficiency of a heat engine is always less than 100% or one. This implies that only a part of heat absorbed at a higher temperature can be converted into work.

Example 1. A carnot engine operating between 500°K and 300°K absorbs 1000 calories from the 500°K reservoir per cycle. Calculate the efficiency of the engine, heat rejected to 300°K reservoir and the work done per cycle.

Solution.

$$\text{Efficiency,} \quad \eta = \frac{T_2 - T_1}{T_2}$$

$$= \frac{500 - 300}{500} = 0.4$$

Work done = heat absorbed × efficiency = 1000 × 0.4 = 400 cal.

Heat rejected = 1000 – 400 = 600 cal.

Example 2. A reversible heat engine works between 49°C and 134°C. Find its efficiency.

Solution. Here we are given that

Temperature of the sink, $T_1 = 49°C = 49 + 273 = 322°A$

and temperature of the source, $T_2 = 134°C = 134 + 273 = 407°A$

Applying the formula,

$$\text{Efficiency,} \quad \eta = \frac{T_2 - T_1}{T_2}, \text{ we get}$$

$$\text{Efficiency,} \quad \eta = \frac{407 - 322}{407} = \frac{85}{407} = 0.2987$$

$$= 29.87\%.$$

Example 3. A reversible heat engine working between 273 K and 373 K absorbs 746 J of heat from the source. Calculate : (*a*) the work done; (*b*) the heat given to the sink; (*c*) the efficiency of the engine; (*d*) what would have been the efficiency if the temperature of the source were 473 K?

Solution. (a) Work done can be calculated by using the equation,

$$\frac{w}{q_2} = \frac{T_2 - T_1}{T_2}$$

or

$$w = \frac{T_2 - T_1}{T_2} \times q_2$$

$$= \frac{373 - 273}{373} \times 746$$

$$= 200 \text{ J.}$$

(b) To calculate heat given to sink, we use

heat given back = Heat absorbed – (heat used in doing work)

$$= 746 - 200$$

$$= 546 \text{ J.}$$

(c) efficiency $\quad \eta = \dfrac{w}{q_2} = \dfrac{T_2 - T_1}{T_2}$

$$= \frac{373 - 273}{373}$$

$$= 0.268.$$

(d) efficiency at 473 K is given by

$$\eta = \frac{w}{q_2} = \frac{T_2 - T_1}{T_2}$$

$$= \frac{473 - 273}{473}$$

$$= 0.423.$$

Example 4. A reversible carnot cycle does work equivalent to 150 kJ per cycle. If heat supplied by cycle is 225 kJ at 227°C per cycle. Find out (*a*) the temperature at which heat is rejected; and (*b*) the thermal efficiency of engine.

Solution. We know

$$\eta = \frac{w}{q}$$

$$= \frac{150}{225}$$

$$= 0.667$$

Also $\quad \eta = \dfrac{Q_1 - Q_2}{Q_1} = \dfrac{T_2 - T_1}{T_2}$

$$T_2 = 273 + 227, \qquad T_1 = ?, \qquad \eta = 0.667$$

$$= 500 \text{ K}$$

Thus, we have

$$0.667 = \frac{500 - T_1}{500}$$

or $\quad T_1 = 166.5$ K.

Second Law of Thermodynamics

To ascertain the direction of a chemical or physical process we need some more information or the knowledge of some other criteria beyond the first law. The principle which would determinee *direction* as also the *extent of a change* is provided by the second law. Like the first law, the second law is also truly a law of experience gathered from observations of nature. Any such experience which would determine the criteria for the direction for the occurrence of a change will form the basis of the second law. These experiences have been expressed in different ways by different workers. That is why we often find a large number of statements for the second law, though all of them lead to the same goal, namely indicate the direction of a process.

It is human experience that when forms of energy including mechanical work, have a natural tendency to be transformed into thermal energy, the latter shows no natural inclination to be transformed into any other form. Only through the introduction of some mechanism, or machine can we convert heat into other forms.

Even then this conversion occurs to a limited extent and not completely. Since heat is readily available, a good deal of human ingenuity is used to find out the conditions and circumstances under which it would be possible to change heat into work. The guidance as to the conditions under which heat changes into the direction of work would obviously come under the Second Law. For this reason, in enunciating the Second Law we often find the statements refer to the conversion of heat into work. It is now known that two conditions must be fulfilled to utilise heat for useful work.

(i) A mechanism, commonly called the thermodynamic engine, is essential. Without the aid of an engine the conversion of heat into work is impossible. Further, the engine must work in a reversible cyclic process.

(ii) The engine must operate between two temperatures. It will take up heat at a higher temperature, convert a portion of it into work and give up the rest of the heat to a body at a lower temperature.

Some important statements of the Second Law are :

(i) It is impossible to convert heat completely into equivalent amount of work without producing some other change in some other part of the system.

(ii) All natural and spontaneous processes take place in one direction only and are thermodynamically irreversible.

(iii) It is impossible, to take heat from a reservoir convert it into work by a cyclic process without transferring it to a cold reservoir.

(iv) Heat cannot pass by itself from cold to a hot body.

(v) It is impossible for a self-acting machine unaided by an external source to convey heat from a cold to a hot body.

It may be noted that various statements of the second law appear to be quite different. However all these state the same fundamental concept i.e., all forms of energy can be converted into heat but no process is available for converting the heat so obtained completely to other forms of energy.

The second law of thermodynamics is based on observations as there is no regorous proof. No deviations from this law are known. This law is applicable to cyclic processes only.

Carnot's Cycle in Reverse Order: Refrigerator

Carnot heat engine is reversible and if operated in reverse order the working substance absorbs heat from the sink at a lower temperature and takes it to a higher temperature source, rejecting a large quantity of heat there. As a result the mechanical energy is converted into heat energy. A machine based on this principle is known as *refrigerator.*

The Refrigerator operates in the direction ADCBA and the net effect of the cycle will be

(i) Quantity $Q_2 = nRT_2 \ln \frac{V_3}{V_4}$ is removed from the sink, and

(ii) a quantity $Q_1 = nRT_1 \ln \frac{V_2}{V_1}$ will be rejected at a higher

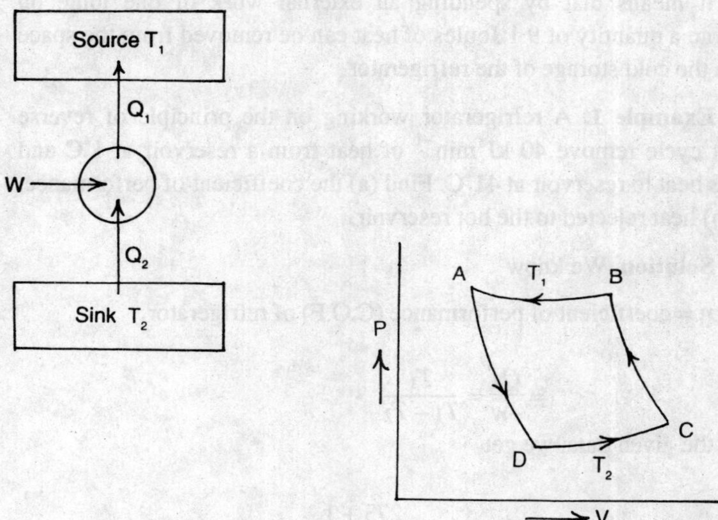

Fig. 4.2. Refrigerator.

temperature. (V_1, V_2, V_3, V_4 are volumes at A, B, C, D respectively).

The difference is the heat $(Q_1 - Q_2)$ = nR $(T_1 - T_2)$ 1n $\frac{V_2}{V_1}$ is produced by converting mechanical energy into heat energy.

Like efficiency in heat engine in case of refrigerators we define a ratio known as *coefficient of performance* (η).

It is the ratio of heat taken from sink to the external work needed to do that

$$\eta = \frac{Q_2}{Q_1 - Q_2} = \frac{Q_2}{w}$$

or
$$\eta = \frac{T_1}{T_1 - T_2}$$

Thus coefficient of performance also depends upon the two temperatures between which the refrigerator works. Generally, a refrigerator works between 0°C and room temperature (30°C) and the coefficient of performance is 9.1.

It means that by spending an external work of one joule on machine a quantity of 9.1 Joules of heat can be removed from the space inside the cold storage of the refrigerator.

Example 1. A refrigerator working on the principle of reverse carnot cycle remove 40 kJ min^{-1} of heat from a reservoir at 1°C and rejects heat to reservoir at 41°C. Find (a) the coefficient of performance, and (b) heat rejected to the hot reservoir.

Solution. We know

η = coefficient of performance (C.O.F) of refrigerator

$$= \frac{Q_2}{-w} = \frac{T_1}{T_1 - T_2}$$

using the given data, we get

$$\eta = \frac{273 + 1}{(273 + 41) - (273 + 1)}$$

$$= \frac{274}{40} = 6.85$$

or $$6.85 = \frac{40}{-(Q_1 + 40)}$$

$$\therefore \quad Q_1 = -\frac{314}{6.85} = -45.84 \text{ kJ}.$$

Example 2. Water is drawn from a well using a heat pump. The temperature of water is 15°C and that of the atmosphere is 40°C. If 10 kg of water is drawn from a depth of 12 m, how much heat is supplied to the well?

Solution. $$\eta = \frac{Q_1}{w}$$

$$= \frac{T_2}{T_2 - T_1}$$

so, $$\eta = \frac{273 + 40}{(273 + 40) - (273 + 15)}$$

$$= \frac{313}{25} = 12.52$$

Work done is given by m.g.h

so $$w = 10 \times 9.80 \times 12$$

$$= 1176 \text{ J} = 1.176 \text{ kJ}.$$

Hence that supplied $= -\eta \times w$

$$= -12.52 \times 1.176 = -14.72 \text{ kJ}.$$

Thermodynamic Scale of Temperature of Kelvin's Scale

In the measurement of temperature we use any property of a substance which depends upon, and changes continuously and uniformly with temperature. The efficiency of a heat engine depends only on the two temperatures between which the engine works and this property can be used to measure the temperatures. Kelvin worked out a theory for a scale of temperature and this is known as Kelvin's or thermodynamic scale of temperature.

Consider a carnot engine working between the temperatures θ_1 (source) and θ_2 (sink). The efficiency of such an engine is given by

$$\eta = \frac{Q_1 - Q_2}{Q_1} = f(\theta_1, \theta_2)$$

where $f(\theta_1, \theta_2)$ is a function of temperatures θ_1 and θ_2.

Simplifying it, we get

$$1 - \frac{Q_2}{Q_1} = f(\theta_1, \theta_2)$$

or $$\frac{Q_2}{Q_1} = 1 - f(\theta_1, \theta_2)$$

$$= F(Q_1, Q_2)$$

where $F(Q_1, Q_2)$ is another function of Q_1, Q_2. Q_1 and Q_2 are the quantities of heat absorbed and heat rejected respectively.

To find exact relationship between the (Q) heat involved and temperature (θ), consider three Carnot engines First working between θ_1 and θ_2, involving Q_1 and Q_2 heats. The second engine works between θ_2 and θ_3 and heats involved are Q_2 and Q_3 respectively. For the third engine temperatures are θ_1 and θ_3 and quantities of heat involved are Q_1 and Q_3 respectively. From these, we have

$$\frac{Q_2}{Q_1} = F(Q_1, Q_2)\,; \qquad \frac{Q_3}{Q_2} = F(Q_2, Q_3)$$

and

$$\frac{Q_3}{Q_1} = F(Q_1, Q_3)$$

The function F is same in all the three cases

$\frac{Q_3}{Q_1}$ can be written as

$$= \frac{Q_3}{Q_2} \times \frac{Q_2}{Q_1}$$

$$\therefore \qquad F(Q_1, Q_2) = F(Q_2, Q_3) \times F(Q_1, Q_2)$$

This relation can be obtained only, if

$$F(Q_1, Q_2) = \frac{\theta_2}{\theta_1} = \frac{Q_2}{Q_1}$$

$$F(Q_2, Q_3) = \frac{\theta_3}{\theta_2} = \frac{Q_2}{Q_2}$$

and

$$F(Q_1, Q_3) = \frac{\theta_3}{\theta_1} = \frac{Q_3}{Q_1}$$

so we conclude that quantity of heat involved α temperature on this scale.

$$Q \propto \theta \quad \text{or} \quad \frac{Q_1}{Q_2} = \frac{\theta_1}{\theta_2}$$

In Kelvin's scale temperature is taken as a quantity which is directly proportional to the quantity of heat absorbed or rejected by the working substance at that temperature.

If Q_1 is determined in the above relationship at triple point of water and take $\theta_1 = 273.16^\circ K$ then this thermodynamic scale is completely defined.

$$\theta_2 = 273.16 \left(\frac{Q_2}{Q_{273.16}} \right)$$

so we can determine θ_2 and the heat exchanged at this temperature.

Zero of Thermodynamic Scale

Zero on this scale is that temperature at which the working substance neither absorbs nor rejects heat during the change of volume or pressure. This is the temperature when the efficiency of carnot engine will be one if this temperature (absolute zero) is taken to be the temperature of the sink. It means that heat engine can convert all the heat it absorbs from the source into mechanical energy, without rejecting any part of heat of the sink.

Fixed Points and the Size of the Degree

On this Kelvin scale, the melting point of ice and the boiling point of water are taken as two fixed temperatures. The difference in these two fixed temperatures is divided into one hundred equal parts and each part is equal to one degree on this scale. So the size of degree on this scale is the same as the size of degree on the celcius scale.

The melting point of ice on this scale is taken as 273.15° and the boiling point of water as 373.15° so the zero of the thermodynamic scale will be equal to –273.15°C.

The triple point of water on this scale is 273.16° and on celcius scale it is .01°C.

Equivalence of Thermodynamic Scale with Absolute Gas Scale

On the absolute gas scale the pressure of an ideal gas becomes zero at zero degrees absolute, if it is cooled to this temperature keeping the volume constant or the volume become zero if it is cooled to this temperature keeping the pressure constant.

So absolute zero is the temperature at which kinetic energy of the molecules of a gas become zero.

In case of carnot's engine the quantity of heat absorbed or rejected at T° in absolute scale is

$$Q = nRT \ \ln \frac{V_2}{V_1} \quad \text{or} \quad Q \propto T$$

when $T = 0, \quad Q = 0$

so absolute zero is the temperature of the sink between which and any other temperature the efficiency of carnot's engine is one.

Again, the size of degrees in absolute gas scale is equal to $\frac{1}{100}$ of the difference in temperature between the melting point of ice and boiling point of water. If Q_1 and Q_2 are heats absorbed or rejected in a carnot's engine at temperature T_1° and T_2° in the absolute gas scale, we have

$$\frac{Q_1}{Q_2} = \frac{T_1}{T_2}$$

By comparison, we can see that the zero of two scales coincide, the size of the degree is same and the quantities of heat involved are in the ratio

$$\frac{Q_1}{Q_2} = \frac{T_1}{T_2} = \frac{\theta_1}{\theta_2}$$

Thus, it can be easily concluded that the two scales are identical, though they are based on different principles.

5

Entropy

Carnot's Theorem. It states that "All reversible heat engines working between the same temperature limits have the same efficiency".

It means that the efficiency of a reversible heat engine depends only on the temperature of sink and source and is independent of the nature of the working substance as well as of the mode of the machine.

Derivation. It can be proved indirectly.

Consider two reversible heat engines A and B working between the same temperature limits T_1 and T_2 and having different efficiencies.

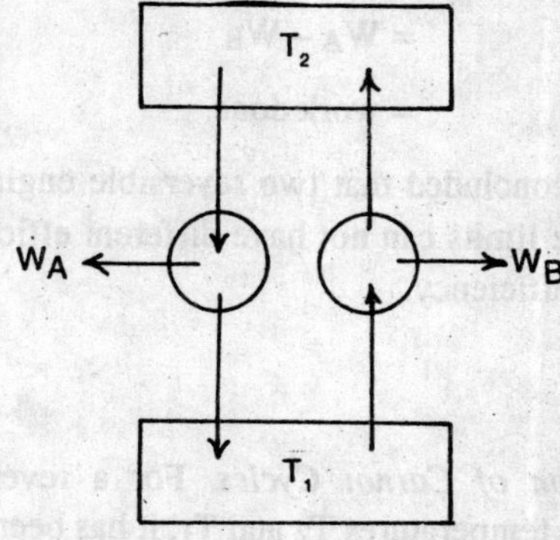

Fig. 5.1.

Engine 'A' takes up the heat Q from the source at T_1 and convert a part of it into work WA and reject the remaining into sink at temperature T_2. Heat rejected $Q_A = Q - W_A$.

In the same way 'B' takes up heat from the source at T_2, convert a part into work W_B and the rest is rejected to the sink at temperature T_1. Heat rejected $Q_B = Q - W_B$.

Suppose efficiency of 'B' is less than the efficiency of 'A'. In that case W_B is less than W_A so Q_B is more than Q_A.

If these two engines are coupled such that 'A' works in forward direction where as 'B' works in reverse direction. This could be achieved since both 'A' and 'B' are reversible. In one complete cycle of such a couple the following heat changes will occur

	'A'	'B'
Heat transferred at T_2	$+Q$	$-Q$
Work done	$+W_A$	$-W_B$
Heat transferred at T_1	$-Q_A$	$+Q_B$

Thus, we find

Heat obtained at $T_1 = Q_B - Q_A$

Work done $= W_A - W_B$

Heat absorbed $= Q - W_B - Q + W_A$

$= W_A - W_B$

= work done.

Thus it is concluded that two reversible engines working within same temperature limits can not have different efficiencies or that they must have same efficiency.

Entropy

Combination of Carnot Cycles. For a reversible carnot cycle working between temperatures T_2 and T_1, it has been seen that

$$\eta \text{ (efficiency)} = \frac{q_2 - q_1}{q_2} = \frac{T_2 - T_1}{T_2}$$

$$1 - \frac{q_1}{q_2} = 1 - \frac{T_1}{T_2} \text{ or } \frac{q_1}{q_2} = \frac{T_1}{T_2} \text{ or } \frac{q_1}{T_1} = \frac{q_2}{T_2}$$

where q_2 is the heat absorbed (+ve) at the temperature T_2 and q_1 is the heat evolved (– ve) at the temperature T_1.

$$\therefore \qquad -\frac{q_1}{T_1} = \frac{q_2}{T_2} \text{ or } \frac{q_1}{T_1} + \frac{q_2}{T_2} = 0.$$

Thus for a Carnot cycle working reversibly between temperature T_1 and T_2, the sum of $\frac{q}{T}$ terms is zero.

or $\Sigma \frac{q}{T} = 0$, where summation [S] stands for two terms of the Carnot cycle.

This will be clear from the discussion below:

Consider any reversible say ABA. It may be regarded as being made up of a number of Carnot cycles drawn across the diagram as shown in Fig. 5.2. Starting at A and going through all the cycles successively from A to B, it can be shown that all paths inside the

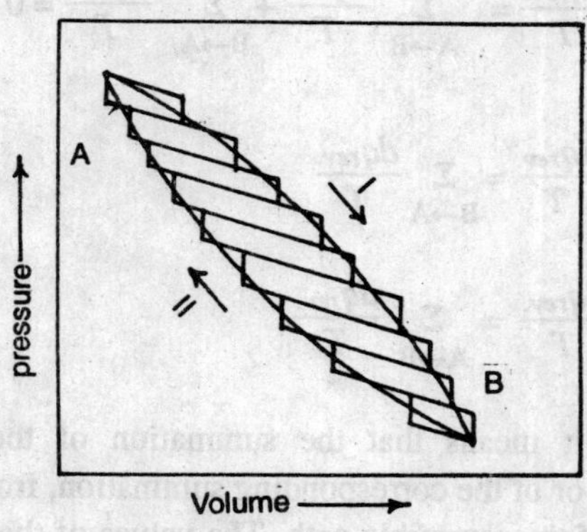

Fig. 5.2. Carnot cycles drawn across the reversible cycle.

closed curve ABA cancel each other leaving only the zig-zag outer path. This zig-zag path is almost the same as the path ABA of the reversible cycle. By making each Carnot cycle smaller and increasing their number, it can be made to correspond more closely to ABA. The reversible cycle can, therefore, be arranged as being made up of an infinite number of Carnot cycles. Hence for all the Carnot cycles, equivalent to the general reversible cycle ABA, it follows that

$$\sum_{\text{cycle}} \frac{q_{rev}}{T} = 0$$

Since an infinite number of small Carnot cycles are required to duplicate the process ABA, it is convenient to write the above equation in the form:

$$\sum_{\text{cycle}} \frac{dq_{rev}}{T} = 0$$

where $dq_{rev.}$ is the infinitesimally small quantity of heat absorbed at temperature T in each of the small isothermal, reversible changes which makes up the reversible cycle ABA.

Definition of Entropy. The summation above, $\Sigma \frac{q_{rev.}}{T} = 0$, applicable to complete (reversible) cycle ABA may be divided in two parts, one for the path A to B and the other back from B to A, Thus:

$$\sum_{\text{cycle}} \frac{dq_{rev.}}{T} = \sum_{A \to B} \frac{dq_{rev.}}{T} + \sum_{B \to A} \frac{dq_{rev.}}{T} = 0$$

or

$$\sum_{A \to B} \frac{dq_{rev.}}{T} = \sum_{B \to A} \frac{dq_{rev.}}{T}$$

or

$$\sum_{A \to B} \frac{dq_{rev.}}{T} = \sum_{A \to B} \frac{dq_{rev.}}{T}$$

Therefore, it means that the summation of the $\frac{dq_{rev.}}{T}$ terms between A and B or of the corresponding summation, from B to A must be independent of the reversible path. The values of these summations are thus determined by states A and B, *i.e.*, by the pressure, volume and temperature and independent of the manner in which the system was brought to these states. It is possible to express the value of each summation in terms of function, S, which depends only on the state of the system. Thus:

$$\sum_{A \to B} \frac{dq_{rev.}}{T} = S_A - S_B = \Delta S$$

or $$\frac{q_{rev.}}{T} = \Delta S$$

where S_A and S_B are the values of the functions at state A and state B respectively.

In the above equation, ΔS is the total change in the functions S, when the state changes from A to B and it is clearly the sum of $\frac{dq_{rev}}{T}$ terms between A and B. Hence for each of these small stages of process, we can write,

$$dS = \frac{dq_{rev.}}{T}$$

where dS is the infinitesimal increase in the function S.

The function or the quantity S is a known as **Entropy.** However the term entropy cannot be easily defined directly and so it is best defined in terms of "increase of entropy" accompanying the process.

"The increase of entropy (ΔS) accompanying a finite reversible process is defined as the quantity of heat ($q_{rev.}$) absorbed isothermally and reversibly divided by the absolute temperature (T) at which heat is absorbed."

Entropy is usually expressed in terms of calories per degree per given amount sometimes called as entropy units. [The quantity *cal. per degree* is called an *entropy unit* (*e.u.*)].

Example 1. Heat transfer takes place from a reservoir at 553 K to another reservoir at 278 K. A quantity of heat 100 kJ is transferred. What will be the change in entropy?

Solution. $T_1 = 553$ K $T_2 = 278$ K

$Q_1 = -100$ kJ $Q_2 = +100$ kJ

According to definition,

$$\Delta S = \frac{Q}{T}$$

$\therefore$ $$\Delta S_1 = \frac{-100}{553}$$

$$= -.181 \text{ kJ/K}$$

and $$\Delta S_2 = \frac{100}{278}$$

$$= .360 \text{ kJ/K}$$

$\therefore$ $$\Delta S_{Total} = \Delta S_1 + \Delta S_2$$

$$= -.181 + .360$$

$$= .179 \text{ kJ/K}$$

Entropy is a State Function

The quantity $q_{rev.}$ used in the definition of entropy is not a state function as it depends on the path adopted to bring about the change. The change can be brought about in a number of ways and the process could be accomplished reversibly in each way. The values of $q_{rev.}$ will however be different in each of these processes. Thus $\delta q_{rev.}$ is not a perfect differential and can not be written as $dq_{rev.}$ But $\frac{\delta q_{rev.}}{T}$ is a state function and is a perfect differential. This was deduced by Clausius who gave this function the name '*Entropy*'. $\frac{\delta q_{rev.}}{T}$ is a perfect differential and a state function can be shown as under.

Suppose 1 mole of an ideal gas is undergoing a reversible expansion, From the first law of thermodynamics,

$$dq_{rev.} = dE + PdV$$

$$= C_v dT + PdV$$

or $$\int dq_{rev.} = \int C_v dT = \int \frac{RT}{V} dV \qquad \left(\because P = \frac{RT}{V} \right)$$

We know that $dq_{rev.}$ is not an exact differential, hence it cannot be integrated. This is confirmed by observing the right-hand side of the above equation where the quantity $\int \frac{RT}{V} dV$ cannot be evaluated unless we specify the path, or in other words, the relation between T and V is known. For different values of T, V will have different values. Thus

$dq_{rev.}$ depends on the path by which the expansion is carried out. Now dividing the above equation by T on both the sides, we get

$$\int \frac{dq_{rev.}}{T} = \int C_v \frac{dT}{T} + \frac{RT}{V} \int \frac{1}{T} dV$$

$$= \int C_v \frac{dT}{T} + \int \frac{RdV}{V}$$

$$= C_v \ln T + R \ln V + \text{constant.}$$

It follows that $\int \frac{dq_{rev.}}{T}$ can be evaluated, hence $\frac{dq_{rev}}{T}$ is an exact differential. It may, therefore, be concluded that entropy change dS is an exact or perfect differential and hence entropy S is a state function.

Entropy Changes and Reversibility of a Process

A. Entropy Changes in Reversible Processes. To evaluate the total entropy changes of the universe, the entropy change of the system as well as of the surroundings should be considered. When a system gains heat, the surroundings lose heat and hence their entropies also change. The total change in entropy will, therefore, be equal to the algebraic sum of entropy changes of the system and the surroundings, i.e., $\Delta S = \Delta S_{sys.} + \Delta S_{sur.}$ Consider an isothermal reversible expansion of an ideal gas. The gas absorbs $q_{rev.}$ heat at temperature T. The entropy change of the gas, *i.e.*, system is given by

$$\Delta S_{sys.} = \frac{q_{rev.}}{T}$$

Since expansion is reversible, the system is equilibrium with its surroundings at all times during the expansion. Hence the surrounding lose an amount of heat equal to $q_{rev.}$ or we can say that the surroundings absorbs $-q_{rev.}$ of heat. The entropy changes of the surroundings is given by

$$\Delta S_{sur.} = -\frac{q_{rev.}}{T}$$

∴ Total entropy change of the universe;

$$\Delta S = \Delta S_{says.} + \Delta S_{sur}$$

$$= \frac{q_{rev.}}{T} - \frac{q_{rev.}}{T}$$

or $\Delta S = 0.$

It is concluded that *for a reversible process the total entropy change of the universe is* **zero.**

B. Entropy Changes in Irreversible Processes. Suppose an irreversible process is occurring at temperature T and let $q_{irrev.}$ be the amount of heat absorbed by the system from its surroundings. The total entropy change of the system $\Delta S_{says.}$ will be greater than the quantity $\frac{q_{irrev.}}{T}$ This can be shown as follows:

Consider a system changes from state 1 to state 2 once reversibly and again irreversibly. From the first law of thermodynamics, the heat changes for the two processes may be written as

$$q_{rev.} = \Delta E + w_{rev.} \quad \text{(i)}$$

$$q_{irrev.} = \Delta E + w_{irrev.} \quad \text{(ii)}$$

Since the initial and final states are the same; subtracting Eqn. (ii) from Eqn. (i), we obtain

$$q_{rev.} - q_{irrev.} = w_{rev.} - w_{irrev.}$$

We know that between the same two states the work done by the system in a reversible process is greater than the work done in irreversible process, i.e.,

$$w_{rev.} > w_{irrev.}, \text{ i.e., } w_{rev.} - w_{irrev.} > 0$$

Hence $\quad q_{rev.} - q_{irrev.} > 0$

or $\quad q_{rev.} > q_{irrev.} > 0$

Dividing both sides by T, we get

$$\frac{q_{rev.}}{T} > \frac{q_{irrev.}}{T}$$

Since $\quad \frac{q_{rev.}}{T} = \Delta S_{sys.}$

$$\therefore \quad \Delta S_{sys.} > \frac{q_{irrev.}}{T}$$

Now we shall calculate the total entropy change of the system in an irreversible process.

When $q_{irrev.}$ amount of heat is absorbed by the system from its surroundings to carry out the process, the entropy increase of the system $\Delta S_{sys.}$ will be same whether the process is carried out reversibly or irreversibly, because the entropy is a state property. ΔS_{sys}, is determined not by the irreversible change and it is as it would have been for a reversible change. Value of $q_{irrev.}$ can very from $q = 0$ to $q = q_{rev.}$ depending on how the actual change occurs.

However, the entropy change of the surrounding does not depend on the change in the state of the system and vary with the process by which the change is brought about. The surroundings have to provide just the quantity q of heat as is needed during expansion of the gaseous system by any specified process. The entropy change of the surroundings may thus be calculated by supposing that the initial state of the surroundings can be restored by supplying $q_{irrev.}$ amount of heat in a reversible manner at temperature T, i.e., by increasing the entropy of the surroundings by the amount $\frac{q_{irrev.}}{T}$.

In the process under consideration the entropy change of the surroundings $\Delta S_{sur.}$ will be numerically equal but opposite in sign of the above value *i.e.*,

$$\Delta S_{sur.} = \frac{q_{irrev.}}{T}.$$

$\therefore$ The total entropy change of the universe.

$$\Delta S = \Delta S_{sys.} + \Delta S_{sur.}$$

$$= S_{sys.} - \frac{q_{irrev.}}{T}$$

Since $\Delta S_{sys.} > \frac{q_{irrev.}}{T}$

$\therefore$ $\Delta S > 0.$

Thus *in an irreversible process, the entropy of the system increases.*

From the above discussion, we conclude that the entropy of the system and its surroundings taken together increases in a thermodynamically irreversible process at constant pressure but it remains constant in a thermodynamically reversible process.

This provides us with a criterion of distinction between an irrversible and a reversible process.

Mathematically,

$\Delta S_{sys.} + \Delta S_{sur.} = 0$ (*for a reversible process*)

and $\Delta S_{sys.} + \Delta S_{sur.} > 0$ (*for irreversible process*)

$$\Delta S_{sys.} + \Delta S_{sur.}\ 0 \geq$$

Hence 'equal to' sign refers to a reversible process and the 'greater than' sign refers to an irreversible process.

These conclusions are quite significant as these help us to predict whether a given process is thermodynamically feasible or not i.e., whether it can take place spontaneously or not.

It is interesting to note that since all the natural spontaneous processes (such as, the transfer of heat from a hot to a colder body) are irreversible and hence they must be associated with a net increase of entropy.

The result about the spontaneous processs is so important that it is sometimes taken as the definition of the **Second law of thermodynamics.**

Hence the second law may be stated as:

"*All spontaneous processes are accompanied by a net increase of entropy.*"

Or

"*The net entropy of the universe tends to increase.*"

Clausius summed up the statements of the first and second laws of thermodynamic as below:

The **energy** *of the universe remains constant; the* **entropy** *of the* *universe tends towards a maximum.*

Physical Significance of Entropy. It is appropriate to refer to an interesting aspect of entropy which throws some light on the physical significance of this apparently theoretical property. An examination of the various processes which take place spontaneously, and which are accompanied by a net increase of entropy, shows that they are associated with an increased randomness of distribution. Entropy has been regarded as a measure of *disorder or randomness of a system.* Thus, when a system goes from a more orderly to less orderly state, there is an increase in its randomness and hence entropy of the system increases. Conversely, if the change is one in which there is an increase in orderliness there is decrease in entropy. For example, when a solid changes to a liquid, an increase in entropy takes place because with the breaking of the orderly arrangement of the molecules in the crystal to the less orderly liquid state, the randomness increases. Conversely, the process of solidification brings about more orderly state and consequently there is a decrease in entropy. The process of vaporisation produces an increase in randomness in the distribution of molecules, hence an increase in entropy. The concept of entropy as a measure of disorderliness or randomness, have led to the conclusion that all *substances in their normal crystalline state at the absolute zero of temperature would be in the condition of maximum orderly arrangement because all motion has essentially ceased at 0°A. In other words, entropy of a substance at 0° A is minimum.*

Evidently expansion of a gas in vacuum and diffusion of a solute from a concentrated to a dilute solution when the two are connected together, are accompanied by increase of disorder (more of randomness) and hence there is increase in entropy.

(1) Entropy and Unavailable Energy.

Suppose a quantity of heat Q be extracted from a reservoir at a temperature T and maximum possible heat is converted into work. If T_o is the temperature of the coldest available reservoir, then

$$W_{max} = Q\left(1 - \frac{T_o}{T}\right)$$

This represents the maximum energy available for work on extracting Q units of heat from reservoir at temperature T. Any energy which resides in the reservoir at T_o can be extracted only in the form of heat and is completely unavailable for work. The potential energy, of a frictionless mechanical device is in a form which is completely available for work. It is desired to establish, *whenever an irreversible process takes place, the effect on the universe is the same as that which would be produced if a certain quantity of energy were converted from a form in which it was completely available for work into a form in which it is completely unavailable for work this amount of energy E is T_o times the entropy change of the universe brought about by the irreversible process*

$$\text{Entropy} = \frac{\text{Unavailable energy}}{\text{Temperature}}$$

$$\text{Unavailable energy} = \text{Entropy} \times \text{Temperature}$$

The following example illustrates the above fact

$$Hg + \tfrac{1}{2}Cl_2 \rightarrow \tfrac{1}{2}Hg_2Cl_2 + 31300 \text{ cal at } 25^\circ C$$

Out of 31300 calories we can utilise a portion for doing useful work. The available energy for useful work computed on the basis of second law of thermodynamics is found to be 25,140 cal. The remaining quantity of 6,160 cal is thus the unavailable energy.

Since unavailable energy is obtained by multiplying decrease of entropy and absolute temperature

$$\therefore \quad \text{Unavailable energy} \propto T$$

or $$\text{Unavailable energy} = \Delta S \times T$$

or $$\Delta S = \frac{\text{Unavailable energy}}{T}$$

$$= \frac{6160}{298} = 20.6 \text{ units.}$$

Sincee in irreversible process some energy becomes unavailable, we conclude that maximum work is obtained when a process takes place reversibly.

Since irreversible processes are continuously going on in nature, therefore, the energy is continually becoming unavailable for work. This is known as *principle of degradation* of energy and first developed by kelvin gives an important physical interpretation of entropy change of the universe. It must be clearly understood that the energy which becomes unavailable for work is not energy which is lost, it is merely transformed from one form to other and thus the first law of thermodynamics is obeyed at all the times. It can be stated that energy is "running down hill".

(2) Entropy and Disorder

In thermodynamics work is a macroscopic concept. Haphazard motion of individual molecules against intermolecular forces do not constitute work. Work involves order or orderly motion. When work dissipates into energy the disorderly motion of the molecules is increased. So during isothermal or adiabate dissipation of work into internal energy the disorderly motion of the molecules of either a reservoir or a system is increased. These processes are thus accompanied by a transition from order to disorder. Two gases when mixed have a higher degree of disorder than when they are separated. On th part of the nature there is a tendency to proceed towards a state of greater disorder. The increase of entropy of the universe during natural processes is an expression of this transition so we can say that the entropy of a system is a measure of the molecular disorder in the system.

Disorder of a system may be calculated by the theory of probability and expressed by a quantity Ω which is known as *thermodynamic probability* the relation between entropy and disorder is

$$S = \text{Constant} \times \ln \Omega$$

We can give a meaning to the entropy of a system in a non-equilibrium state using this equation. The non-equilibrium state corresponds to a certain degree of disorder and therefore to a definite entropy.

Out of the three state of matter the solid state is most ordered and has the least entropy where as gaseous state is most disorderly and has the least entropy where as gaseous state is most disorderly and has maximum entropy.

(3) Entropy and Probability

We have studied in the previous article that molecular disorder and entropy go together and this relationship between them is given by simple relationship

$$S = k \ln \Omega$$

where Ω is thermodynamic probability and is a measure of disorder.

Boltzmann gave a relationship between entropy(s) and thermodynamic probability (Ω) in 1896.

$$S = k \ln \Omega + C$$

Plank, in 1912, suggested C = O in above equation, hence

$$S = k \ln \Omega$$

The k in this equation is known as Boltzmann's constant and this equation is known as Boltzmann-Plank equation.

Entropy Calculations

Entropy is a state function and 1 the change in entropy for a given process is independent of the path between the initial and the final states. The process may occur by any path reversible or irreversible, but the entropy change will be the same. To compute the value of ΔS, a path for which a calculation is as simple as possible is chosen. The most covenient path frequently chosen is the reversible path, because $dE = \frac{dq_{rev.}}{T}$ is applicable to reversible processes and employed change calculations.

Entropy change for an ideal gas. Consider a system consisting of n moles of an ideal gas occupying a volume V at a pressure P and temperature T. If $dq_{rev.}$ amount of heat is absorbed by the system reversibly, then the increase in entropy ds is given by $ds = \frac{dq_{rev.}}{T}$.

From the first law of thermodynamics, for a reversible process in which only pressure-volume work is involved.

$$dq_{rev.} = dE + PdV \qquad (1)$$

Substituting the value of $dq_{rev.}$ from equation (1)

$$ds = \frac{dE + PdV}{T} \qquad (2)$$

For *n* moles of an ideal gas, we know that

$$P = \frac{nRT}{V}$$

and $\quad dE = n\,C_v\,dT \qquad [\because C_v = \frac{dV}{dT} \text{ for one mole }]$

where C_v is the molar heat capacity at constant volume.

Substituting these values of P and dE in eqn. (2), we get

$$dS = \frac{nC_v dT + (nRT/V)\,dV}{T}$$

or $$dS = nC_v \frac{dT}{T} + nR\frac{dV}{V} \qquad (3)$$

If the entropy, volume and temperature of the system in the initial states are S_1, V_1 and T_1 and in the final state are S_2, V_2 and T_2 respectively; then integrating eq. (3) between these limits, we get

$$\int_{S_1}^{S_2} dS = \int_{T_1}^{T_2} nC_v\,\frac{dT}{T} + \int_{V_1}^{V_2} nR\,\frac{dV}{V}$$

Assuming C_v to be independent of temperature over the temperature range considered, we obtain

$$\Delta S = S - S_1 = n\,C_v \ln\frac{T_2}{T_1} + nR\ln\frac{V_2}{V_1} \qquad (4)$$

For 1 mole of an ideal gas

$$\Delta S = C_v \ln\frac{T_2}{T_1} + R\ln\frac{V_2}{V_1} \qquad (5)$$

An alternate form of eqn. (4) involving pressure term can be obtained in the following manner:

If P_1 is the pressure of the system in the initial state and P_2 in the final state, then from the gas equation,

$$\frac{P_1V_1}{T_1}=\frac{P_2V_2}{T_2}$$

or
$$\frac{V_2}{V_1}=\frac{P_1T_2}{P_2T_1}$$

Substituting in eqn. (4), we obtain

$$DS = n\,C_v \ln\frac{T_2}{T_1} + n\,R \ln\frac{P_1T_2}{P_2T_1}$$
$$= n\,C_v \ln\frac{T_2}{T_1} + n\,R\,\ln\frac{T_1}{T_2} + n\,R\,\ln\frac{P_1}{P_2}n$$
$$= n(C_v + R)\,\ln\frac{T_2}{T_1} + n\,R\,\ln\frac{P_1}{P_2}$$

Since $C_v + R = C_p$, hence

$$\Delta S = C_p \ln\frac{T_2}{T_1} + nR \ln\frac{P_1}{P_2} \quad (6)$$

$$= C_p \ln\frac{T_2}{T_1} - nR \ln\frac{P_2}{P_1} \quad (7)$$

From equations (4) and (6) entropy changes for an ideal gas can be calculated.

These equations take different forms under different conditions.

(i) **At Constant Temperature for an Isothermal Process** *i.e.*, when $T_1 = T_2$. For such processes equations (5) and (6) take the form.

$$\Delta S_T = nR \ln\frac{V_2}{V_1} \quad (8)$$

$$= nR \ln\frac{P_1}{P_2} = -\,nR \ln\frac{P_2}{P_1}$$

The subscript T in ΔS_T indicate that the process is occurring at constant temperature. In an expansion $V_2 > V_1$ or $P_1 > P_2$, hence ΔS_T is positive but in contraction $V_1 > V_2$ or $P_2 > P_1$ and hence S_T is negative. From this it is concluded that in an isothermal expansion entropy increases and in contraction it decreases.

(ii) **At Constant pressure for an isobaric process**, i.e., when $P_1=P_2$. For such processes the equation (6) takes the form

$$\Delta S_p = n\, C_p \ln \frac{T_2}{T_1} \quad (9)$$

(iii) **At Constant Volume for an Isochoric Process** i.e., when $V_1=V_2$ the equation (5) takes the form

$$\Delta S_v = C_v \ln \frac{T_2}{T_1} \quad (10)$$

(iv) **Adiabatic Processes are Isoentropic.** In case of adiabatic processes taking place reversibly, no heat enters or leaves the system, $q_{rev.}$ is zero.

$$\therefore \quad \Delta S = \frac{q_{rev.}}{T} = 0.$$

Those processes in which *there is no change in entropy of a system are known as* **isoentropic processes.** So all the reversible adiabatic processes are isoentropic.

It may be noted that

(i) These equations are perfectly general and apply to all types of substances viz., solid, liquid or gas.

(ii) Since entropy changes depend only upon the initial and final states of system, and does not depend on the path by which the changes are brought about, the equations are applicable to the reversible as well irreversible processes.

(iii) Strictly speaking the heat capacities C_p and C_v are not independent of temperature. But while evaluating the integral we have assumed them to be constant, *i.e.*, independent of temperature. Hence it is assumed that only the mean values of the heat capacities over the temperature range T_1 and T_2 are used.

Entropy Change in Heating a Solid or Liquid. When a solid or a liquid is heated without producing change in state, the heat change is given by

Quantity of heat absorbed = mass × specific heat × temperature rise = No. of moles × molar heat × Temperature rise.

Thus is *m* gm of a solid or liquid of specific heat '*s*' is heated reversibly through temperature *d*T, the amount of heat absorbed is given by $dq_{rev} = m.s.dT$

$$\frac{dq_{rev.}}{T} = \frac{msdT}{T} \qquad \text{(... dividing by T)}$$

or

$$= \int_{T_1}^{T_2} \frac{msdT}{T} \qquad \cdots \left(\frac{dq_{rev.}}{T} = dS\right)$$

Assuming that *s* remains constant within the temperature limits T_1 and T_2.

On integration the above equation gives $\Delta S = ns \ln \frac{T_2}{T_1}$

$$= 2.303\ ms \log \frac{T_2}{T_1}.$$

Entropy Change accompanying Change of Phase. A process of change of state, *e.g.*, melting of solid or vaporisation of liquid may be carried out at constant temperature reversibly as the two phases are in equilibrium for all times during the change. Suppose the process of change of state of 1 mole of substance is carried out reversibly, the amount of heat absorbed will be equal to the molar heat of fusion or vaporisation and the temperature will be the melting point or the boiling point.

Thus the *molar entropy change* accompanying a fusion process is given by :

$$\Delta S_f = \frac{\Delta H_f}{T_f}$$

where ΔS_f is the molar entropy change accompanying fusion process, ΔH_f is the molar heat of fusion and T_f is the freezing or melting point on the absolute scale.

Similarly the molar entropy change ΔS_v accompaying a vaporisation process is given by:

$$\Delta S_v = \frac{\Delta H_v}{T_b}$$

where ΔH_v is the molar heat of vaporization and T_b is the boiling point on the absolute scale.

Since ΔH_f and ΔH_v are both positive, an increase in entropy takes place during the processes of fusion and vaporisation.

In the change of state from vapour to liquid or from liquid to solid, ΔH_v and ΔH_f will be negative and during processes of condensation of vapour or freezing of a liquid, decrease in entropy takes place.

Entropy Increase during Isothermal Mixing

It has already been proved that

$$\Delta S = C_p \ln \frac{T_2}{T_1} - R \ln \frac{P_2}{P_1} \quad \text{[see Equation 7]}$$

So for one mole of an ideal gas increase of entropy at constant temperature is

$$\Delta S = S_2 - S_1 = -R \ln \frac{P_2}{P_1}$$

$$= R \ln \frac{P_1}{P_2}$$

So for n moles this will be

$$\Delta S = nR \ln \frac{P_1}{P_2}$$

or $$S_2 - S_1 = nR \ln \frac{P_1}{P_2}$$

$$\therefore \quad S_2 = S_1 + nR \ln \frac{P_1}{P_2}$$

$$= (S_1 + nR \ln P_1) - nR \ln P_2$$

or $$S = n\,(S^\circ - R \ln P)$$

where S° is the molar entropy of one mole of a gas in standard state.

If two gases A and B, kept in different vessels under the same conditions of temperature and pressure and have the entropies S_A and S_B respectively. Then

$$S_A = n_A\,(S_A^\circ - R \ln P)$$

$$S_B = n_B (S_B^\circ - R \ln P)$$

where n_A and n_B are the number of moles of gas A and B respectively and S_A° and S_B° are molar entropies of A and B respectively.

If these gases are mixed at same temperature and pressure the total entropy will be given by

$$S_T = n_A (S_A^\circ - R \ln P_A) + n_B (S_B^\circ - R \ln P_B)$$

P_A, P_B are the partial pressure of gas A and B in the mixture. The value of P_A and P_B can be calculated using the Dalton's law of partial pressures

$$P_A = \frac{n_A}{n_A + n_B} P = X_A P$$

$$P_B = \frac{n_B}{n_A + n_B} P = X_B P$$

where X_A and X_B are the mole fractions of A and B respectively, the change in entropy (ΔS) is given by

$$\Delta S = S_T - (S_A + S_B)$$

$$= n_A (S_A^\circ - R \ln P_A) + n_B (S_B^\circ - R \ln P_B)$$

$$- [n_A (S_A^\circ - R \ln P) + n_B (S_B^\circ - R \ln P)]$$

$$= -n_A R \ln \frac{P_A}{P} - n_B R \ln \frac{P_B}{P}$$

$$= -n_A R \ln X_A - n_B \ln X_B$$

In general $\Delta S_m = -R\Sigma n_m \ln X_m$

Problems

Example 1. One mole of Neon is mixed with two moles of Helium, both at the same temperature and pressure, calculate ΔS if the total volume is unchanged.

Solution. From the given data, we have

$$V_{He} = 2V_{Ne}$$

After mixing total volume will be

$$V_T = V_{Ne} + V_{He} = 3V_{Ne}$$

Each of the two gases will show a change in entropy due to volume change

$$\Delta S_{Ne} = nR \ 1n \frac{V_2}{V_1}$$

$$= 1 \times 1.987 \times 2.303 \log \frac{3V_{Ne}}{V_{Ne}} = 2.183 \text{ e.u.}$$

$$= \Delta S_{He} = nR \ 1n \frac{V_2}{V_1}$$

$$= 2 \times 1.987 \times 2.303 \log \frac{3V_{Ne}}{2V_{Ne}} = 1.612 \text{ e.u.}$$

Thus $\Delta S_{mixing} = \Delta S_{He} + \Delta S_{Ne}$

$$= 1.612 + 2.183 = 3.795 \text{ e.u.}$$

Example 2. Calculate the entropy change involved in expanding 1 mole of an ideal gas from 15 litres at 2 atmosphere pressure to 50 litres as 1 atmosphere pressure (C_p = 7.4 cal/°/ mole)

Solution. We know that

$$\Delta S = nC_p \ 1n \frac{T_2}{T_1} + nR \ 1n \frac{P_1}{P_2}$$

$$= 2.303 \ nC_p \log \frac{T_2}{T_1} + 2.303 \ nR \log \frac{P_1}{P_2}$$

Here $n = 1$ mole, $C_p = 7.4$ cal. deg.$^{-1}$ mole^{-1}

$R = 1.987$ cal. deg.$^{-1}$ mole^{-1}

$P_1 = 2$ atm., $P_2 = 1$ atm.

For an ideal gas $\dfrac{P_1V_1}{T_1} = \dfrac{P_2V_2}{T_2}$

or

$$\frac{T_2}{T_1} = \frac{P_2V_2}{P_1V_1} = \frac{50 \times 1}{15 \times 2} = \frac{5}{3}$$

$$\therefore \quad \Delta S = 2.303 \times 1 \times 7.4 \times \log \frac{5}{3} + 2.303 \times 1 \times 1.987 \times \log \frac{2}{1}$$

$$= 3.779 + 1.377 = 5.156 \text{ cal. deg}^{-1}$$

Example 3. Calculate the entropy change when 2 moles of an ideal gas are allowed to expand from a volume of 1 litre to a volume of 10 litres at 27°C.

Solution. We know that for an isothermal expansion of an ideal gas

$$\Delta S_T = nR \ 1n \frac{V_2}{V_1} = 2.303 \ nR \log \frac{V_2}{V_1}$$

Here $\quad V_1 = 1$ litre, $V_2 = 10$ litres, $n = 2$ moles

$R = 1.987$ cal. deg^{-1} mole^{-1}

$$\therefore \quad \Delta S_T = 2.303 \times 2 \times 1.987 \log \frac{10}{1}$$

$$= 9.152 \text{ cal deg}^{-1}.$$

Example 4. Calculate the change in entropy when 2 moles of lead are heated from 27°C to 28°C. The specific heat of lead over this temperature range is 0.03 cal. gm^{-1} and the atomic weight of lead is 207.

Solution. We know that

$$\Delta S = 2.303 \ ms \log \frac{T_2}{T_1}$$

Here $\quad m = 2 \times 207 = 414$ gm

$s = 0.03$ cal gm.$^{-1}$, $T_1 = 273 + 27 = 300$°K

$T_2 = 273 + 28 = 301$ K.

$$\therefore \quad \Delta S = 2.303 \times 414 \times 0.03 \times \log \frac{301}{300}$$

$$= 0.0428 \text{ cal. deg}^{-1}$$

Example 5. Calculate the entropy change when one mole of an ideal gas is allowed to expand from a temp. of 20°C to 40°C (C_p=6.189).

Solution. We know that for an isobaric process

$$\Delta S_p = nC_p \ 1n \frac{T_2}{T_1} = 2.303 \ nC_p \log \frac{T_2}{T_1}$$

Here $n = 1$ mole, $C_P = 6.189$ cal. deg.$^{-1}$

$$T_1 = 273 + 20 = 293°K\ ;\ T_2 = 273 + 40 = 313°K$$

$$\therefore \quad \Delta S_P = 2.303 \times 1 \times 6.189 \times \log \frac{313}{293}$$

$$= 0.4077 \text{ cal deg}^{-1}$$

Example 6. Calculate the entropy change when 5 moles of an ideal gas are allowed to expand isothermally at 20°C from a pressure of 10 atm. to a pressure of 2 atm.

Solution. We know that for an isothermal expansion involving pressure changes.

$$\Delta S_T = nR \ 1n \frac{P_1}{P_2} = 2.303 \ nR \ 1n \frac{P_1}{P_2}$$

Here $P_1 = 10$ atm, $P_2 = 2$ atm; $n = 5$ moles.

$$R = 1.987 \text{ cal deg}^{-1} \text{ mole}^{-1}$$

$$\therefore \quad \Delta S_T = 2.303 \times 5 \times 1.987 \times \log \frac{10}{2}$$

$$= 16 \text{ cal deg}^{-1}.$$

Example 7. Calculate entropy change when 3 moles of an ideal gas ($C_v = 7.88$ cal. deg^{-1} mole^{-1}) are heated from a volume of 200 litres at 50°C to a volume of 300 litres at 150°C.

Solution. We know that

$$DS = nCv \ 1n \frac{T_2}{T_1} + nR \ 1n \frac{V_2}{V_1}$$

$$= 2.303 \ nC \log \frac{T_2}{T_1} + 2.303 \ nR \log \frac{V_2}{V_1}$$

Here $n = 3$ moles

$C_v = 7.88$ cal deg^{-1} mole^{-1},

$R = 1.987$ cal deg^{-1} mole^{-1},

$T_1 = 273 + 50 = 323°K$, $T_2 = 273 + 150 = 423°K$

$V_1 = 200$ lit., $V_2 = 300$ lit.

$$\Delta S = 2.303 \times 2 \times 7.88 \times \log \frac{423}{323} + 2.303 \times 1.987 \times 3$$

$$\times \log \frac{300}{200}$$

$$= 6.375 + 2.417 = 8.792 \text{ cal deg}^{-1}$$

Example 8. Calculate the entropy change when 1 mole of ethanol is evaporated at 350°K. The molar heat of vaporisationof ethanol is 9522 cal mole^{-1}.

Solution. We know that

$$\Delta S_v = \frac{\Delta H_v}{T_b}$$

Here $\Delta H_v = 9522$ cal mole^{-1}

$T_b = 350$°K

$$\therefore \quad \Delta S_v = \frac{9522}{350} = 2.178 \text{ deg.}^{-1} \text{ mole}^{-1}.$$

Example 9. Calculate the total entropy change when 5 moles of ice are converted into water at 0°C. The latent heat of fusion of ice is 80 cal gm^{-1}.

Solution. We know that molar entropy change

$$\Delta S_f + \frac{\Delta H_f}{T_f}$$

Here latent heat of ice = 80 cal gm^{-1}

$\therefore$ Molar heat of fusion $\Delta H_f = 80 \times 18$ cal. mole^{-1}

$T_f = 273$°K.

$$\therefore \quad \Delta S_f = \frac{80 \times 18}{273} = 5.247 \text{ cal deg}^{-1} \text{ mole}^{-1}.$$

The entropy change per mole

$= 5.274$ cal deg^{-1}

$\therefore$ The total entropy change for 5 moles $= 5 \times 5.274$

$= 26.37$ cal. deg^{-1}.

Example 10. 1 mole of an ideal gas an (C_v = 25.12 and C_p = 33.44 J/mol K) expands adiabatically from a initial state at 340°K and

500 K Pa to double its volume. Find the final temperature of the gas, the work done and the entropy change of the gas for (i) reversible expansion (ii) free expansion.

Solution. (*i*) For reversible adiabatic expansion of an ideal gas with constant heat capacities, we have

$$T_2 = T_1 \left(\frac{V_1}{V_2}\right)^{\gamma-1}$$

Here

$$\gamma = \frac{C_p}{C_u} = 1.331$$

and

$$\frac{V_2}{V_1} = .5$$

$$\therefore \quad T_2 = 340 \times .5^{.331} = 270.3 \text{ K}$$

Applying first law,

$$W = -C_u \Delta T$$

$$= -25.12 \times (270.3 - 340) = 1751 \text{ J}$$

In accordance with second law of thermodynamics the entropy change for a reversible adiabatic process is zero.

(ii) Free expansion of a gas produces no work. For an adiabatic process insuch a case both W and Q are zero. Hence $C_v \Delta T$ should be zero to satisfy the requirement of first law that $\Delta E = 0$. Hence T_2 must be equal to T_1.

Entropy change is given by

$$\Delta S = R \ 1n \frac{V_1}{V_1}$$

$$= 8.314 \times 1n\ 2 = 5.7628 \text{ J/K}$$

The value of ΔS for the gas is also ΔS_{Total} because $\Delta_{surr.} = 0$ for an adiabatic process. So we get a positive value of ΔS.

Example 11. Calculate the change in entropy suffered by two moles of an ideal gas being heated from a volume of 100 litres at 323 K to a volume of 150 litres at 423 K (C_p = 32.97 JK^{-1} mol^{-1}, R = 8.314

$JK^{-1}\ mol^{-1}$) [*Bombay 1973*]

Solution. We know

$$\Delta S = 2.303\ n\ C_v \log \frac{T_2}{T_1} + 2.303\ n\ R \log \frac{V_2}{V_1}$$

In this exercise, we are given

$$n = 2,\ C_v = 32.97\ JK^{-1}\ mol^{-1},\ R = 8.314\ JK^{-1}\ mol^{-1}$$

$V_1 = 100$ litres, $V_2 = 150$ litres

$T_1 = 323$ K, $T_2 = 423$ K.

Making these substitution, we get

$$\Delta S = 2 \times 2.303 \times 32.97\ 1n \frac{423}{323} + 8.314\ 1n \frac{150}{100}$$

$$= 24.50\ JK{-}1$$

Example 12. Find out the change in entropy when 2 moles of an ideal gas on being heated from a pressure of 5 atmosphere at 323 to a pressure of 10 atmospheres at 373 K.

Given $C_p = 41.34\ JK^{-1}\ mol^{-1}$.

Solution. We know,

$$\Delta S = 2.303\ n\ \ C_p \log \frac{T_2}{T_1} + 2.303\ n\ R \log \frac{P_1}{P_2}$$

Given, $n = 2$, $T_1 = 323$ K, $T_2 = 373$ K

$P_1 = 5$ atm, $P_2 = 100$ atm

$R = 8.314\ JK^{-1}\ mol^{-1}$

$C_p = 41.34\ JK^{-1}\ mol^{-1}$

Substituting these values, we get

$$\Delta S = 2 \times 2.303 \left[41.34\ \ 1n \frac{373}{323} + 8.314\ 1n\ \frac{5}{10} \right]$$

$$= 0.374\ JK^{-1}\ mol^{-1}.$$

Example 13. Calculate the entropy change for fusion of 1 mole of a solid which melts at 300 K. The latent heat of fusion is 2510.4 J mol^{-1}. [*G.N.D. Uni. 1979*]

Solution. $$\Delta S = \frac{\Delta H_f}{T_f}$$

$$= \frac{2510.4}{300} = 8.368\ JK^{-1}$$

Example 14. The Entropy of a liquid A at 0.05 atmo, and 32.0°C is 0.52 kJ/deg/kg. Assuming vapours of the liquid A to be ideal. Find out the entropy of the vapours at 32°C and 0.05 atmo. Heat of vapourisation under the conditions specified above is 2422.12 kJ kg^{-1}.

Solution. $$\Delta S = \frac{\Delta H_{vap.}}{T_b}$$

$$= \frac{2422.12}{305}$$
$$= 7.94\ kJ\ kg^{-1}\ deg^{-1}$$

$\therefore$ $$S_{Vapour} = 0.52 + 7.94 = 8.46\ kJ\ kg^{-1}\ deg^{-1}.$$

TROUTON'S RULE

For liquids which slow ideal behaviour, this rule can be stated mathematically, as

$$\Delta S_{vap} = 21\ cal\ mol^{-1}\ deg^{-1}$$

$$= 87.822\ J\ mol^{-1}\ deg^{-1}.$$

where T_b is the boiling point of the liquid.

The rule can thus be used to find out either ΔH_{vap} or T_b if the other is known.

Deviations from this rule occur in case of those liquids which have strong intermolecular interactions.

Example 1. Given the boiling points (T_b) for CCl_4, $CHCl_3$ and H_2S as 76.7°C, 61.5°C and –59°C respectively. Calculate the molar heat of vapourisation if these liquids obey Trouton's rule.

Solution. For CCl_4

$$\Delta H_{vap} = -S_{vap} \times T_b = 87.822 \times 349.7$$
$$= 30.71 \text{ kJ mol}^{-1}.$$

For $CHCl_3$
$$\Delta H_{vap} = 87.822 \times 334.5$$
$$= 29.87 \text{ kJ mol}^{-1}$$

For H_2S
$$\Delta H_{vap} = 87.822 \times 213.4$$
$$= 18.75 \text{ mol}^{-1}.$$

Example 2. What will be the vapour pressure of toluene at 77°C. The boiling point of toluene is 110.6°C.

[Assume that toluene behaves ideally]

Solution.
$$\Delta S = \frac{\Delta H_{vap}}{T_b} = 87.822 \qquad \text{[Trouton's Rule]}$$

Since toluene behaves ideally so it obeys Trouton's rule.

Hence
$$\Delta H_{vap} = 87.822 \times 383.6$$
$$= 33688.5 \text{ J mol}^{-1}$$

Since at boiling point vapour pressure is equal to atmospheric pressure (760 mm). If ΔH_{vap} is independent of temperature, we get

$$\log \frac{P_2}{P_1} = \frac{\Delta H_{vap}}{2.303R}\left[\frac{1}{T_2} - \frac{1}{T_1}\right]$$

or

$$\log P_2 = \log 760 - \frac{33688.5}{2.303 \times 8.314}\left[\frac{1}{350} - \frac{1}{383.6}\right]$$

or
$$P_2 = 275.7 \text{ mm}.$$

Characteristics of Entropy. Entropy is a state function like internal energy, enthalpy etc.

It is defined as under

$$dS = \frac{dQ_{rev}}{T} \qquad (1)$$

where dQ_{rev} is the amount of heat reversibly absorbed and T is the temperature at which this heat absorption takes place.

If in this change system also performs some work of expansion in addition to its increase of internal energy, then in accordance with first law of thermodynamics, we have

$$dQ_{rev} = dE + PdV \qquad \text{(ii)}$$

Dividing by T, we obtain

$$\frac{dQ_{rev}}{T} = \frac{dE + PdV}{T}$$

or

$$dS = \frac{1}{T} dE + \frac{P}{T} dV \qquad \text{(iii)}$$

Equation (iii) gives the variation of entropy as the internal energy and volume of a system vary. The term $\frac{1}{T}$ and $\frac{P}{T}$ occurring in this relation are positive so dS is positive or entropy of a system increases when (i) internal energy increases at constant volume and/or (ii) volume increases at constant internal energy.

These characteristics of the entropy function can be stated mathematically by the equation as follows:

$$dS = \left(\frac{\partial S}{\partial E}\right)_V dE + \left(\frac{\partial S}{\partial V}\right)_E dV \qquad \text{(iv)}$$

From a comparison of (iii) and (iv), we get

$$\left(\frac{\partial S}{\partial E}\right)_V = \frac{1}{T} \text{ and } \left(\frac{\partial S}{\partial V}\right)_E = \frac{P}{T}$$

Substituting these values in (iii), we get

$$dE = TdS - PdV$$

*d*E being an exact differential, Euler's theorem in applicable. So we get

$$\left(\frac{\partial T}{\partial V}\right)_S = -\left(\frac{\partial P}{\partial S}\right)_V$$

This one of the Maxwells relations. Other relations have been derived in a separate section.

Entropy as a Function of T and V. This can be stated as

$$S = f(T, V)$$

differentiating, we get

$$dS = \left(\frac{\partial S}{\partial T}\right)_V dT + \left(\frac{\partial S}{\partial V}\right)_T dV \qquad (1)$$

We have already calculated in previous article

$$dS = \frac{1}{T}\ dE + \frac{P}{T}\ dV \qquad (2)$$

Again E is a function of T, V it can be stated as

$$E = f(T, V)$$

$$\therefore \qquad dE = \left(\frac{\partial E}{\partial T}\right)_V dT + \left(\frac{\partial E}{\partial V}\right)_T dV \qquad (3)$$

Substituting the value of dE from Eq (3) into equation (2), we get

$$dS = \frac{1}{T}\left(\frac{\partial E}{\partial T}\right)_V dT + \frac{1}{T}\left[P + \left(\frac{\partial E}{\partial V}\right)_T\right] dV \qquad (4)$$

By comparing equations (1) and (4),

$$\left(\frac{\partial S}{\partial T}\right)_V = \frac{1}{T}\left(\frac{\partial E}{\partial V}\right)_V \qquad (5)$$

and

$$\left(\frac{\partial S}{\partial V}\right)_T = \frac{1}{T}\left[P + \left(\frac{\partial E}{\partial V}\right)_T\right] \qquad (6)$$

Equation (5) gives the change of entropy of the system with respect to temperature when its volume is kept constant and Equation

(6) given the change of entropy of the system with respect to volume when its temperature is kept constant.

For a finite change of temperature at constant volume equation (5) can be integrated to give

$$\Delta S = \int_{T_1}^{T_2} n \frac{C_v}{T} dT \qquad \text{(for } n \text{ moles)}$$

if C_v is assumed to be constant in this temperature range, we get

$$\Delta S = n\, C_v \ln \frac{T_2}{T_1} \qquad \text{(for } n \text{ moles)}$$

If C_v is dependent on temperature then for an ideal gas

$$C_p - C_v = R$$

and C_p is given by

$$C_p = a + bT + cT^2 + \ldots$$

$$\therefore \quad C_v = a + bT + cT^2 + \quad -R$$

$$\therefore \quad \Delta S = n \int_{T_1}^{T_2} \frac{a + bT + cT^2 + \quad -R}{T} dT$$

$$\Delta S = n\left[(a - R) \ln \frac{T_2}{T_1} + b(T_2 - T_1) + \frac{C}{2}(T_2^2 - T_1^2) + \ldots\right] \qquad (7)$$

Another important relation can be derived from equation (4) if we use the internal energy function.

By definition, we know

$$C_v = \frac{dE}{dT} \qquad \text{or} \qquad dE = C_v dT \qquad (8)$$

and

$$C_p = \frac{dH}{dT}$$

Substituting the value of C_v from (8) into equation (4), we get

$$dS = \frac{C_v}{T} dT + \frac{1}{T}\left[P + \left(\frac{\partial E}{\partial v}\right)_V\right] dV \qquad (9)$$

From equation (5), we have

$$\left(\frac{\partial S}{\partial T}\right)_V = \frac{1}{T}\left(\frac{\partial E}{\partial T}\right)_V$$

$$= \frac{C_v}{T} \qquad \left[\because \left(\frac{\partial E}{\partial T}\right)_V = C_v\right] \tag{10}$$

and from equation (6), we get

$$\left(\frac{\partial S}{\partial V}\right)_T = \frac{1}{T}\left[P + \left(\frac{\partial E}{\partial V}\right)_T\right]$$

or

$$\left(\frac{\partial E}{\partial V}\right)_T = T\left(\frac{\partial S}{\partial V}\right)_T - \text{P} \tag{11}$$

Differentiating equation (10) with respect to volum at constant temperature

$$\frac{\partial^2 S}{\partial V \partial T} = \frac{\partial}{\partial V}\left(\frac{C_v}{T}\right)_T$$

$$= \frac{1}{T}\frac{\partial}{\partial V}\left(C_v\right)_T$$

$$= \frac{1}{T}\frac{\partial}{\partial T}\left[\left(\frac{\partial E}{\partial T}\right)_V\right]_T \left[C_v = \left(\frac{\partial E}{\partial T}\right)_V\right] \tag{12}$$

Again differentiating quation (11) with respect to temperature at constant volume, we get

$$\frac{\partial^2 S}{\partial T \partial V} = \frac{\partial}{\partial T}\left[\frac{1}{T}\left\{P + \left(\frac{\partial E}{\partial T}\right)_V\right\}\right]_T$$

$$= \frac{P}{T^2} - \frac{1}{T^2}\left(\frac{\partial E}{\partial V}\right)_T + \frac{1}{T}\left(\frac{\partial P}{\partial T}\right)_V$$

$$+ \frac{1}{T}\frac{\partial^2 E}{\partial T \partial V} \tag{13}$$

Since entropy is a state function, so we get in accordance with

Euler's theorem

$$\frac{\partial^2 S}{\partial T \partial V} = \frac{\partial^2 S}{\partial v \partial T}$$

Comparing equations (12) and (13), we get

$$\frac{1}{T}\,\frac{\partial^2 E}{\partial V \partial T} = -\frac{P}{T^2} - \frac{1}{T^2}\left(\frac{\partial E}{\partial V}\right)_T + \frac{1}{T}\left(\frac{\partial P}{\partial V}\right)_V$$

$$+\frac{1}{T}\,\frac{\partial^2 E}{\partial T \partial V} \qquad (14)$$

Since E is a state function so we have

$$\frac{\partial^2 E}{\partial V \partial T} = \frac{\partial E}{\partial T \partial V}$$

Thus from equation (14) we have

$$\left(\frac{\partial E}{\partial V}\right)_T = \mathrm{T}\left(\frac{\partial P}{\partial T}\right)_V - \mathrm{P} \qquad (15)$$

This is an important expression and is known as *thermodynamic equation of state.*

This equation is found to be true for all substancs under all conditions.

Since it relates various variables of state, it can be compared with other equation.

On comparison with equation (6), we get

$$\left(\frac{\partial S}{\partial V}\right)_T = \left(\frac{\partial P}{\partial T}\right)_V \qquad (16)$$

Again, we get

$$\left(\frac{\partial S}{\partial V}\right)_T = \frac{\alpha}{\beta} \qquad \left[\because \left(\frac{\partial E}{\partial V}\right)_T = \frac{\alpha}{\beta}\right]$$

For most cases α is positive and β is always positive at constant volume. Therefore, entropy increases with an increase in volume at constant temperature. In case of water, in the temperature range 0°C to 4°C, α has a negative value, so we get a negative value of $\left(\frac{\partial S}{\partial V}\right)_T$.

In case of condensed system $\frac{\alpha}{\beta}$ is neglible and the dependence of entropy of volume can be ignored except in case of gases.

Application of thermodynamics equation of state to (i) 'Ideal gas (ii) Van der Waals' gas.

(i) **Ideal Gas.** For *n* moles of an ideal gas, we have

$$PV = nRT \qquad \text{(Gas equation)}$$

or

$$\left(\frac{\partial P}{\partial T}\right)_V = \frac{nR}{V} = \frac{P}{T}$$

Substituting these values in the thermodynamic equation of state, we get

$$\left(\frac{\partial E}{\partial V}\right)_T = T\left(\frac{\partial P}{\partial T}\right)_V - P$$

$$= T \times \frac{P}{T} - P = 0$$

Thus for an ideal gas $\left(\frac{\partial E}{\partial V}\right)_T$ is zero.

(ii) Van der Waal's Gas. The equation for n moles of a gas obeying.

Van der Waals' equation is

$$\left(P + \frac{n^2 a}{V^2}\right)(V - nb) = nRT$$

On differentiating this equation w.r.t. T, we get

$$\left(\frac{\partial P}{\partial T}\right)_V = \frac{nR}{V - nb}$$

Substituting this value in the thermodynamic equation of state, we get

$$\left(\frac{\partial E}{\partial T}\right)_V = \frac{nRT}{V - nb} - P$$

$$= \frac{nRT}{V - nb} - \left[\frac{nRT}{V - nb} - \frac{an^2}{V^2}\right]$$

or

$$\left(\frac{\partial E}{\partial T}\right)_V = \frac{an^2}{V^2}$$

For one mole, we have

$$\left(\frac{\partial E}{\partial V}\right)_T = \frac{a}{V^2}$$

or $$dE = \frac{a}{V^2}\,\mathrm{dV}$$

or $$E = E_o - \frac{a}{V} \qquad \text{[on integrating]}$$

Entropy as a Function of Temperature and Pressure

The dependence of entropy on temperature and pressure can be mathematically expressed as

$$\mathrm{S = f\,(T, P)} \qquad (1)$$

Differentiating, we get

$$dS = \left(\frac{\partial S}{\partial T}\right)_P \mathrm{dT} + \left(\frac{\partial S}{\partial P}\right)_T dP \qquad (2)$$

From the definition of entropy

$$\mathrm{TdS = dQ}_{rev}$$

$$= \mathrm{dE + PdV} \qquad [\,dQ = dE + PdV]$$

or $$dS = \frac{1}{T}dE + \frac{P}{T}dV \qquad \text{[Dividing by T]} \quad (3)$$

Again, from the definition of enthalpy

$$H = E + PV$$

or $$E = H - PV$$

$$\therefore \qquad dE = dH - PdV - VdP \qquad (4)$$

Since H is a function of temperature and pressure

so $$\mathrm{H = f\,(T, P)}$$

Hence $dE = \left\{\left(\frac{\partial H}{\partial T}\right)_P dT + \left(\frac{\partial H}{\partial P}\right)_T dP\right\} - PdV - VdP$ (5)

Making substitutions from equation (5) into equation (3), we get

$$dS = \frac{1}{T}\left\{\left(\frac{\partial H}{\partial T}\right)_P dT + \left(\frac{\partial H}{\partial P}\right)_T dP\right\} - \frac{P}{T}dV$$

$$- \frac{V}{P}dP + \frac{P}{T}dV$$

$$= \frac{1}{T}\left(\frac{\partial H}{\partial T}\right)_P dT + \left\{\frac{1}{T}\left(\frac{\partial H}{\partial P}\right)_T - V\right\} dP \qquad (6)$$

By comparing equation (2) and (6),

$$\left(\frac{\partial S}{\partial T}\right)_P = \frac{1}{T}\left(\frac{\partial H}{\partial T}\right)_P dT$$

$$= \frac{C_P}{T} \qquad \left[\frac{1}{T}\left(\frac{\partial H}{\partial T}\right)_P = C_P\right] \qquad (7)$$

and

$$\left(\frac{\partial S}{\partial P}\right)_T = \frac{1}{T}\left\{\left(\frac{\partial H}{\partial P}\right)_T - V\right\} \qquad (8)$$

Equation (8) can be simplified to

$$\left(\frac{\partial H}{\partial P}\right)_T = T\left(\frac{\partial S}{\partial P}\right)_T + V \qquad (9)$$

Applying Euler's theorem, it can be easily deduced, that

$$\left(\frac{\partial S}{\partial P}\right)_T = -\left(\frac{\partial V}{\partial T}\right)_P$$

and

$$\left(\frac{\partial H}{\partial P}\right)_T = \left\{V - T\left(\frac{\partial V}{\partial T}\right)_P\right\}$$

Change in Entropy on Simultaneously Heating and Expanding of a Van der Waals' Gas

If n moles of a Van der Waals' gas at temperature T_1 and pressure

P_1 are brought to the temperature T_2 and pressure P_2 by simultaneously heating and expanding the gas then from the first and second laws of thermodynamics, we have

$$TdS = Sq_{rev} = dE + PdV \quad (1)$$

Since $\quad E = f(T, V)$

$$\therefore \quad dE = \left(\frac{\partial E}{\partial T}\right)_V dT + \left(\frac{\partial E}{\partial V}\right)_T dV$$

$$= C_v\, dT + \left(\frac{\partial E}{\partial V}\right)_T dV \qquad \left[\frac{\partial E}{\partial T} = C_v\right] \quad (2)$$

From equations (1) and (2), we get

$$dS = \frac{C_v}{T}\, dT + \frac{1}{T}\left\{\left(\frac{\partial E}{\partial V}\right)_T + P\right\} dV \quad (3)$$

For one mole of a Van der Walls' gas, we have

$$\left\{P + \frac{a}{V^2}\right)(V - b) = RT$$

and

$$\left(\frac{\partial E}{\partial V}\right)_T = \frac{a}{V^2}$$

Substituting in equation (3), we get

$$dS = \frac{C_v}{T}\, dT + \frac{R}{V - b}\, dV$$

Integrating, we get

$$S_2 - S_1 = \Delta S = C_v \ln \frac{T_2}{T_1} + R \ln\left(\frac{V_2 - b}{V_1 - b}\right) \quad (4)$$

Entropy Change when a Gas Obeys Virial Equation of State

For one mole of a gas obeying the virial equation of state, we have

$$\frac{PV}{RT} = 1 + \frac{B(T)}{V} \qquad (1)$$

From the Relations,

$$dS = \frac{C_v}{T}\,dT + \frac{1}{T}\left[\left(\frac{\partial E}{\partial V}\right)_T + P\right] dV \qquad \text{[See Eq. (3)]}$$

and

$$\left(\frac{\partial E}{\partial V}\right)_T = T\left(\frac{\partial P}{\partial T}\right)_V - P \qquad \text{[See Eq. (15) Page 203]}$$

we get

$$dS = \frac{C_v\,dT}{T} + \left(\frac{\partial P}{\partial T}\right)_V dV \qquad (2)$$

Differentiating Eq. (1) w.r.t. temp. at constant volume, we get

$$\left(\frac{\partial P}{\partial T}\right)_V = \frac{R}{V} + \frac{R}{V^2}\;B(T) + \frac{RT}{V^2}\left(\frac{dB}{dT}\right)_V \qquad (3)$$

Making substitution from (3) into (2), we get

$$dS = C_v\frac{dT}{T} + \left[\frac{R}{V} + \frac{R}{V^2}\,B(T) + \frac{RT}{V^2}\left(\frac{dB}{dT}\right)_V\right] dV$$

Integrating this equation, we get

$$\Delta S = S_2 - S_1 = C_v\,\ln\frac{T_2}{T_1} + R\,\ln\frac{V_2}{V_1} - RB\left[\frac{1}{V_2} - \frac{1}{V_1}\right]$$

$$- RT\frac{dB}{dT}\left[\frac{1}{V_2} - \frac{1}{V_1}\right]$$

or

$$dS = C_v\,\ln\frac{T_2}{T_1} + R\,\ln\frac{V_2}{V_1} + R\left[\left(B + \frac{TdB}{dT}\right)\left(\frac{1}{V_2} - \frac{1}{V_1}\right)\right]$$

From this the entropy change can be calculated.

SOLVED EXAMPLES

Example 1. Pure liquid water can be cooled to temperature below 0°C. If a mass of water is cooled to –5°C and a small crystal of ice is added to it and the subsequent change of state occurs adiabatically and at constant pressure what fraction of the system solidifies? What is the entropy change of the system?

Given that

Latent heat of fusion of water at 0°C = 334 kJ/kg.

The heat capacity of water between

0 and –5°C = 4.22 kJ/kg.

Solution. The final state of the system can be presumed to be the equilibrium state between ice and water. This state exists at 0°C or 273.15°K. This change of the state takes place in a closed system at constant pressure so $Q = \Delta H^t = m.\Delta H$. Since the process is adiabatic $\Delta H = 0$.

ΔH is a state property and is independent of path adopted so we consider an arbitrary path made up of the following two steps:

(i) The sub-cooled liquid is warmed from –5°C to 0°C in liquid state.

(ii) The heat added in above step is withdrawn at 0°C causing a part of liquid to freeze.

The amounts of heat involved in these two steps in such that $Q = \Delta H = 0$ for each kilogram of water.

Making these calculation for step (i),

$$\Delta H_1 = C_p \, \Delta T$$

$$= 4.22 \times 5 = 21.2 \text{ kJ/kg}$$

For step (ii), $\Delta H_2 = -Z\Delta\, H_{fusion} = -333.4\ Z$ kJ/kg

where Z is the fraction of the system that freezes.

Hence –333.42 + 21.1 = 0 or Z = .0633

Thus 6.33% of the system solidifies.

Since the process as it occurs is irreversible so ΔS is positive. Calculating entropy along the same path, we get

For step (i),

$$\Delta S_1 = C_p \ln \frac{T_2}{T_1}$$

$$= 4.22 \times \ln \frac{273.15}{268.15}$$

$$= .07796 \text{ kJ/kg K.}$$

For step (ii), Temperature is constant (0°C)

$$\Delta S_2 = \frac{Q_2}{T_2} = \frac{-21.2}{273.15}$$

$$= -.07725 \text{ kJ/kg K}$$

$$\Delta S_{Total} = \Delta S_1 + \Delta S_2 = +.00071 \text{ kJ/kg K.}$$

Example 2. Find out ΔS for 5 moles of nitrogen heated at constant pressure from 26°C to 100°C. C_p for nitrogen varies as

$$C_p = 6.524 + 1.25 \times 10^{-3}\, T - .01 \times 10^{-7}\, T^2.$$

Solution. For one mole of a gas, we have

$$\left(\frac{\partial S}{\partial T}\right)_p = \frac{C_p}{T}$$

$$\therefore \quad \left(\frac{\partial S}{\partial T}\right)_p = \frac{6.524}{T} + 1.25 \times 10^{-3} - .01 \times 10^{-7}\, T$$

or

$$dS = \left(\frac{6.524}{T} + 1.25 \times 10^{-3} - .01 \times 10^{-7} T\right) dT$$

Integrating, we get

$$DS = \int_{298}^{373} \left(\frac{6.524}{T} + 1.25 \times 10^{-3} - .01 \times 10^{-7} T\right) dT$$

$$= 6.524 \times 2.303 \ln \frac{373}{298} + 1.25 \times 10^{-3} \times 75$$

$$- .005 \times 10^{-7} (373^2 - 298^2)$$

$$\Delta S = 1.465 + .0937 + 2.516 \times 10^{-5}$$

$$= 1.559 \text{ e.u.}$$

For 5 moles, we get

$$\Delta S = 5 \times 1.559 \text{ e.u.} = 7.795 \text{ e.u.}$$

Example 3. Calculate the change in entropy for 10 moles of an ideal gas when heated from 400 K to 800 K: (i) at constant pressure; and (ii) at constant volume.

$$\left(\text{For ideal gas } C_P = \frac{5}{2} R\right)$$

Solution. At constant pressure, we have

(i) $$\Delta S_P = n\, C_p \ln \frac{T_2}{T_1}$$

$\therefore$ $$\Delta Sp = 10 \times \frac{5}{2} \times 1.987 \times 2.303 \log \frac{400}{800}$$

$$= 34.33 \text{ e.u.}$$

(ii) At constant volume, we have

$$\Delta S_v = n\, C_v \ln \frac{T_2}{T_1}$$

$$= 10 \times \frac{3}{2} \times 1.987 \times 2.303 \log \frac{400}{800}$$

$$= 20.66 \text{ e.u.}$$

Example 4. What will be the change in entropy when 2 moles of N_2, 3 moles of H_2 and 2 moles of NH_3 are mixed isothermally.

[Assume that they do not have any chemical reaction).

Solution. ΔS for isothermal mixing for different systems is given by

$$\Delta S_{says} = -R\,\Sigma ni\ 1n\ xi$$

where ni = number of moles,

xi = mole fraction of ith component.

We have

$$X_{N_2} = \frac{2}{7},\ X_{H_2} = \frac{3}{7},\ X_{NH_3} = \frac{2}{7}$$

Hence $$\Delta S = -8.314 \times \left[2\ 1n\frac{2}{7} + 3\ 1n\frac{3}{7} + 2\ 1n\frac{2}{7}\right]$$

$$= 62.80\ JK^{-1}.$$

6

Equations of State for Real Gases

1. Introduction

Based on the postulates of kinetic theory of gases, it is possible to deduce an equation for the pressure of an ideal gas. This is known as *kinetic equation for gases.*

$$PV = \frac{1}{3}mne^2$$

P and V are pressure and volume respectively of *n* molecules of a gas, each having mass *m, c* is the root mean square velocity of the molecules.

The kinetic gas equation is derived from kinetic theory. However in case of an ideal gas the gas equation called *equation of state* is used. It is

$$PV = RT$$

This equation is based on various gas laws and is thus an experimental formulation. However their equation makes use of two basic assumptions of kinetic theory i.e.

(i) The gaseous molecules are point masses, and

(ii) there are no intermolecular forces between molecules.

The above equation (i.e. PV = RT) has been found to be valid to a large extent at ordinary conditions of temperature and pressure.

However at high pressure and low temperature, the real gases show a deviation from ideal behaviour.

Deviations from Ideal Gas Equation

To explain the deviation of real gases from ideal behaviour under the conditions of Law temperature and high pressure Van der Waals found fault with two basic postulates of kinetic theory of gases on which the ideal gas equation is based. He observed that

(i) Read gases do have molecular volumes under ordinary conditions, free space between the molecules is much larger in comparison to the molecular volume and therefore, neglecting the volume occupied by these gaseous molecules does not involve any measurable error. But at high pressures and low temperatures, when molecules of gas are relatively close together, molecular volume constitutes a significant fraction of the total volume. Thus the volume of the ideal gas should be lower than that of real gas.

(ii) Moreover under conditions of high pressures and low temperatures, when molecules are brought quite close together, some intermolecular attractions come into play. The pressure exerted by the gas get reduced and therefore, ideal pressure should be higher than the real pressure.

Taking into account these factors Van der Waals suggested a modified form of gas equation for real gases.

Van der Waals Equation

Van der Waals in 1873 proposed an equation of state. For one mole of a gas it can be written as

$$\left(P + \frac{a}{V^2}\right) (V - b) = Rt$$

'a' and 'b' are Van der Waals constants and theory depend upon the nature of the gas.

'b' denotes the volume excluded from molecular motion for each mole of the gas having a molal volume υ. The increased impact rate of the molecules in the restricted volume (V – b) will cause the kinetic pressure of the gas due to kinetic energy to increase to

$$\left(\frac{RT}{V}\right)\left(\frac{V}{V-b}\right) = \frac{PT}{V-b}$$

The cohesive forces between molecules may thus be considered to decrease rapidly with distance between the molecules and so this force is appreciable only at small distances.

A molecule at a large distance from the wall of the container is attracted by its neighbouring molecules from all sides and thus the net effect is zero. However for a molecule near the wall of the container, there is an average net force normal to the wall and this tends to pull the molecule from the wall and thus the pressure on the wall is reduced. The decrease in pressure is proportional to the number of molecules being pulled and is also proportional to the number of molecules that perform the pulling. Thus the cohesive pressure because of potential energy must be proportional to the square of the density of the gas and can be written as $\frac{a}{V^2}$ where 'a' is a constant.

Thus the net pressure p is given by

$$P = \frac{RT}{V-b} - \frac{a}{V}$$

Rearranging we can write

$$\left(P + \frac{a}{V^2}\right)(V-b) = RT.$$

This is Van der Waals equation of state.

Liquifaction of Gases

Higher pressure bring the gas molecules together, the effect of attractive forces become appreciable. The attractive forces are opposed by motion of the gas molecules. At low temperature average K.E. of the molecules is low. Thus at extremes of these conditions gases liquify. Higher the temperature higher the pressure that is required to be applied to affect liquifaction. However for each gas there is a temperature above which it is impossible to liquify the gas no matter low high is the applied pressure. This temperature is called the *critical temperature of the gas.* The corresponding pressure required to be applied at critical temperature to liquify a gas is called *critical pressure.*

The Vander Waals equation is represented graphically as a family of isotherms

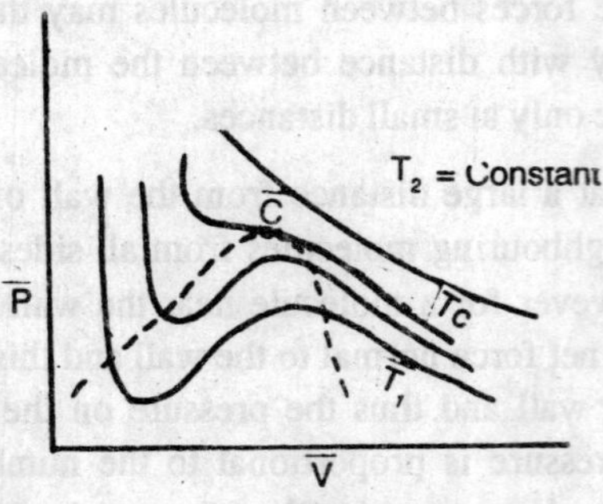

Fig. 6.1.

Since the Van der Waals equation is cubic equation in V so V has three roots for any given value of P and T. At high temperature such as at T_2 only one root is real for all pressures. At lower temperatures such as T_1, all three roots are real over a certain pressure range. Thus a maximum and minimum are formed on the isotherm in the region where experimentally the phenomenon of condensation or vapourisation occurs. For some values of P and T the three real roots became identical and the isotherm is labelled as T_C which has a point of inflection at C with a horizontal tangent. This point is called the *critical point.*

Since the critical isotherm, has a point of inflection at the critical point, we have at this point

$$\left(\frac{\partial P}{\partial V}\right)_{T_c} = 0 \qquad \text{and} \qquad \left(\frac{\partial^2 P}{\partial v^2}\right)_{T_c} = 0$$

Differentiating the Vander Waals equation and applying the results to critical point, we have

$$\left(\frac{\partial P}{\partial V}\right)_{T_c} = -\frac{RT_c}{(V_c - b)^2} + \frac{2a}{V_c^3} = 0$$

and

$$\left(\frac{\partial^2 P}{\partial V^2}\right)_{T_c} = \frac{2RT_c}{(V-b)^3} - \frac{6a}{V_c^4} = 0$$

Solving these two equations together with the Van der Waals equation as applied to the critical point, we obtain the following expression for the critical properties P_c, T_c, V_c in terms of Vander Waals constants a and b.

$$P_c = \frac{a}{27b^2} \tag{1}$$

$$T_c = \frac{8a}{27Rb} \tag{2}$$

$$V_c = 3b \tag{3}$$

Conversely, the Van der Waals constants a and b can be expressed in terms of the critical properties. Thus the simultaneous solution of the above equations gives (1, 2 and 3)

$$a = \frac{27}{64}\frac{(RT_c)^2}{P_c} \tag{4}$$

$$b = \frac{RT_c}{8P_c} \tag{5}$$

Though Van der Waals equation describes the general behaviour of real gases, it holds only qualitatively and any attempt to test the equation quantitatively is doomed to failure one definite and clear deficiency is seen in the value of the ratio of $\frac{P_cV_c}{RT_c}$, which according to Vander Waals equation is $\frac{P_cV_c}{RT_c} = \frac{\left(\frac{a}{27b^2}\right)(3b)}{R\left(\frac{8a}{27Rb}\right)} = \frac{3}{8} = 0.375$

for all gases. But actual values of $\frac{P_cV_c}{RT_c}$ for different gases formed experimentally are different. In most cases they are much smaller that predicted value.

However this equation for real gas is not expected to hold good at high densities such as the critical region due to Van der Waals simplified reasoning. Testing of the equation for lower densities should give better correlation.

Other Equations of State

A large number of equations of state have been suggested to account for the behaviour of real gases. Some are based on theoretical arguments where as some are strictly empirical.

One such equation proposed was BEATTIE-BRIDGEMAN Equation

The equation can be written as

$$P = \frac{RT\left(1 - \dfrac{C}{VT^3}\right)}{V^2}\left[V + B_0\left(1 - \frac{b}{V}\right)\right] - \frac{A_0}{V^2}\left(1 - \frac{a}{V}\right) \qquad (6)$$

In addition to the gas constant R, in this equation, we find five specific constants (A_o, a, B_o, b, c) for a gas. Values of these constants for a large number of gases have been found experimentally.

This equation is reasonably accurate when the volumes involved in calculations are greater than twice the critical volumes. In the critical region it is inaccurate.

Benedict-Webb Rubin Equation

It is entirely empirical equation and was obtained by putting experimental data of light hydrocarbon. It may be written as

$$P = \frac{RT}{V} + \left(B_o RT - A_0 - \frac{C_0}{T^2}\right)\frac{1}{V^2} + (bRT - a)\frac{1}{V^3} + \frac{\alpha}{V^6}$$

$$+ \frac{C\left(1 + \dfrac{\gamma}{\varsigma^2}\right)}{T^2} \cdot \frac{1}{V^3}\, e^{-\gamma/v^2} \qquad (7)$$

In this equation there are eight parameters B_o, A_o, C_o, b, a, c, α and γ are numerical constants which are different for different gases.

This equation is sufficiently accurate in describing P–V–T relation of real gases upto densities of 1.8 times the critical density.

Martin and Hou Equation

$$P = \frac{RT}{V-b} + \frac{A_2 + B_2T + C_2\, e^{-5.475 \frac{T}{T_c}}}{(V-b)^2} + \frac{A_3 + B_3T + C_3 e^{-5.475 \frac{T}{T_c}}}{(V-b)^3}$$

$$+ \frac{A_4}{(V-b)^4} + \frac{B_5T}{(V-b)^5} \tag{8}$$

In this equation we have nine specific constants (i.e. b, A_2, B_2, C_2, A_3, B_3, C_3, A_4 and B_5).

Later on this equation has been modified by adding to it one more term

$$\frac{A_5 + C_5\, e^{-5.475 \frac{T}{T_c}}}{(V-b)^5}$$

By addition of this terms the number of specific constants has increased to 11 from 9 the new specific constants being A_5 and C_5.

The values of various specific constants have been found for a number of gases.

This equation is valid for both polar and non-polar gases upto densities of 1.5 times the critical density.

In general, the more specific constants used in an equation of state, the more accurate the equation can be. However, it is not always so.

One of the simpler equations that is quite accurate over a considerable wide range has only two specific constants. This equation is **REDLICH-KWONG** Equation.

It can be written as

$$P = \frac{RT}{(V-b)} - \frac{a}{T^{1/2}\ V(V+b)} \quad (9)$$

It has only two constants (i.e. a, b). These constants can be evaluated from critical constants. These are

$$a = \frac{0.42748\, R^2\, T_c^{2.5}}{P_c}$$

and

$$b = \frac{0.08664 RT_c}{P_c}$$

This equation given satisfactory results above the critical temperature for any pressure.

DIETERICIS Equation

$$P(V-b) = RT\, e^{\frac{a}{RTV}} \quad (10)$$

CLAUSIUS Equation

$$\left[P + \frac{a}{T(V+C)^2}\right][V-b] = RT \quad (11)$$

In this equation C is another constant.

BERTHOLET Equation

$$\left(P + \frac{a}{TV^2}\right)(V-b) = RT \quad (12)$$

VIRIAL Equation of State

$$PV = A + BP + CP^2 + DP^3 + \ldots \quad (13)$$

In this equation A must be equal to RT to satisfy the ideal gas equation at zero density.

So $$\frac{PV}{RT} = 1 + \frac{B}{V} + \frac{C}{V^2} + \frac{D}{V^3} \ldots.$$

In this B, C, D etc. are called second, third, fourth etc. One called *virial coefficients.*

These coefficients are functions of temperature and also of gas.

The virial equation can also be expressed in terms of a power series of pressure as

$$\frac{PV}{RT} = 1 + B'P + C'P + D'P^3 + ..$$

where the coefficients B′, C′, D′ etc. are functions of temperature of the gas.

Any of the equations of state can be expanded in the virial form e.g. we can expand the Vander Waals equation in the virial form as follows.

$$PV = RT\left(1 - \frac{b}{V}\right)^{-1} - \frac{a}{V} \qquad \text{[Van der Waals Equ.]}$$

using Binomial theorem, we get

$$PV = RT\left(1 + \frac{b}{V} + \frac{b^2}{V^2} + \frac{b^3}{V^3} + \ldots\right) - \frac{a}{V}$$

or

$$PV = 1 + \frac{b - \left(\frac{a}{RT}\right)}{V} + \frac{b^2}{V^2} + \frac{b^3}{V^3} + \ldots$$

This is the Van der Waals equation in the virial form in terms of density. The virial coefficients are

$$B = b - \frac{a}{RT}\ ; \quad C = b^2\ ; \ D = b^3 \text{ etc.}$$

Boyles' Temperatures (T_B): The temperature at which a gas shows ideal gas behaviour over a considerable range of pressure is known as Boyes temperature.

Relation between Boyles temperature (T_B) and Van der Waals Constants (*a, b*)

$$T_b = \frac{a}{b \cdot R}$$

Equation of State and Second Virial Coefficient

We can write the Van der Waals equation as

$$PV = \frac{RT}{\left(1 + \frac{a}{PV^2}\right)\left(1 - \frac{b}{V}\right)}$$

The correction terms (i.e. $\frac{a}{PV^2}$ and $\frac{b}{V}$) are both small, of the gas is not too much compressed these terms can be neglected.

Using the binomial theorem and neglecting squares and products of correction terms, the equation becomes

$$PV = RT\left(1 + \frac{b}{V} - \frac{a}{PV^2}\right)$$

$$= RT + \frac{RT}{V}b - \frac{RT}{PV^2} \cdot a$$

The second and third terms of R.H.S are small correction terms. Thus by approximation, we can write the equation, using the approximation as $PV = RT$ is above equation, it becomes

$$PV = RT + \left(b - \frac{a}{RT}\right)P$$

According to Vander Waals equation the second virial coefficient B is given by

$$B = b - \frac{a}{RT}$$

The *Boyles temperature* (T_B) is given by

$$B = 0$$

Thus $\quad T_B = \frac{a}{Rb}$

We also known, $T_C = \frac{8a}{27R.b}$ $\left[T_c = \text{ critical temp.}\right]$

Thus according to Van der Waals equation,

$$\frac{T_B}{T_C} = \frac{27}{8} = 3.375$$

If we now consider Bertholet equation, which may be written as

$$PV = RT + \frac{9}{128} \cdot \frac{RT_c}{P_c}\left(1 - \frac{8T_c^2}{T^2}\right)P$$

According to this equation the second virial coefficient B is given by

$$B = \frac{9}{128} \; \frac{RT_c}{P_c}\left(1 - \frac{6T_c^2}{T^2}\right)$$

and the Boyles Temperature is given by

$$\frac{T_B}{T_C} = \sqrt{6} = 2.45$$

This value is found to be in better agreement with experimental results as compared to the ratio of $\frac{T_B}{T_C}$ predicted by Van der Waals equation.

Principle of Corresponding States

We can define reduced temperature (θ), reduced pressure (π) and reduced volume (φ) as

$$\theta = \frac{T}{T_c}$$

$$\pi = \frac{P}{P_c}$$

$$\phi = \frac{V}{V_c}$$

where T_c, P_c and V_c are critical temperature, critical pressure and critical volume.

Thus θ is the ratio of the temperature of the gas to its critical temperture, π is the ratio of pressure of gas to its critical pressure and ϕ is the ratio of volume of gas to its critical volume.

From above ratios, we get

$$T = \theta \, . \, T_c$$

$$P = \pi \, . \, P_c$$

$$V = \phi \, . \, V_c$$

Substituting these values of P, T, V and also the values of T_c, P_c, Vc in the Van der Waals equation, we get

$$\left(\pi + \frac{3}{\phi^2}\right)(3\phi - 1) = 8\,\theta \qquad \text{(i)}$$

Equation (i) is called the ***reduced equation of state*** of Van der Waals.

In this equation we find that there are no constants that depend on the nature of the substance under consideration. Thus if for any two substances any two of the quantities θ, π and ϕ are same, then the third will also be same. This is called *law of corresponding states.*

This principle of corresponding states is valid, if

(i) The effect of quantization of the translational degrees of freedom is negligible.

(ii) The molecules are spherically symmetrical either actually or by virtue of free rotation.

(iii) The inter molecular degrees of freedom are independent of the density.

(iv) The potential energy of a pair of molecules is of the form given by de Boer.

The (i) condition excludes all the isotopes of the lightest molecules hydrogen and helium, whereas the polar molecules are ruled out by conditions (ii) to (iv).

However it is not possible to divide substances into those for which principle holds and those for which at does not hold good i.e. Different substances accord with the principle to a greater or lesser degree. Heavy inert elements follow the principle to a great extent whereas highly polar molecules, molecules forming hydrogen bonds and chemically reactive molecules show a large deviation from the principle of corresponding states.

Suggestion has applied the principle of corresponding states to examine.

(i) The critical parameters

(ii) The Virial coefficients and the Boyles temperature

(iii) The densities of coexistent phases

(iv) The vapour pressure, heat capacity, coefficient of thermal expansion and surface tension of the liquid phase.

(v) The temperature and the pressure at the triple point.

The following gases follow the principle:

Argon, krypton, Xenon, Neon (less accuracy).

The following gases follow the principle with fair accuracy:

Nitrogen, oxygen, carbon dioxide, methane.

de Boer discussed this law from the point of view of quantum theory. He introduced new parameters, characteristic energy (ε) and characteristics length (σ) for each molecular species.

The macroscopic quantities temperature, pressure and volume in terms of molecular units are $\frac{\varepsilon}{\kappa}$, $\frac{\varepsilon}{\delta^3}$ and $N\sigma^3$ respectively where K is Boltzmann's constant and N is Avagadro's number of particles.

According to law of corresponding states then there should be a unique relation between pressure, volume and temperature expressed in molecular units. It has been found that the relation which is obtained depends also on a parameter.

$$\Lambda^* = h/\sigma \ . \ \sqrt{m\varepsilon}$$

which has a characteristics value for each substance considered.

The classical law of corresponding states is obtained by the limiting value $\Lambda^* \to 0$.

Various properties of gaseous elements have been examined as this basis and it has been formed that, for higher elements, the molar volume and internal energy depart from the classical limit. This derivation can be related to zero-point energy.

The solid-liquid transition has also been considered and an explanation has been provided of the fact that a helium isotope of mass 4 cannot be solidified under its equilibrium vapour pressure.

de Boer made calculations of the molar volume, as a function of temperature, of the helium isotope of mass 3 in the liquid and the gaseous states and the equilibrium vapour pressure of the liquid. The vapour pressure curve agrees fairly accurately with the predictions made by de Boers. More recently various other measurements such as those of density function, viscosity etc. have been made for this isotope of helium and the values are found to agree with the predicted value at 4.2 K. The values are about 12% lower at 1.3 K.

The law of corresponding states has now been extended to solids at low temperatures, considering the frozen rare gases. This method has been used to calculate the potential energy and the interatomic distances of Ne, Ar, Kr and Xe. The values are found to be in agreement with experimental values.

7

Kinetic Theory of Gases

Kinetic Theory of Matter. In this theory the two important concepts that have been combined are

(i) heat is a form of energy, and

(ii) matter is molecular in nature.

The essential features of the theory are

(a) Every substance is made up of a very large number of tiny particles called *molecules.*

(b) Molecules are in a state of continuous motion with all possible velocities in all possible directions. The velocity of molecules increase with an increase in temperature.

(c) The molecules attract one another with a force which rapidly decreases as the distance between molecules increases.

(d) The energy of molecules in motion constitutes the heat energy of the matter. Any increase in the quantity of heat in a body increases the energy of its molecules. In case of gases the heat energy is mostly in the form of kinetic energy of the molecules. However in solids it is partly in the form of potential energy and partly in the form of kinetic energy.

Kinetic Theory of Gases

The *Kinetic molecular theory of gases,* developed during

eighteenth and nineteenth century by various scientists provides an explanation for the behaviour of gases.

This theory was developed for *ideal gas*. Though ideal gas does not exist, all gases approach the ideal gas behaviour under moderate conditions of temperature and pressure.

The important postulates of this theory are

(i) A gas is made up of particles (atoms or molecules) whose dimensions are very small compared with the very large average distance between the particles.

(ii) These particles are in constant random motions, moving in straight lines unless they collide with another particle or the walls of the container.

(iii) Attraction or repulsion between gas particles is negligible.

(iv) The average kinetic energy (the energy of motion) of these particles is indirectly proportional to the Kelvin temperature.

Although the (iii) postulate states that attraction between gas particles are negligible, attractions must exist or gases would never condense into the liquid state. Attractive forces are responsible for holding the liquid state together. Clearly the assumption that these forces are absent in gases is a simplification. Nevertheless, attractive forces between gas particles are negligible under moderate conditions of temperature and pressure.

According to (iv) postulate, the average kinetic energy of gas particles is directly proportional to the Kelvin temperature. The term directly proportional means that the average kinetic energy will double when Kelvin temperature is doubled. The motion of gas particles became more chaotic and violent as the temperature increases and more restrained as the temperature decreases.

These postulates explain the behaviour of gases as follows:

Gases fill a container completely and uniformly because the particle move constantly and rapidly and virtually have no attraction to one another. Gases are easily compressed and have low densities because of very large average distances between the particles. In fact most of the space in any gas sample, under moderate conditions is empty space (without any gas particles). Finally gases exert a pressure on any surface that they touch because the gas particles collide with the surface and push against it.

Pressure Exerted by a Gas

When a certain mass of a gas is enclosed in a vessel, the molecules collide with the walls of the container and are being reflected from them. During the collision each molecule suffers a change in momentum.

According to Newton's second law of motion, the rate of change of momentum is equal to the force exerted. This pressure can be formed out using the postulates of kinetic theory of gases.

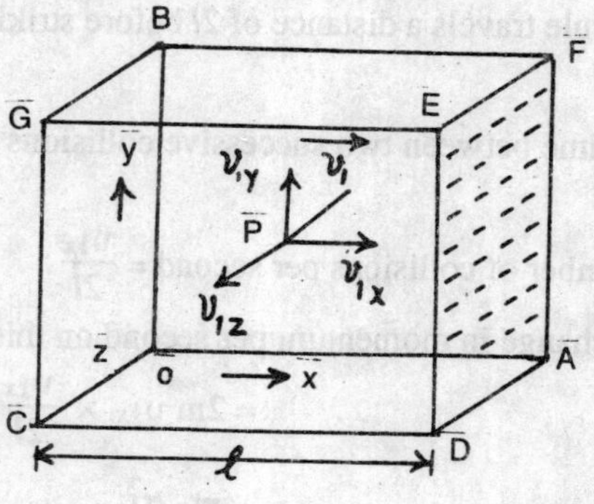

Fig.

Consider a perfect gas enclosed in a cubical vessel of side 'l' having perfectly elastic walls. the number of gas molecules enclosed in the vessel be '*n*' and the mass of each molecule be '*m*'. Let the sides of the cube represent mutually perpendicular X, Y, Z axes.

Consider a molecular P moving in random direction with a velocity υ_1. Let $\upsilon_{1\xi}$, υ_{1y} and v_{1z} be the components of υ_1 along three mutually perpendicular axes x, y and z, respectively. Then

$$\upsilon_1^2 = \upsilon_{1x}^2 + \upsilon_{1y}^2 + \upsilon_{1z}^2$$

The component of velocity with which the molecule P will strike the face DFEA is υ_{1x} and the momentum of the molecule along x-direction before collision is mx υ_{1x}.

Since the collision is perfectly elastic so during collision there is no change in the speed of the molecule and only the direction of velocity changes. Hence

the momentum after collision, along x-axis is

$$= m(-\upsilon_{1x})$$

$$= -m\,\upsilon_{1x}$$

The change in momentum $= m\,\upsilon_{1x} - (-m\upsilon_{1x})$

$$= -2m\upsilon_{1x}$$

For simplicity we treat each component of velocity separately.

The molecule travels a distance of $2l$ before striking the same face again.

Thus the time between two successive collisions on the same face in $\frac{2l}{\upsilon_{1x}}$

Hence number of collisions per second $= \frac{\upsilon_{1x}}{2l}$

Thus the change in momentum per second on this face

$$= 2\text{m}\,\upsilon_{1x} \times \frac{\upsilon_{1x}}{2l}$$

$$= \frac{m\,.\,\sigma_{1x}^2}{l}$$

Since all the molecules of a gas are identical and the time spent in collision is negligible, we can safety assume that the original molecule continues to move with its initial speed. Considering all the n molecules striking the some face, we get total change in momentum.

$$\upsilon_{1x}^2 = \frac{m\,.\,\upsilon_{1x}^2}{l} + \frac{m\,.\,\upsilon_{2x}^2}{l} + \frac{m\,.\,\upsilon_{3x}^2}{l} + \ldots\ldots \frac{m\,.\,\upsilon_{nx}^2}{l}$$

$$= \frac{m}{l}\sum_i \upsilon_{ix}^2$$

where $\upsilon_{1x}, \upsilon_{2x}, \ldots \upsilon_{nx}$ are x-components of the velocities $\upsilon_1, \upsilon_2 \ldots$ $.\upsilon_n$ of the n molecules.

Since rate of change of momentum is equal to the force exerted by the gas, so the net force exerted by the gas on this face is equal to

$$= \frac{m}{l} \sum_i \upsilon_{ix}^2$$

Hence Pressure P is given by

$$P_x = \frac{\text{Force}}{\text{Area}}$$

$$= \frac{m}{l \cdot l^2} \sum_i \upsilon_{ix}^2$$

or

$$P_x = \frac{m}{V} \sum \upsilon_{ix}^2$$

Similarly we can get

$$P_y = \frac{m}{V} \sum \upsilon_{iy}^2$$

and

$$P_z = \frac{m}{V} \sum \upsilon_{iz}^2$$

When the cube in small, the gravity does not play any role and the pressure on all faces may be taken as equal in magnitude

i.e. $\quad P_x = P_y = P_z$

By adding we get

$$3P = P_x + P_y + P_z = \frac{m}{V} \sum_i \left(\upsilon_{ix}^2 + \upsilon_{iy}^2 + \upsilon_{iz}^2\right)$$

or

$$P = \frac{1}{3} \frac{m}{V} \sum_i \left(\upsilon_{ix}^2 + \upsilon_{iy}^2 + \upsilon_{iz}^2\right)$$

$$= \frac{1}{3} \frac{m}{V} \sum_i \upsilon_i^2$$

Now if υ^2 is the mean average speed of all the molecules, then

$$\upsilon^2 = \frac{V_1^2 + V_2^2 + \ldots\ldots V_n^2}{n} = \frac{\sum_i \upsilon_i^2}{n}$$

or

$$\sum \upsilon_1^2 = n\,\upsilon^2$$

Thus

$$P = \frac{1}{3}\,\frac{m\,n\,\overline{\upsilon^2}}{V}$$

But $m\,n$ = M = Total mas of the gas

So $\frac{m\,n}{V} = l$ = density of the gas

Therefore pressure of the gas becomes

$$P = \frac{1}{3}\,l\,\upsilon^2$$

or

$$P = \frac{2}{3}\,.\,\frac{1}{2}\,l\,\overline{\upsilon^2} = \frac{2}{3}\,E$$

where

$$E = \frac{1}{2}\,l\,\overline{\upsilon^2} = \text{K.E. per unit volume.}$$

Thus *the pressure of a gas is equal to two-third of total translational k.E. of the molecules per unit volume.*

Some deduction from Expression for Pressure

If 1 gram molecule of a gas are contained in a Volume V at Temperature T absolute, then

$$P = \frac{1}{3}\,\frac{m\,N}{V}\,\overline{\upsilon^2}$$

where

N = Avagadro's Number.

Kinetic Interpretation of Temperature

From the above equation, we have for 1 mole of a gas

$$PV = \frac{1}{3} m N \overline{\upsilon^2}$$

According to ideal gas equation, for 1 mole.

$$PV = RT$$

Comparing these two equations, we get

$$\frac{1}{3} m N \overline{\upsilon^2} = RT$$

or

$$\frac{1}{2} m \overline{\upsilon^2} = \frac{3}{2} \frac{R}{N} T$$

$$= \frac{3}{2} k T$$

where $k = \frac{R}{N}$ = Boltzmann's constant = 1.38×10^{-23} J/K

From this equation, we observe that *Average kinetic energy of translation of the molecules is directly proportional to the absolute temperature of the gas.*

Thus average kinetic energy of the molecules can be considered as a measure of temperature.

Moreover at absolute zero, the K.E. of gas molecules must be zero i.e. the molecular motion of a gas ceases. Hence **absolute zero** is that temperature at which all molecular motion of a gas ceases.

From the above equation, we also have $\overline{\upsilon^2} \propto T$

i.e. *the mean square velocity of molecules is directly proportional to absolute temperature.*

Boyles Law

The kinetic molecular theory of gases provides an explanation for Boyles' law. According to the (ii) postulate of the theory, gas molecules are in constant, random motion, moving in straight line, until they collide with another particle or the walls of the container. Since average kinetic enregy is directly proportional to Kelvin temperature, so the average energy of collision when one gas molecule hits the wall of the container will not change when volume decreases at constant temperature. However the pressure will increase when the volume deoreases because there will be more collisions with the walls of the container per unit area. This is Boyles' law.

Also from kinetic equation of gas, we have

$$PV = \frac{1}{3} mN\overline{v^2}$$

Since

$$\frac{1}{2} m\overline{v^2} \alpha T$$

So at constant T, we have

$$PV = \frac{2}{3}N \cdot \frac{1}{2}m\ v^2 = \frac{2}{3} N \alpha\ T$$

This is Boyles Law.

Charles's Law

The kinetic molecular theory explains Charles law.

Consider a gas contained in an elastic container. On cooling the pressure inside the container will not change because the pressure of the gas inside an elastic container is always identical to atmospheric pressure. However due to decrease in temp. the average kinetic energy of the gas particles will decrease. Less powerful collisions with the walls of container will occur since the gas particles exert less pressure on the walls of the container and to keep the pressure equal to atmospheric pressure the area of the walls of container must decrease if the collision force decreases, hence the container must contract i.e. its volume must decrease. This is Charles law.

From the kinetic gas equation

$$PV = \frac{1}{3} mN\overline{v^2}$$

We know, K.E. $= \frac{1}{2} m\overline{v^2}\ \alpha\ T$

$$\therefore \qquad PV = \frac{2}{3} N \alpha T$$

V α T for a constant value of P. This is Charles law.

Gas Equation

$$PV = \frac{1}{3} m N \overrightarrow{\upsilon^2}$$

Also $\quad \frac{1}{2} m \overrightarrow{\upsilon^2} \; \alpha \, T$

or

$$PV = \frac{2}{3} \text{N}. \frac{1}{2} m \overrightarrow{\upsilon^2} = \frac{2}{3} N \alpha T$$

or

$$\text{PV} = \text{RT}$$

Where $\quad \text{R} = \frac{2N\alpha}{3} = \text{N} . \text{k} = \text{R}.$

A Vagadro's Hypothesis

Avagadro's law is explained by (i) postulate of kinetic molecular theory, which says that dimensions of gas particle are very small compared with the very large average distances between them.

Consider two gases A and B at a pressure P and each having a volume V. Let m_1 and m_2 be the masses of each molecule of A and B respectively and n1 and n2 be the number of molecules of A and B respectively. If $\overrightarrow{\upsilon_1^2}$ and $\overrightarrow{\upsilon_2^2}$ be the mean square speeds of A, B respectively. Then

Pressure of A is $\text{P} = \frac{1}{3} \; \frac{m_1 n_1 \overrightarrow{\upsilon_1^2}}{V}$

and Pressure of B is $\quad \text{P} = \frac{1}{3} \frac{m_2 n_2 \overrightarrow{\upsilon_2^2}}{V}$

From these two equation, we have

$$m_1 n_1 \overrightarrow{\upsilon_1^2} = m_2 n_2 \overrightarrow{\upsilon_2^2} \qquad \text{(i)}$$

If the two gases are at the same temperature (T), they must nave the same kinetic energy.

i.e. $$\frac{1}{2} m_1 \overrightarrow{v_1^2} = \frac{1}{2} m_2 \overrightarrow{v_2^2}$$

or

$$m_1 \overrightarrow{v_1^2} = m_2 \overrightarrow{v_2^2} \qquad \text{(ii)}$$

Thus $n_1 = n_2$ [Comparing (i) and (ii)]

This is Avagadro's hypothesis.

Graham's Law of Diffusion

In accordance with kinetic theory, we have

$$P = \frac{1}{3}\, e \,.\, \overrightarrow{v^2}$$

or

$$v_{ims} = \sqrt{\overrightarrow{v^2}} = \sqrt{\frac{3P}{e}}$$

i.e. root mean square velocity of the molecules of a gas is universely proportional to the square root of its density.

Consider two gases with root mean square velocities v_1 and v_2 respectively.

Then $$v_1 = \sqrt{\frac{3P}{e_1}}$$

$$v_2 = \sqrt{\frac{3P}{e_2}}$$

or

$$\frac{v_1}{v_2} = \sqrt{\frac{e_2}{e_1}}$$

As rate of diffusion of gas is proportional to the root mean square speed of the molecules, thus from above equation it follows that the rate of diffusion of a gas is inversely proportional to the square root of its density.

This is Graham's law of diffusion.

Dalton's Law of Partial Pressures

Let p_1 and p_2 be the partial pressures of the individual gases in a mixture of volume V_1 Let n_1 molecules of gas with partial pressure p_1 be present in the mixture. Let the average square velocity of these molecules be $\overrightarrow{\upsilon_1^2}$ with each having a mass m_1. Let the mixtur contain n_2 molecules of the gas with partial pressure p_2. Let the mass of its molecules be m_2 and their mean square velocity be $\overrightarrow{\upsilon_2^2}$.

As each gas occupies the whole volume V at its own pressure, therefore

$$p_1 V = \frac{1}{3} m_1 n_1 \overrightarrow{\upsilon_1^2}$$

$$p_2 V = \frac{1}{3} m_2 n_2 \overrightarrow{\upsilon_2^2}$$

$$\text{So that } (p_1 + p_2 + \ldots.)\, V = \frac{1}{3}\left(m_1 n_1 \overrightarrow{\upsilon_1^2} + \frac{1}{2} m_2 \overrightarrow{\upsilon_2^2} + \ldots \right)$$

Since the gaseous mixture is at the same temperature so the kinetic energy of translation per molecule of each gas is the same i.e.

$$\frac{1}{2} m_1 \overrightarrow{\upsilon_1^2} = \frac{1}{2} m_2 \overrightarrow{\upsilon_2^2} + \ldots..$$

$$= \frac{1}{2} m \overrightarrow{\upsilon^2}$$

From the above two equations, we get

$$(p_1 + p_2 + \ldots.)\, V = \frac{1}{3}(n_1 + n_2 + \ldots.)\, m\upsilon^2$$

But $(n_1 + n_2 + \ldots\ldots)$ is the total number of molecules in the mixture, Hence net pressure P is given by

$$PV = \frac{1}{3}(n_1 + n_2 + \ldots.)\, m\, \overrightarrow{\upsilon^2}$$

or $$P = p_1 + p_2 + \ldots\ldots \text{ (By comparing above equation).}$$

i.e. The net pressure due to a mixture of non-reacting gases contained in the same vessel is equal to the sum of the pressures which each gas would exert if it were alone contained in the vessel. This is Dalton's law of partial pressures.

The **Reason for upper atmosphere being Cool** inspite of the fact that its kinetic temperature is of the order of 1000K.

In the upper atmosphere the heat content of an individual molecule increases by about 3.5 times as compared to that at the earth's surface. Kinetic temperature on earth's surface is about 300K whereas in the upper atmosphere it is of the order of 1000K. But the pressure and hence the density is very low in upper atmosphere i.e. in the upper atmosphere the number of molecules per C.C. is very small. Hence the total heat content in the upper atmosphere is very small as compared to the heat content on the earth in the same volume of the gas. This results in very low temperature of the upper atmosphere.

Distribution of Molecular Speeds and Energy Maxwell's Distribution of Speeds (Velocities)

According to assumptions of kinetic theory of gases, the molecules of a gas are moving continuously in all directions with different velocities and are colliding with one another. The velocity changes after each collision. Since collisions are perfectly elastic no loss of k.E. occurs. Thus after collision some molecules will be speeded up and some will be slowed down. It means that even at constant temperature molecular chaos exists.

In 1860, Maxwell pointed out that inspite of this molecular choas there is one regularity. He suggested that at any *particular temperature the fraction of molecules possessing particular velocity remains almost constant*. This is known as *steady state*. Considering the probability laws he calculated the fraction of molecules having particular velocity at a particular temperature. On plotting the fraction of molecules against the corresponding velocities the curve obtained was always the same as shown in Fig. 7.2 (Maxwells distribution of velocities)

From the curve it can be easily observed.

(i) The fraction of molecules with very low or very high velocities are very small.

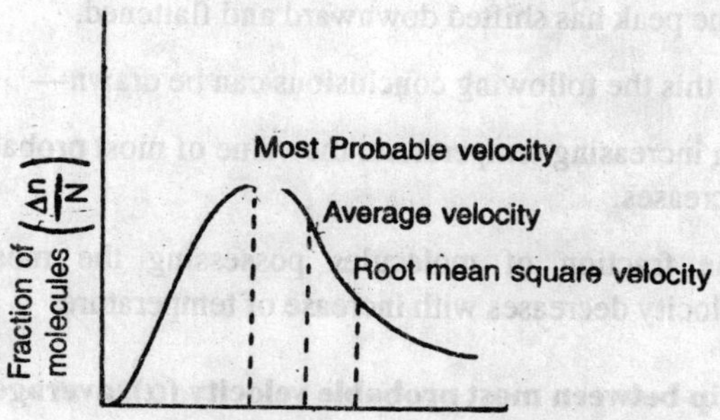

Fig. 7.2 Maxwells distribution of velocities.

(ii) The fraction of molecules possessing higher and higher velocities goes on increasing till the peak is reached. After the peak is reached this fraction starts decreasing.

(iii) The maximum fraction of molecules posses a velocity corresponding to the peak in the curve is known as most probable velocity (α).

(iv) The area under the curve represent total number of molecules.

Effects of Temperature on Maxwells Distribution of Velocities

On increasing the temperature curve shifts as shown in Fig. The important changes that can be noticed are.

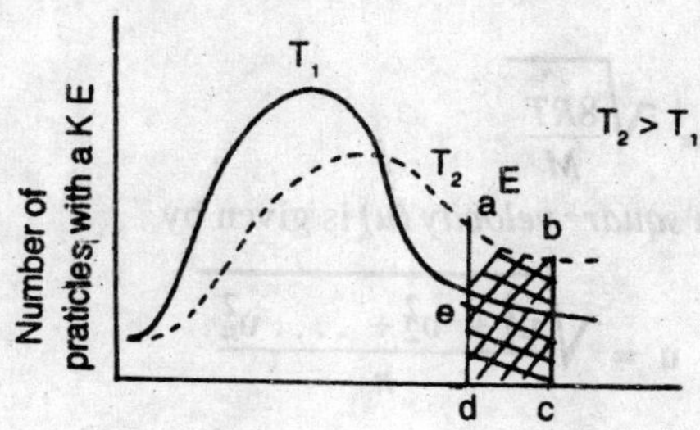

Fig. 7.3

(i) The peak has shifted forward.

(ii) The peak has shifted downward and flattened.

From this the following conclusions can be drawn—

(i) On increasing temperature, the value of most probable velocity increases.

(ii) The fraction of molecules possessing the most probable velocity decreases with increase of temperature.

Relationship between most probable velocity (α), average velocity ($\overline{\upsilon}$) and the root mean square velocity (u).

From the graph of Maxwells' distribution of velocities (Fig. 7.2) we can see that the maximum of the curve indicates the *most probable velocity* (α). i.e. the velocity possessed by the largest fraction of the molecules.

Thus $$\alpha = \sqrt{\frac{2RT}{M}}$$

The *average velocity* ($\overline{\upsilon}$) is given by

$$\overline{\upsilon} = \frac{\upsilon_1 + \upsilon_2 + \upsilon_3 + \ldots . \upsilon_n}{n}$$

where υ_1, υ_2, $\upsilon_3 \ldots . \upsilon_n$ are the actual velocities possessed by the individual molecules at any time and n is the total number of gas molecules.

or

$$\overline{\upsilon} = \sqrt{\frac{8RT}{M}}$$

Root mean square velocity (u) is given by

$$u = \sqrt{\frac{\upsilon_1^2 + \upsilon_2^2 + \ldots . \upsilon_n^2}{n}}$$

$$= \sqrt{\frac{3RT}{M}}$$

From graph and from calculations, we can get

$$\alpha = 0.8164 \times u_{rms} \quad = \sqrt{\frac{2}{3} \times u_{rms}}$$

$$\bar{\upsilon} = 1.128 \ddot{o}\, \alpha$$

Hence $u_{rms} : \bar{\upsilon} : \alpha = 1 : 0.9213 : 0.816$

Also $u_{rms} > \bar{\upsilon} > \alpha$.

Some important results

(i) $u_{rms} = 1.58 \times 10^4 \sqrt{\frac{T}{M}}$ cm s^{-1}

(ii) When the absolute temperature of the gas is *quadrupled*, the values of its various types of velocities gets *doubled*

(iii) For two different gases at same temperature

$$\frac{\upsilon_1}{\upsilon_2} = \sqrt{\frac{M_2}{M_1}}$$

(iv) For the same gas at two different temperatures

$$\frac{\upsilon_1}{\upsilon_2} = \sqrt{\frac{t_2}{t_1}}$$

Collision Properties

Collision diameter: The distance between the centres of molecules when they are close to undergo collision is known as **Collision diameter** (σ).

It can be calculated as follows:

(i) $\sigma = \left[\frac{m\bar{\upsilon}}{3\sqrt{2}\pi\eta}\right]^{\frac{1}{2}}$

Where $\bar{\upsilon}$ = average velocity

m = man of the molecule

η = viscosity

(ii) $$\sigma = \frac{9}{4}\,\eta\,.\,b\sqrt{\frac{\pi}{PTM}}$$

where η = viscosity

b = Van der Waals Constant

R = gas constant

M = molecular man

Mean Free Path. The path described by any molecule between two successive collisions may be called the *free path* and the average of such free paths as the *mean free path.* If $\alpha_1, \alpha_2 \ldots . \alpha_n$ are the successive free paths traversed in total time t, then

$$\alpha_1, + \alpha_2 + \ldots . \alpha_n = \bar{c}\,.\,t$$

where $\bar{c}$ is the average speed of the molecules

If n is the total number of collisions in time t, then

$$\alpha = \frac{\alpha_1 + \alpha_2 + \alpha_3 + \ldots . \alpha_n}{n}$$

$$= \frac{\bar{c}\,.\,t}{n} \qquad (\alpha_1 + \alpha_2 + \ldots \alpha_n = \bar{c}\,.\,t)$$

α can also be calculated using the relation

$$\alpha = \frac{1}{\sqrt{2}\,\pi\,\sigma^2 n}$$

where σ = collision diameter

N = Avagadro's number of molecules/cm^3

Also at constant P, mean free path α T

and at constant T, mean free path $\alpha \frac{1}{P}$

Also mean free path of a molecule is given by

$$\text{Mean free path} = \frac{3\eta}{e\ \upsilon_{rms}}$$

where η = coefficient of viscosity

e = density of gas

υ_{rms} = root mean square speed of molecules.

Collision Number. It is the number of collisions which a molecule registers with other molecules per second.

It can be calculated using the relation,

$$\text{Collision number} = \sqrt{2}\pi\,\sigma^2\,\bar{\upsilon}\,N$$

where $\bar{\upsilon}$ = average velocity

σ = collision diameter

N = number of molecules/cm^3

Collision Frequency (Z). It is the total number of collision taking place in a unit volume of the gas under particular conditions in one second.

It can be calculated using the relation

$$z = \frac{1}{\sqrt{2}}\sigma^2\,\bar{\upsilon}\,N^2$$

where σ = collision diameter

$\bar{\upsilon}$ = average velocity

N = number of molecules/cm^3

Also at constant pressure,

collision frequency α $T^{2/3}$

and at constant temperature

collision frequency α P^2

Kinetic Energy of a Gas molecule

For 1 mole of a gas we have

$$PV = RT = \frac{1}{3} M\,\upsilon_{rms}^2$$

Kinetic energy for 1 gram molecule (mole) of a mono atomic gas is

$$\frac{1}{2} M v_{rms}^2 = \frac{3}{2} RT$$

$$\text{k.E.} = \frac{3}{2} \text{RT}$$

$$\text{k.E.} = \frac{3}{2} \text{P}^{\text{V}} \qquad (\because \text{PV} = \text{RT})$$

If the mass of each molecule is m and there are N numbers (Avagadro's number) of molecules then k.E. of each molecule,

$$\text{K.E. of each molecule} = \frac{3}{2} \text{kT}$$

Where K is Boltzmann's constant

Boltzmann's Constant. The ratio of molar gas constant (R) to the Avagadro's number N is called Boltzmann's constant.

i.e. $$k = \frac{R}{N} = \frac{8.314 \times 10^7}{6.023 \times 10^{23}}$$

$$= 1.38 \times 10^{-23} \text{ Joules/degree}$$

$$= 8.167 \times 10^{-5} \text{ ev/k.}$$

$\therefore$ k.E. of a molecule is directly proportional to absolute temperature.

Equipartition of Energy. According to the principle of equipartition of energy, the average energy of each molecule (whether monoatomic, diatomic or polyatomic) in a gas associated with each degree of freedom is $\frac{1}{2}$ K.T.

Degree of Freedom

The total independent terms required to express the total energy of the molecule is called the *degree of freedom.*

(a) **Monoatomic gas**. The energy of each molecule in a monoatomic gas is expressed as

$$E = \frac{1}{2} m v_x^2 + \frac{1}{2} m v_y^2 + \frac{1}{2} m v_z^2$$

Here three independent variables v_x, v_y, v_z are required to express its energy.

So the degree of freedom of a monoatomic gas molecule is three.

(b) **Diatomic gas.** The energy of each molecule of a diatomic gas is expressed as

$$E = \frac{1}{2} m\,v_x^2 + \frac{1}{2} m\,v_y^2 + \frac{1}{2} m\,v_z^2 + \frac{1}{2} I_x\,\omega_x^2 + \frac{1}{2} I_2\,\omega_z^2$$

Here the total indpendent terms required to explain its energy is five.

So the degree of freedom of a diatomic gas is five.

When the vibrational energy is taken into account the degree of freedom becomes seven.

(c) **Polyatomic gas.** The energy of each molecule of a polyatomic gas is expressed as

$$E = \frac{1}{2} m\,v_x^2 + \frac{1}{2} m\,v_y^2 + \frac{1}{2} m\,v_z^2 + \frac{1}{2} I_x\omega_x^2 + \frac{1}{2} I_y\,\omega_y^2 + \frac{1}{2} I_2\,\omega_z^2$$

Here the total independent terms required is six.

So the degree of freedom of a polyatomic molecule will be six.

Internal Energy of an Ideal Gas

Monoatomic gases like He, Ne etc. only possess the translational motion i.e. they possess only linear motion.

But if we consider diatomic or polyatomic gases like O_2, H_2, CO_2, NH_3, CH_4 etc. their molecules possess both translational and rotational energy. So the internal energy of the gas is shared between translational and rotational motion of molecules.

The internal energy of an ideal gas does not depend on its volume. It depends only on its temperature. The internal energy associated with a monoatomic gas is

Fig. 7.5 Monoatomic molecule (only translatory motion).

$$\text{Internal Energy (U)} = \text{n N}\left(\frac{3}{2}kT\right) = \text{n} \cdot \frac{3}{2}\text{RT}\left[\because k = \frac{R}{N}\right]$$

Similarly internal energy associated with a diatomic molecule is

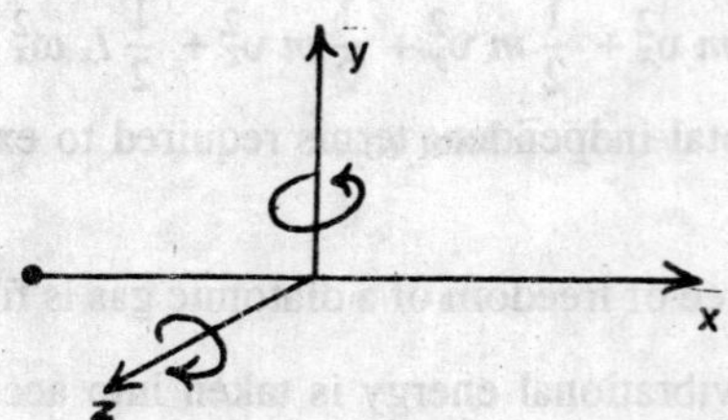

Fig. 7.6 Diatomic molecule (Translatory & rotatory motion).

$$U = n\,\text{N}\left(\frac{5}{2}kT\right)$$

$$= n.\frac{5}{2}RT$$

where n = amount of gas in moles

N = Avagadro's number

R = Universal gas constant

Average energy per molecule of

a monoatomic gas $= \frac{3}{2}$ kT (Translational)

Average energy per molecule of a diatomic gas (if the molecules do not vibrate)

$= \frac{3}{2}$ kT (translational)

$+ \frac{2}{2}$ kT (rotational)

$= \frac{5}{2}$ kT

If the molecules do vibrate, then the average energy per molecule

$$= \frac{3}{2} kT \text{ (translational)}$$

$$+ \frac{2}{2} kT \text{ (rotational)}$$

$$+ \frac{5}{2} kT$$

If the molecules do vibrate, then the average energy per molecule

$$= \frac{3}{2} \text{kT (translational)}$$

$$+ \frac{2}{2} \text{kT (rotational)}$$

$$+ \frac{2}{2} \text{kT (vibrational)}$$

$$= \frac{7}{2} \text{kT.}$$

Numericals

Example. 1. Calculate the RMS velocity of oxygen molecule at 1 atm and 0°C (the density of oxygen is 1.42 g per litre).

Solution. $C = \sqrt{\frac{3p}{d}}$

Given $P = 1$ atmo = 76 cm of Hg

$= 76 \times 13.6 \times 981$ dynes/cm^2

d = 1.429 g/l = 0.001429 g/cc.

Substituting these values, we get

$$C = \sqrt{\frac{3p}{d}} = \sqrt{\frac{3 \times 76 \times 13.6 \times 981}{0.001429}}$$

$$= 46120 \text{ cm/s.}$$

Example. 2. Calculate the r.m.s and average velocity of hydrogen molecules at 0°C.

Solution. For 1 mole of a gas, we have

$$PV = RT = \frac{1}{3} m\,Nc^2 \qquad (N = \text{Avagadro's number})$$

$$= \frac{1}{3} Mc^2 \qquad (M = \text{Mol. wt})$$

or

$$C = \sqrt{\frac{3RT}{M}}$$

Substituting the given values, we get

$$C = \sqrt{\frac{3 \times 8.314 \times 10^7 \times 273}{2.016}} \qquad [\because R = 8.31 \times 10^7]$$

$$= 18.4 \times 10^4 \text{ cm/s.}$$

Example. 3. Calculate the K.E. in Joules of 1 mole of CO_2 at 27°C, assuming ideal behaviour. What is the average kinetic energy of a single molecule?

Solution. $E = \frac{3}{2} RT$

$$= \frac{3}{2} \times 8.314 \times 10^7 \times 300$$

$$= 3741 \times 10^7 \text{ ergs.}$$

$$= 3741 \text{ J.}$$

Average K.E. of a molecule $= \frac{3}{2} \frac{RT}{N_0}$

$$= \frac{3741 \times 10^7}{6.02 \times 10^{23}}$$

$$= 6.214 \times 10^{-14} \text{ ergs.}$$

Example. 4. Calculate the total kinetic energy, in Joules, of the molecules in 8g of methane at 27°C.

Solution. K.E $= \frac{3}{2} RT = 3741$ J (See previous example)

Total energy of 8g of methane $= \frac{1}{2}$ mole of methane

$$= \frac{1}{2} \times 3741$$

$$= 1870.5 \text{ J.}$$

(Note: Average K.E. per molecule will be same at the same temperature, irrespective of the nature of the gas).

Example. 5. Calculate in ergs the k.E. of a molecule of monoatomic gas moving with a velocity of 5×10^4 cm sec^{-1}. Given the molecular weight as 20.2.

Solution.

Weight of 1 molecule of gas $= \dfrac{20.2}{6.02 \times 10^{23}}$

$$= 3.355 \times 10^{-23} \text{ g/mole cule}$$

K.E. of a single molecule $= \frac{1}{2} m c^2$

$$= \frac{1}{2} \times 3.355 \times 10^{-23} \times (5 \times 10^4)^2$$

$$= 4.194 \times 10^{-14} \text{ ergs.}$$

8

Maxwell's Thermodynamic Relations

In the previous chapters we discussed the first and second laws of thermodynamics. We have derived certain mathematical equations involving the thermodynamics properties to define the state of a system. To define the state of a system. any pair of quantities P, V, T, S, E, H, A, G may be used. We have established that the functions E, H, S, A, G. These are state functions and so their differentials are exact. All these have been shown to be extensive properties.

To solve a thermodynamical problem a suitable pair is chosen so that the quantities constituting the pair are independent variables.

From the first law of thermodynamics, we have

$$dE = dQ - PdV \quad (1)$$

and

$$dH = dQ + VdP \quad (2)$$

From the combination of first and second law, we have

$$dE = TdS - PdV \quad (3)$$

$$dH = TdS + VdP \quad (4)$$

$$dA = -SdT - PdV \quad (5)$$

$$dG = -SdT + VdP \quad (6)$$

These equations give us independent variables which may be used for each of the four functions. These are

$$E = f(S, V)$$

$$H = f(S, P)$$

$$A = f(T, V)$$

$$G = f(T, P)$$

Since dE, dH, dA and dG are exact differentials, therefore

$$\left(\frac{\partial E}{\partial S}\right)_V = T \text{ and } \left(\frac{\partial E}{\partial V}\right)_S = -P \tag{7}$$

$$\left(\frac{\partial H}{\partial S}\right)_P = T, \quad \left(\frac{\partial H}{\partial P}\right)_S = V \tag{8}$$

$$\left(\frac{\partial A}{\partial T}\right)_V = -S, \quad \left(\frac{\partial A}{\partial V}\right)_T = -P \tag{9}$$

$$\left(\frac{\partial G}{\partial T}\right)_P = -S, \quad \left(\frac{\partial G}{\partial P}\right)_T = V \tag{10}$$

From these equations we can see the change in one dependent variable with a change in one independent variable, keeping the appropriate variable constant.

Since these are exact differentials so Euler's theorem is applicable. Hence we get the following relations known as Maxwells relations.

Applying Euler's reciprocity theorem to equations 3, 4, 5, 6 we get

$$\left(\frac{\partial T}{\partial V}\right)_S = -\left(\frac{\partial P}{\partial S}\right)_V \tag{11}$$

$$\left(\frac{\partial T}{\partial P}\right)_S = \left(\frac{\partial V}{\partial S}\right)_P \tag{12}$$

and

$$\left(\frac{\partial S}{\partial V}\right)_T = \left(\frac{\partial P}{\partial T}\right)_V \tag{13}$$

$$\left(\frac{\partial S}{\partial P}\right)_T = -\left(\frac{\partial V}{\partial T}\right)_P \qquad (14)$$

By reversing the equations (11) and (12), we get

$$\left(\frac{\partial S}{\partial P}\right)_V = -\left(\frac{\partial V}{\partial T}\right)_S$$

and $$\left(\frac{\partial S}{\partial V}\right)_P = \left(\frac{\partial P}{\partial T}\right)_S$$

All these are important as they relate the rate of change of a particular quantity to P, V and T.

The given diagram known as thermodynamics square was proposed by Max Born and F.O. Koeing, which can be used for a system of constant mass and composition which does only the work of expansion.

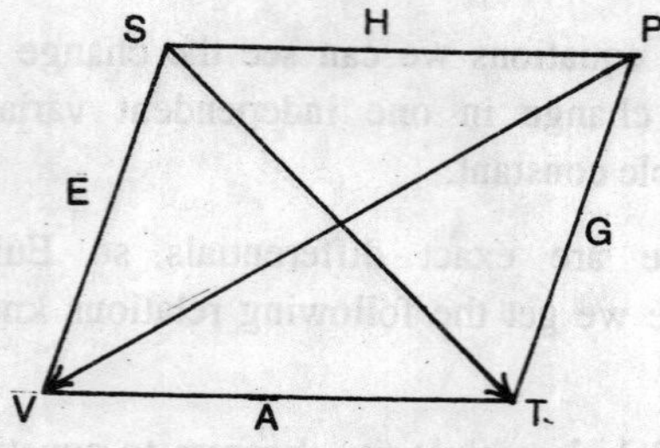

Fig.

In this diagram the functions E, H, G, A are placed on the sides of a square and the corners of the sides contain the respective independent variables.

i.e., $E = f(S, V), H = f(S, P), G = f(T, P), A = f(T, V)$

To read this square remember that the coefficient of the differential of the independent variables is placed diagonally opposite to this variable and the arrow pointing away from a natural variable is indicative of a positive sign. So we can write

$$dE = TdS - PdV$$

$$dH = TdS + VdP$$

$$dA = -SdT - PdV$$

$$dG = -SdT + VdP$$

These relations can also be written as under:

(*i*) **Relation I.** Taking T and V as independent variables. we have

$$\left(\frac{\partial S}{\partial V}\right)_T = \left(\frac{\partial P}{\partial T}\right)_V \qquad \text{[equation 13]}$$

But

$$dS = \frac{dH}{T}$$

$$\therefore \quad \left(\frac{\partial H}{\partial V}\right)_T = \mathrm{T}\left(\frac{\partial P}{\partial T}\right)_V \qquad (15)$$

(*ii*) **Relation II.** Taking T and P as independent variables,

$$\left(\frac{\partial S}{\partial P}\right)_T = -\left(\frac{\partial V}{\partial T}\right)_P \qquad \text{[equation 14]}$$

But

$$dS = \frac{dH}{T}$$

$$\therefore \quad \left(\frac{\partial H}{\partial P}\right)_T = -\left(\frac{\partial V}{\partial T}\right)_P \qquad (16)$$

(*iii*) **Relation III.** Taking S, V as independent variables,

$$\left(\frac{\partial T}{\partial V}\right)_S = -\left(\frac{\partial V}{\partial S}\right)_V \qquad \text{[equation 11]}$$

But

$$dS = \frac{dH}{T}$$

$$\therefore \quad \left(\frac{\partial T}{\partial P}\right)_S = -\mathrm{T}\left(\frac{\partial P}{\partial H}\right)_V \qquad (17)$$

(iv) *Relation IV*. Taking S, V as independent

$$\left(\frac{\partial T}{\partial P}\right)_S = \left(\frac{\partial V}{\partial S}\right)_P \qquad \text{[equation 12]}$$

But

$$dS = \frac{dH}{T}$$

$$\therefore \quad \left(\frac{\partial T}{\partial P}\right)_S = T\left(\frac{\partial V}{\partial H}\right)_P \qquad (18)$$

(v) **Relation V.** Taking P, V as independent we can derive the following relation by making suitable substitutions.

$$\left(\frac{\partial T}{\partial P}\right)_S \left(\frac{\partial S}{\partial V}\right)_P - \left(\frac{\partial T}{\partial V}\right)_P \left(\frac{\partial S}{\partial P}\right)_V = 1 \qquad (19)$$

(vi) **Relation VI**

and

$$\left(\frac{\partial P}{\partial T}\right)_S \left(\frac{\partial V}{\partial S}\right)_T - \left(\frac{\partial P}{\partial S}\right)_T \left(\frac{\partial V}{\partial T}\right)_S = 1 \qquad (20)$$

The relations 15 to 20 are important relations and are used in solving problems.

Applications of Maxwell's Relations

(i) To derive relation between C_p and C_v for an ideal gas,

We have $\quad C_p = \left(\frac{\partial H}{\partial T}\right)_P$

$$= T\left(\frac{\partial S}{\partial T}\right)_V \qquad \left[dS = \frac{dH}{T}\right]$$

and

$$C_v = \left(\frac{\partial H}{\partial T}\right)_V = T\left(\frac{\partial S}{\partial T}\right)_V$$

Taking S as a function of T and V,

$$dS = \left(\frac{\partial S}{\partial T}\right)_V dT + \left(\frac{\partial S}{\partial V}\right)_T dV$$

or

$$\left(\frac{\partial S}{\partial T}\right)_P = \left(\frac{\partial S}{\partial T}\right)_V \left(\frac{\partial T}{\partial T}\right)_V + \left(\frac{\partial S}{\partial V}\right)_T \left(\frac{\partial V}{\partial T}\right)_P$$

We also know that

$$\left(\frac{\partial S}{\partial V}\right)_T = \left(\frac{\partial P}{\partial T}\right)_V \qquad \text{[Maxwell relation]}$$

So we get, by substitution,

$$\left(\frac{\partial S}{\partial T}\right)_P = \left(\frac{\partial S}{\partial T}\right)_V + \left(\frac{\partial P}{\partial T}\right)_V \left(\frac{\partial V}{\partial T}\right)_P \tag{21}$$

Multiply by T,

$$T = \left(\frac{\partial S}{\partial T}\right)_P = T\left(\frac{\partial S}{\partial T}\right)_V + T\left(\frac{\partial P}{\partial T}\right)_V \left(\frac{\partial V}{\partial T}\right)_P \tag{22}$$

Hence

$$C_P = C_V + T\left(\frac{\partial P}{\partial T}\right)_V \left(\frac{\partial V}{\partial T}\right)_P$$

or

$$C_P - C_V = T\left(\frac{\partial P}{\partial T}\right)_V \left(\frac{\partial V}{\partial T}\right)_P \tag{23}$$

(ii) *To prove that for an ideal gas* $C_P - C_V = R$.

For an ideal gas, we have

$$PV = RT \qquad \text{(one mole)}$$

$$\left(\frac{\partial P}{\partial T}\right)_V = \frac{R}{V} \text{ and } \left(\frac{\partial V}{\partial T}\right)_P = \frac{R}{P}$$

But $C_P - C_V = T\left(\frac{\partial P}{\partial T}\right)_V\left(\frac{\partial V}{\partial T}\right)_P$ [use eq. 23]

$$\therefore \quad C_P - C_V = T \times \frac{R}{V} \times \frac{R}{P}$$

$$= \frac{TR^2}{VP} = \frac{TR^2}{RT} = R$$

Hence $C_P - C_V = R.$

(iii) To derive a relation between C_P and C_V in case of a van der Waals' gas.

For one mole of such a gas,

$$\left(P + \frac{a}{V^2}\right)(V-b) = RT$$

or

$$\left(P + \frac{a}{V^2}\right) = \frac{RT}{V-b}$$

Differentiating, we get

$$\left(\frac{\partial P}{\partial T}\right)_V = \frac{R}{V-b}$$

Again differentiating *w.r.t.* T at constant P, we get

$$-\frac{2a}{V^3}\left(\frac{\partial V}{\partial T}\right)_P = \frac{R}{V-b} - \frac{RT}{(V-b)^2}\left(\frac{\partial V}{\partial T}\right)_P$$

or

$$\left(\frac{\partial V}{\partial T}\right)_P\left[\frac{RT}{(V-b)^2} - \frac{2a}{V^3}\right] = \frac{R}{V-b}$$

We have already derived,

$C_P - C_V$ = Substituting, we get $\left(\frac{\partial P}{\partial T}\right)_V\left(\frac{\partial V}{\partial T}\right)_P$ [eq.23]

$$C_P - C_V = \frac{T\left(\frac{R}{V-b}\right)\left(\frac{R}{V-b}\right)}{\left[\left(\frac{RT}{V-b}\right)^2 \frac{2a}{V^3}\right]}$$

$$= \frac{RT(V-b)^2}{RT(V-b)^2}\left[\frac{R}{1-\dfrac{2a(V-b)^2}{V^3RT}}\right]$$

b can be neglected as compared to V.

$$\therefore \quad C_p - C_v = \frac{R}{1-\dfrac{2a}{VRT}}$$

$$= R\left(1-\frac{2a}{VRT}\right)^{-1}$$

Applying binormal theorem, we get

$$C_p - C_v = R\left[1+\frac{2a}{RTV}\right]$$

[$\therefore$ *a* is small neglect higher terms]

(iv) Clapeyron's heat equation

From Maxwell's relations, we know

$$\left(\frac{\partial H}{\partial V}\right)_T = T\left(\frac{\partial P}{\partial T}\right)_V \qquad \text{[Eq. 15]} \qquad (24)$$

The $\left(\frac{\partial H}{\partial V}\right)_T$ is the latent heat as it represents the heat absorbed per unit increase in volume at constant temperature. If for a change in volume from V_1 to V_2 of a unit mass of a substance ûH is the latent heat (L), then

$$\partial H = L \qquad \text{and} \qquad \partial V = V_2 - V_1$$

$$\therefore \quad \frac{L}{V_2 - V_1} = T\left(\frac{\partial P}{\partial T}\right)_V \qquad \text{[From eq 24]}$$

or

$$\frac{L}{V_2 - V_1} = T\frac{dP}{dT}$$

or

$$\frac{dP}{dT} = \frac{L}{T(V_2 - V_1)} \qquad (25)$$

This is known as Clapeyron's Latent heat equation.

(v) *First TdS equation.* Taking entropy as a function of T, V,

$$dS = \left(\frac{\partial S}{\partial T}\right)_V dT + \left(\frac{\partial S}{\partial V}\right)_T dV \tag{25}$$

Multiply by T,

$$TdS = T\left(\frac{\partial S}{\partial T}\right)_V dT + T\left(\frac{\partial S}{\partial V}\right)_T dV$$

But $\quad C_v = \left(\frac{\partial H}{\partial T}\right)_V \quad$ (definition)

and $\quad dS = \frac{dH}{T}$

$\therefore \quad C_v = T\left(\frac{\partial S}{\partial T}\right)_V$

Again from Maxwell's relations, we know

$$\left(\frac{\partial S}{\partial V}\right)_T = \left(\frac{\partial P}{\partial T}\right)_V \quad \text{[Equation 13]}$$

By substitution in (25), we get

$$TdS = Cv\, dT + T\left(\frac{\partial P}{\partial T}\right)_V dV \tag{26}$$

This is known as the first TdS equation.

(*vi*) **Second *TdS* equation.** Taking entropy as a function of T, P. We have

$$dS = \left(\frac{\partial S}{\partial T}\right)_P dT + \left(\frac{\partial S}{\partial P}\right)_T dP$$

or

$$TdS = T\left(\frac{\partial S}{\partial T}\right)_P dT + T\left(\frac{\partial S}{\partial P}\right)_T dP \tag{27}$$

But

$$C_v = \left(\frac{\partial H}{\partial T}\right)_V$$

and

$$dS = \frac{dH}{T}$$

so

$$C_P = T\left(\frac{\partial S}{\partial T}\right)_P$$

We know from Maxwell's relations,

$$\left(\frac{\partial S}{\partial P}\right)_T = -\left(\frac{\partial V}{\partial T}\right)_P$$

Thus by substitution in (27), we get

$$TdS = C_P\, dT - T\left(\frac{\partial V}{\partial T}\right)_P dP \qquad (28)$$

This is known as second TdS equation.

(*vii*) *Energy Equation.* The energy equations for ideal gas and gas obeying van der Waal's gas equation have already been derived. These can also be derived as under:

For an infinitesimal change in a reversible process. ΔE is given by

$$dE = TdS - PdV$$

But

$$TdS = C_V\, dT + T\left(\frac{\partial P}{\partial T}\right)_V dV$$

From these two equations, we get

$$dE = C_V\, dT + \left[T\left(\frac{\partial P}{\partial T}\right)_V - P\right] dV \qquad (29)$$

Since E is a function of T and V.

So
$$dE = \left(\frac{\partial E}{\partial T}\right)_V dT + \left(\frac{\partial E}{\partial V}\right)_T dV \qquad (30)$$

Comparing (29) and (30),

$$\left(\frac{\partial E}{\partial V}\right)_T dV = \left[T\left(\frac{\partial P}{\partial T}\right)_V - P\right] dV$$

or

$$\left(\frac{\partial E}{\partial V}\right)_T = \left[T\left(\frac{\partial P}{\partial T}\right)_V - P\right] \quad (31)$$

This is known as energy equation. Internal energy of any system, whose equation of state is known, can be calculated with the help of this equation.

(i) *Ideal gas.* We know for ideal gas,

$$P = \frac{nRT}{V} \quad (32)$$

or

$$\left(\frac{\partial P}{\partial T}\right)_V = \frac{nR}{V} \quad (33)$$

From (31) and (33), we get

$$\left(\frac{\partial E}{\partial V}\right)_T = \left[T\,\frac{nR}{V} - P\right]$$

or

$$\left(\frac{\partial E}{\partial V}\right)_T = \mathrm{P} - \mathrm{P} = 0$$

So the internal energy of an ideal gas is independent of its volume at constant temperature.

(*ii*) *For a gas obeying van der Waals equation*

We have

$$\left(P + \frac{a}{V^2}\right)(V-b) = RT$$

$$P = \frac{RT}{V-b} - \frac{a}{V^2}$$

$$\therefore \quad \left(\frac{\partial P}{\partial T}\right)_V = \frac{R}{V-b}$$

Substituting, we get

$$\left(\frac{\partial E}{\partial V}\right)_T = \left[T\left(\frac{\partial P}{\partial T}\right)_V - P\right] \tag{34}$$

$$= \left[\frac{TR}{V-b} - P\right]$$

On substitution, we get

$$\left(\frac{\partial E}{\partial V}\right)_T = \left[\frac{RT}{V-b} - \frac{RT}{V-b} + \frac{a}{V^2}\right]$$

$$= \frac{a}{V^2} \tag{35}$$

It means that for actual gases the internal energy increases with increase in volume at constant temperature.

Example 1. Use Clausius Clapeyron equation $\frac{dP}{dT} = \frac{L}{T(V_2 - V_1)}$ to calculate the change in boiling point of water where the pressure is increased from 1.0 to 1.2 atmosphere. Given specific volume of steam = 1677 cc/gm, latent heat of stream = 540 cal/gm, boiling point of water at one atmospheric pressure = 1.0×10^6 dynes/cm^2.

[Agra, 1970]

Solution. Using Clapeyron-Clausius equation, we have

$$dT = \frac{T(V_2 - V_1)\,dP}{L}$$

Substituting these values, we get

$$dT = \frac{373\,(1677-1)\,(1.2-1)}{540 \times 4.2 \times 10^7}$$

$$= \mathbf{5.5\ K} \qquad \text{or} \qquad \mathbf{5.5^\circ C.}$$

Example 2. Calculate the depression in (freezing point) melting point of ice by one atmospheric increase of pressure. Given that the latent heat of fusion of ice L = 3.4×10^5 J/kg, specific volume of 1 kg of ice and water at 0°C are 1.091×10^{-3} m^3 and 1.000×10^{-3} m^3 respectively. [1 atmo. pressure = 10^5 N/m^2] [*Rohilkhand 1979*]

Solution. Using Clapeyron-Clausius equation, we have

$$dT = \frac{dP \times T(V_2 - V_1)}{L}$$

Substituting these values, we get

$$dT = \frac{10^5 \times 273 \times (-0.091 \times 10^{-3})}{3.4 \times 10^5}$$

$$= -0.0073 \text{ K or } -0.0073\text{°C}.$$

Example 3. Calculate the pressure required to make ice freeze at –1°C. Change of specific volume when 1 g of water freezes into ice = 0.091 cc, J = 4.2×10^7 ergs/cal, 1 atmosphere = 10^6 dynes/cm^2 and the latent heat of ice = 80 cal/gm. (*Delhi 1973, Agra 1963, 65*)

Solution. Using Clapeyron's Latent heat equation,

$$\frac{dP}{dT} = \frac{L}{T(V_2 - V_1)}$$

$$dP = \frac{L}{T(V_2 - V_1)} \times dT$$

$$= \frac{80 \times 4.2 \times 10^7 \times (-1)}{273 \times (-0.091)} \text{ dynes/cm}^2$$

$$= 135.2 \text{ atm.}$$

Hence to lower the melting point of ice by 1°C, the pressure must be raised by 135.2 atmospheres.

Thus the required pressure is 135.2 + 1 = **136.2 atmospheres.**

Example 4. Calculate the change in vapour pressure of water as the boiling temperature changes from 100°C to 103°C. Given latent heat of steam = 540 cal/gm and specific volume of steam = 1670 cc/gm.

[*Agra 1976*]

Solution. Using Clapeyron's equation

$$\frac{dP}{dT} = \frac{L}{T(V_2 - V_1)}, \text{ we get}$$

$$dP = \frac{L \times dT}{T(V_2 - V_1)}$$

Substituting various values,

$$dP = \frac{540 \times 4.2 \times 10^7 \times 3}{373 \times (1670 - 1)}$$

$$= \mathbf{0.17091 \text{ atmospheres.}}$$

Example 5. Calculate the specific heat of saturated steam at 100°C from the following data.

L at 90°C = 545.25 cal

L at 100°C = 539.30 cal

L at 110°C = 533.17 cal

Specific heat of water at 100°C = 1.013 cal/gm.

[*Delhi 1975; Kanpur 1977*)

Solution. We have

$$\frac{dL}{dT} = \frac{L \text{ at } 110°C - L \text{ at } 90°C}{20}$$

$$= \frac{533.17 - 545.25}{20}$$

$$= -\frac{12.08}{20}$$

$$= -0.604 \text{ cal/gm.}$$

Hence from

$$C_2 - C_1 = \frac{dL}{dT} - \frac{L}{T}$$

We have

$$C_2 = C_1 + \frac{dL}{dT} - \frac{L}{T}$$

or

$$C_2 = 1.013 - 0.604 - \frac{539.0}{373}$$

$= -1.037$ cal/gm.

Example 6. If L = 800 – 0.705 T, show that the specific heat of steam is negative.

Solution. Given that L = 800 – 0.705 T

$$\therefore \quad \frac{dL}{dT} = -705$$

Now at normal boiling point i.e., at 373 K

We have $\quad L = 800 - 0.705 \times 373$

$$= 537$$

Specific heat of liquid water (C_1) is 1 thus the second specific heat (C_2) is given by

$$C_2 = C_1 + \frac{dL}{dT} - \frac{L}{T}$$

$$= 1 + (-0.705) - \frac{537}{373}$$

$$= -1.14.$$

Thus we find that the specific heat of saturated water vapour at 100°C is negative.

Example 7. Calculate the pressure at which water would boil at 150°C, given that the change in specific volume when 1 g of water changes into steam is 1676 cc, J = 4.2×10^7 erg/cal, 1 atmo. = 10^6 dynes/cm^2 and Latent heat of vapourisation of steam = 540 cal/gm.

Solution. Using Maxwell's relation,

We have

$$\left(\frac{\partial S}{\partial V}\right)_T = \left(\frac{\partial P}{\partial T}\right)_V$$

$$\left(\frac{\partial S}{\partial V}\right)_T = T\left(\frac{\partial P}{\partial T}\right)_V$$

or

$$\left(\frac{\partial Q}{\partial V}\right)_T = T\left(\frac{\partial P}{\partial T}\right)_V \qquad [\because T \cdot \partial T = \partial Q]$$

Substituting various values, we get

$$\frac{540 \times 4.2 \times 10^7}{1676} = 373 \times \left(\frac{\partial P}{50}\right)$$

or

$$\partial P = \frac{50 \times 540 \times 4.2 \times 10^7}{373 \times 1676}$$

$$= 1.815 \times 10^6 \text{ dynes/cm}^2$$

$$= \mathbf{1.815\ atmospheres.}$$

So the pressure at which water will boil at 150°C is

$$1 + 1.815 = \mathbf{2.185\ Atmo.}$$

9

Free Energy Work Functions

We have tried in the previous chapters to find a criterion for the spontaneity of a process. We have tried to establish that in case entropy is kept constant the system. Tries to attain a state of minimum energy and in case the energy is kept constant it tries to attain a state of maximum entropy. But in most of the processes neither of the two i.e., entropy of internal energy remain constant so we introduce two new functions based on internal energy and entropy. These are:

(i) The Helmholtz free energy function or work function (A)

(ii) Gibb's free energy function or thermodynamic potential (G)

These are defined as

$$A = E - TS$$

and

$$G = H - TS$$

We have already studied that TS represent the amount of unavailable energy which can not be converted into useful work (or unavailable energy). Thus A represents isothermally available internal energy and G represents isothermally available enthalpy.

Like internal energy and Enthalpy A and G are also state functions and thus these also depend only on the state of the system and are independent of the path adopted.

The work function 'A'

'A' has been defined as $A = E - TS$

Since A is a state function so dA must be an exact differential. It depends on extensive properties E and S so it should also be an extensive property. It is more convenient to use this function for constant volume processes.

For a reversible process, we have

$$dW_{max} = -dA$$

or

$$W_{max} = -\Delta A$$

In an irreversible process $W_{irr} < W_{max}$ but ΔA has the same value. This is known as *work function* or *Helm holtz free energy* function as it represents the maximum amount of available energy which can be converted into useful work.

In the only work done by the system is PV work then under constant volume conditions dW = 0 and for an isothermal process, we get

$$dA_{T,V} < 0$$

So we conclude that at equilibrium

$$dA_{T,V} = 0 \text{ for any process.}$$

Another relationship between A and S can be obtained as follows:

For a system changing from state (i) to final state (ii), we have

$$A_1 = E_1 - T_1S_1$$

and

$$A_2 = E_2 - T_2S_2$$

$$\therefore \quad \Delta A = A_2 - A_1$$

$$= (E_2 - E_1) - (T_2S_2 - T_1S_1)$$

$$= \Delta E - (T_2S_2 - T_2S_1)$$

For an isothermal pocess

$$T_1 = T_2$$

$$\Delta A = \Delta E - T(S_2 - S_1)$$
$$= \Delta E - T\Delta S$$

Relation between A and S for Small Changes

$$A = E - TS$$
$$dA = dE - TdS - SdT$$
$$= dQ_{rev} - PdV$$
$$- TdS - SdT \qquad [\because dE = dQ_{rev} - PdV]$$
$$= - PdV - SdT \qquad [dQ_{rev} = TdS]$$

Variation of work function with Temperature and Volume

For an infinitesimal change for an isothermal process, which is reversible and in which the only work done is work of expansion, then

$$dS = \frac{dq_{rev}}{T}$$

or $$dS = \frac{dE + PdV}{T} \qquad [\because dq_{rev} = dE + PdV]$$

or $$TdS = dE + PdV$$

But $A = E - TS$

$$\therefore \quad dA = dE - TdS - SdT$$
$$= -PdV - SdT \qquad \text{[See previous Art]}$$

So at constant volume,

$$dA_v = -SdT_v \qquad [\because dV = 0]$$

or $$\left(\frac{\partial A}{\partial T}\right)_V = -S \qquad \text{(i)}$$

Again at constant temperature,

$$dA_T = - PdV_T \qquad [\because dT = 0]$$

or $$\left(\frac{\partial A}{\partial V}\right)_T = -P \qquad \text{(ii)}$$

Equations (i) and (ii) gives the variations of the work function with temperature and volume respectively.

Isothermal change in Work Function

From the definition of work function, we have

$$A = E - TS$$

$$\therefore \quad dA = dE - TdS - SdT$$

But $\quad TdS = dE + PdV$

$$\therefore \quad dA = -PdV - SdT$$

For an isothermal process $dT = 0$

Hence $dA = -PdV$

For an appreciable isothermal change the change in work function ΔA can be obtained by integrating the above equation between the limits of the initial state and final state

$$\therefore \quad \Delta A = -\int_{V_1}^{V_2} PdV$$

For one mole of an ideal gas, $\quad PV = RT$

or $$V = \frac{RT}{P}$$

or $$P = \frac{RT}{V}$$

Substituting the value of P, we get

$$\Delta A_T = -\int_{V_1}^{V_2} \frac{RT}{V} dV$$

$$= -RT\int_{V_1}^{V_2} \frac{dV}{V}$$

$$= -RT \ln \frac{V_2}{V_1} = RT \ln \frac{V_1}{V_2}$$

This is in agreement with the relationship obtained between the change in work function in an isothermal process and the reversible work obtainable.

It should be noted that this equation is applicable to isothermal physical changes only.

Gibb's Free Energy Work Function, 'G'

It is defined by the equation.

$$G = H - TS$$

It is more useful and is derived from entropy.

As H, T, S are state functions so 'G' is also a state function and depends only on the state of the system and is independent of the path adopted.

Thus dG is an exact differential. Like G, S, E, H the Gibbs free energy 'G' is also an extensive property.

Thus for a change of state from 1 to 2 at constant temperature, we get

$$G_2 - G_1 = (H_2 - H_1) - T(S_2 - S_1)$$

$$\Delta G = \Delta H - T\Delta S$$

But $$\Delta H = \Delta E + P\Delta V \qquad \text{(i)}$$

$$\therefore \quad \Delta G = \Delta E - T\Delta S + P\Delta V = \Delta + P\Delta V$$

$$[\quad \Delta A = \Delta E - T\Delta S] \qquad \text{(ii)}$$

(at T, P constant)

From (i) and (ii), we observe that A is related to G in the same manner as E to H.

For a process at constant pressure from (ii), we have

$$\Delta G_P = \Delta A_P + P\Delta V$$

If this process is isothermal also then

$$\Delta A = -W_{rev}$$

$$\therefore \qquad \Delta G_{P,T} = -W_{rev} + P\Delta V$$

or
$$-\Delta G_{P,T} = W_{rev} - P\Delta V$$

$$\begin{bmatrix}\text{Decrease is}\\ \text{free energy}\end{bmatrix} = \begin{bmatrix}\text{Max work}\\ \text{obtainable}\end{bmatrix} - \begin{bmatrix}\text{work of}\\ \text{expansion}\end{bmatrix}$$

= Maximum useful work
[non-mechanical work]

W_{rev} is the total reversible work obtainable in the given change so $(W_{rev} - P\Delta V)$ represents the net work and may be represented by W'_{rev}.
Then
$$-\Delta G_{P,T} = W'_{rev}$$

Therefore, the decrease in free energy of a process that takes place at constant pressure and temperature is equal to the maximum work obtainable from the process. Since 'G' is a measure of useful work it is known as *free energy*. It is also referred to as *'thermodynamic potential'* and as *'available energy'*.

It is only when T, P are constant that the free energy change is equal to the maximum net work available for the given change in state.

Variation of free Energy with T, P

From the definition, $G = H - TS$

$$= E + PV - TS \qquad [H = E + PV]$$

Differentiating, we get

$$dG = dE + PdV + VdP - TdS - SdT$$

For an infinitesimal change in reversible process involving only work of expansion, we get

$$dG = dE - (dE + PdV) - SdT + PdV + VdP$$

or
$$dG = VdP - SdT$$

At constant pressure, $dP = 0$

$$\therefore \qquad dG_P = -SdT_P$$

or $$\left(\frac{\partial G}{\partial T}\right)_P = -S \qquad (1)$$

A constant temperature $dT = 0$

$\therefore$ $$dG_T = VdP_T$$

or $$\left(\frac{\partial G}{\partial P}\right)_T = V \qquad (2)$$

These equations (1) and (2) gives the variation of free energy with temperature and pressure.

These results are applicable to closed systems.

Isothermal Change in Free Energy

By definition, $G = H - TS = E + PV - TS$

Differentiating,

$$dG_T = dE + PdV + VdP - TdS - SdT$$

$$= dq + VdP - TdS - SdT \quad [dq = dE + PdV]$$

$$= TdS + VdP - TdS - SdT \quad [dq = TdS]$$

$$= VdP - SdT$$

For an isothermal process, $dT = 0$

$\therefore$ $$dG_T = VdP$$

If G_1 and G_2 are the values of free energy in the initial and final state of the system. Then at constant temperature the change in free energy is given by

$$\Delta G_T = G_2 - G_1 = \int_{P_1}^{P_2} VdP$$

where P_1, P_2 are initial and final pressure respectively.

For 1 mole of an ideal gas,

$$PV = RT$$

or $$V = \frac{RT}{P}$$

$$\therefore \qquad \Delta G_T = RT\int_{P_1}^{P_2}\frac{dP}{P}$$

$$= \text{RT } 1\text{n}\,\frac{P_2}{P_1}$$

and for *n* moles of the gas,

$$\Delta G_T = nRT\,1\text{n}\,\frac{P_2}{P_1}$$

But for 1 mole of an ideal gas,

$$P_1V_1 = P_2V_2$$

$$\therefore \qquad \frac{P_2}{P_1} = \frac{V_1}{V_2}$$

Hence $\Delta G_T = \text{RT } 1\text{n}\,\frac{V_1}{V_2}$

Since for an ideal gas $\frac{P_2}{P_1}$ is equal to $\frac{V_1}{V_2}$ at constant temperature so ΔG_T is equal to ΔA_T. It should be noted that this equality is applicable only to an isothermal process with an *ideal gas*, but it is not true generally.

For condensed phase it is given by

$$G_2 - G_1 = V(P_2 - P_1)$$

Example 1. Find out the change in work function (ΔA) for the process

2 He (g, 300 K, 1 atmo) → 2 He (g, 300 K, 5 atmo)

[Assume the process as reversible].

Solution. The above process when considered as a reversible process becomes a case of reversible isothermal process.

For a reversible isothermal process

$$\Delta A = -w_{max} = -\text{ (work of compression)}$$

$$= -\text{nRT } 1\text{n}\,\frac{P_1}{P_2}$$

$$= -(2)(8.314)(300)\ \ln \frac{1}{5}$$

$$= 8.029 \text{ kJ.}$$

Example 2. What will be the value of ΔA for the change represented by

$$C_8H_{18}(g) + \frac{25}{2} O_2 (g) \rightarrow 8CO_2 (g) + 9H_2O (l)$$

Given $\Delta H^\circ = -5.93 \times 10^3$ kJ

$$S^\circ C_8H_{18} = 463.65 \text{ e.u.}$$

$$S^\circ O_2 = 205.05 \text{ e.u.}$$

$$S^\circ CO_2 = 213.65 \text{ e.u.}$$

$$S^\circ H_2O = 69.95 \text{ e.u.}$$

The reaction is carried out in a Bumb calorimeter. The reactants were at 1 atmo and 25°C to start with and the reaction is complete.

Solution. We know

$$\Delta A = \Delta E - T\Delta S$$

or

$$\Delta E = \Delta H - \Delta(pV)$$

$$= \Delta H - \Delta nRT$$

$$= -5.93 \times 10^3 - \left(8 - \frac{25}{2} - 1\right)(8.314)(298)$$

$$= -5916.65 \text{ kJ.}$$

It is not possible to calculate ΔS simply by the difference in entropies of products and reactants because there is a change in the number of mole when products are formed which will result in a change in pressure. The final pressure is given by

$$P_{final} = P_{initial} \cdot \frac{nf}{ni}$$

$$= 1 \times \frac{8}{13.5} = 0.595 \text{ atmo.}$$

There will be no change in entropies of liquids.

There will a change in entropy of CO_2 gas. It can be calculated as under:

$$S = S^\circ + R \ln \frac{P_1}{P_2}$$

$$= 213.65 + 8.314 \frac{1}{.595}$$

$$= 218 \text{ e.u.}$$

Thus $\Delta S = (8 \times 218 + 9 \times 69.95) - (463.65 + 12.5 \times 205.05)$

$$= -653.0 \text{ e.u.}$$

Again using the relation,

$$\Delta A = \Delta E - T\Delta S$$

$$= -5916.65 - 298 \times (-653)\,(10^{-3})$$

$$= -5722.0 \text{ kJ}$$

Example 3. During a reversible isothermal compression of one mole of an ideal gas its pressure increases from 1 atmo. to 100 atmo. The compression was carried out at 300 K. What is the change in entropy and change in free energy of the gas ($R = 8.314 \text{ J K}^{-1} \text{ mol}^{-1}$).

Solution. Since the change is isothermal so $\Delta E = 0$.

Now $\Delta G = 23.03\, nRT \log \frac{P_2}{P_1}$

Substituting various values, we get

$$\Delta G = 2.303 \times 1 \times 8.314 \times 300 \log \frac{100}{1}$$

$$= 11486.20$$

Now $\Delta G = \Delta H - T\Delta S$

Again for an isothermal process $\Delta H = 0$

$\therefore$ $\Delta G = -TDS$

$\therefore$ $\Delta S = -\frac{\Delta G}{T}$

$$= -\frac{11486.20}{300}$$

$$= -38.29 \text{ J K}^{-1} \text{ mol}^{-1}.$$

Driving Force of a Process including Chemical Reactions

It has been pointed out earlier that the process at constant pressure is spontaneous or feasible if ΔG has a negative value. Thus the value of $-\Delta G$ is a *measure of driving force.*

We know $$\Delta G = \Delta H - T\Delta S$$

or $$-\Delta G = -\Delta H + T\Delta S$$

Thus $-\Delta G$ will be most negative when ΔH is negative and ΔS is positive. So in a natural process there is a tendency to acquire lowest enthalpy (or energy) and the highest entropy.

If ΔH is sufficiently negative then changes with small decrease in entropy may be spontaneous. Even an endothermic change where ΔH is positive can occur if $T\Delta S$ is sufficiently positive to counteract the positive contribution of ΔH to ΔG. Most of the natural spontoneous chemical processes taking place at room temperature and pressure are so highly exothermic that $T\Delta S$ is not quite important but $T\Delta S$ factor becomes quite important in endothermic spontaneous processes. The following four cases can be distinguished which depends on the values of ΔH and ΔS.

(*i*) **When ΔH is negative and ΔS is positive**. Such reactions can be carried out at any temperature as the value of ΔG in such cases will always be negative, e.g.,

$$2H_2O_{2(g)} \rightarrow 2H_2O_{(g)} + O_{2(g)}$$

(*ii*) **When ΔH is Negative and ΔS is also Negative.** Such reactions can be carried out only if $T\Delta S$ is less than ΔH. It is only under these conditions that we can get negative values for ΔH. Such reactions must be strongly exothermic. At high temperature $T\Delta S$ becomes larger than ΔH and then such reactions can no longer proceed, e.g., $3H_{2(g)} + N_{2(g)} \rightarrow 2NH_{2(g)}$

(*iii*) **When ΔH is positive and ΔS is also positive.** Such reactions can take place only if $T\Delta S > \Delta H$. Generally these reactions become spontaneous at high temperature, e.g.,

$$N_2O_{4(g)} \rightarrow 2NO_{2(g)}$$

(*iv*) **When ΔH is positive and ΔS is negative.** In such cases it would not be possible to get negative values of ΔG so these are prohibited at all temperatures, e.g.,

$$N_{2(g)} + O_{2(g)} \rightarrow 2NO_{2(g)}$$

Relationship Between ΔG and ΔA

From definition

$$\Delta G = \Delta H - T\Delta S$$

and $$\Delta A = \Delta E - T\Delta S$$

$\therefore$ $$\Delta G - \Delta A = \Delta H - \Delta E$$

$$= \Delta E + P\Delta V - \Delta E \qquad [\because \Delta E = \Delta E + P\Delta V]$$

$$= P\Delta V$$

or $$\Delta G = \Delta A + P\Delta V$$

At constant volume, $\Delta V = 0$

$\therefore$ $$\Delta G_V = \Delta A_V.$$

SOLVED EXAMPLES

Example 1. Two moles of an ideal gas expand isothermally from one litre to 10 litres at 300°K. Find the change in free energy of the Gas.

Solution. In case of isothermal change for an ideal gas,

$$\Delta E = 0$$

$$q_{rev} = w_{rev} \; 2.303 \, n\text{RT} \log \frac{V_2}{V_1}$$

Here we have

$n = 2,$ R = 1.987 cal/°/mole T = 300°K,

V_1 = litre, V_2 = 10 litres.

$$\therefore \quad q_{rev} = 2.303 \times 2 \times 1.987 \times 300 \log \frac{10}{1}$$

$$= 2740 \text{ cal.}$$

$$\Delta S = \frac{q_{rev}}{T} \quad \text{or} \quad T\Delta S = q_{rev}$$

For an ideal gas at constant temperature

$$\Delta(PV) = 0$$

Now $$\Delta H = \Delta E + \Delta PV$$

$$= 0 + 0 = 0$$

$\therefore$ $$\Delta G = \Delta H - T\Delta S$$

$$= 0 - 2740 = -2740 \text{ cal.}$$

Example 2. Vapour pressure of liquid mercury at 160°C is 4.19 mm Hg. Find the free energy change for expansion of 2 moles of mercury in equilibrium with liquid mercury at 160°C to a pressure of 1 atmosphere isothermally. Mercury vapours behave as monoatomic ideal gas.

Solution. At equilibrium,

$$\Delta G = 0$$

$$\Delta G = nRT \, 1n \frac{P_2}{P_1}$$

$$= 2 \times 8.314 \times 10^7 \times 433 \times 2.303 \log 183.1$$

$$= 3.752 \times 0^{11} \text{ ergs.}$$

Gibb's Helmholtz Equations

Two equations, one in terms of changes in free energy and enthalpy or heat content and the other in terms of change in internal energy and work function, derived below and known as Gibb's-Helmholtz equations.

The one in terms of changes in free energy and enthalpy is more frequently used. This is applicable to chemical as well as physical processes provided they take place in a closed system i.e., mass is constant and is in equilibrium with external pressure.

These equations were derived by J.W. Gibbs (1875) and H. Von

Helmholtz (1882).

(i) **Gibbs-Helmholtz equation in terms of change in free energy and enthalpy.**

Free energy is given by

$$G = H - TS$$

and $$H = E + PV$$

$$\therefore \quad G = E + PV - TS$$

Differentiating,

$$dG = dE + PdV + VdP - TdS - SdT$$

$$= dq + VdP - TdS - SdT \qquad [dq = dE + PdV].$$

For a reversible process

$$dS = \frac{dq_{rev}}{T} \text{ or } TdS = dq$$

$$\therefore \quad dG = dq + VdP - dq - SdT$$

$$= VdP - SdT$$

In case pressure is constant dP = 0

$$\therefore \quad dG = -SdT$$

Let G_1 be the free energy of the system in initial state at temperature T and $G_1 + dG_1$ be the free energy of the system at T + dT where dT is infinitesimally small and pressure is constant.

Then $$dG_1 = -S_1 dT \qquad (1)$$

(where S_1 is the entropy in initial state) The free energy in final state of the system at temperature T be G_2 and $G_2 + dG_2$ is the free energy of the system at T + dP in the final state. Then

$$dG_2 = -S_2 dT \qquad (2)$$

(where S_2 is entropy of the system in final state).

Subtracting (1) from (2), we get

$$dG_2 - dG_1 = -(S_2 - S_1)\, dT \qquad (3)$$

or $$d(\Delta G) = -\Delta S dT$$

So at constant pressure,

$$\left(\frac{\partial(\Delta G)}{\partial T}\right)_P = -\Delta S \qquad (4)$$

But we know $\Delta G = \Delta H - T\Delta S$

or $$-\Delta S = \frac{\Delta G - \Delta H}{T} \qquad (5)$$

From (4) and (5), we get

$$\frac{\Delta G - \Delta H}{T} = \left[\frac{\partial}{\partial T}(\Delta G)\right]_P$$

or

$$\Delta G = \Delta H = T\left[\frac{\partial}{\partial T}(\Delta G)\right]_P \qquad (6)$$

Equation (6) is known as Gibbs-Helmholtz equation in terms of free energy and enthalpy changes at constant pressure.

(ii) Gibbs-Helmholtz Equation in Terms of Internal Energy and Work Function.

The work function 'A' is given by

$$A = E - TS \qquad (7)$$

or $$\Delta A = \Delta E - T\Delta S$$

or $$-\Delta S = \frac{\Delta A - \Delta E}{T} \qquad (8)$$

On differentiating (7), we get

$$dA = dE - TdS - SdT \qquad (9)$$

From the 1st law of thermodynamics,

$$dq = dE + PdV$$

At constant volume $dV = 0$

$\therefore$ $$dq = dE$$

and for a reversible process,

$$dq = TdS$$

Thus for a reversible process at constant volume, we have

$$dq = dE \tag{10}$$

From (9) and (10), we get

$$dA = -SdT \tag{11}$$

Let A_1 be the work function of the system in initial state at temperature T and $A_1 + dA_1$ is the work function of the system in initial state at temperature T + dT volume being kept constant.

A_2 be the work function of the system in final state at temperature T and $A_2 + dA_2$ is the work function of the system in final state at temperature T + dT, volume being kept constant.

$$\therefore \quad dA_1 = -S_1\, dT \tag{12}$$

and

$$dA_2 = -S_2\, dT \tag{13}$$

where S_1 and S_2 are the entropy values of the system in the initial and final state respectively.

Subtracting (12) from (13), we get

$$dA_2 - dA_1 = -(S_2 - S_1)\, dT$$

$$d(\Delta A) = -\Delta S dT$$

At constant volume,

$$\left(\frac{\partial A}{\partial T}\right)_V = -\Delta S \tag{14}$$

From equation (8) and (14), we get

$$\frac{\Delta A - \Delta E}{T} = \left[\frac{\partial}{\partial T}(\Delta A)\right]_V$$

or

$$\Delta A = \Delta E + T\left[\frac{\partial}{\partial T}(\Delta A)\right]_V \tag{15}$$

Equation (15) is known as Gibbs-Helmholtz equation in terms of changes in internal energy and work function at constant volume.

From these we can deduce another most frequently used integrated form of Gibbs-Helmholtz equation. With its help the change in free energy at a particular temperature can be evaluated if the free energy change and heat of reaction at any other temperature are known for the same process.

Consider the Gibbs-Helmholtz equation,

$$\Delta G = \Delta H + T\left(\frac{\partial}{\partial T}\Delta G\right)_P$$

Divide the equation by T^2 and rearrange

$$\frac{1}{T}\left[\frac{\partial}{\partial T}(\Delta G)\right]_P - \frac{\Delta G}{T^2} = -\frac{\Delta H}{T^2}$$

or

$$\frac{\partial}{\partial T}\left[\frac{\Delta G}{T}\right] = -\frac{\Delta H}{T^2} \tag{16}$$

Let the temperature limits be T_1 and T_2 and corresponding values of free energy changes be ΔG_1 and ΔG_2

Integrating (16), we get

$$\frac{\Delta G_2}{T^2} - \frac{\Delta G_1}{T_1} = -\int_{T_1}^{T_2} \frac{\Delta H}{T^2}\, dT \tag{17}$$

Similarly, we can obtain

$$\frac{\Delta A_2}{T_2} - \frac{\Delta A_1}{T_1} = -\int_{T_1}^{T_2} \frac{\Delta E}{T^2}\, dT \tag{18}$$

Equations (17) and (18) are known as integrated forms of Gibb's, Helmholtz equation.

Example 1. For the reaction

$$N_2(g) + 3H_2(g) \rightarrow 2NH_3(g)$$

the free energy changes at 25°C and 35°C are –16.65 kJ and –14.1 kJ respectively. Calculate the heat of reaction.

Solution. We know $\Delta G = H + T\left(\frac{\partial \Delta G}{\partial T}\right)_P$

Given $G_1 = -16.65$ kJ $T_1 = 273 + 25 = 298$ K

$G_2 = -14.1$ kJ $T_2 = 273 + 35 = 308$ K

$$\left(\frac{\partial}{\partial T}(\Delta G)\right)_P = \frac{-14.1 - (-16.65)}{308 - 298}$$

$$= \frac{2.55}{10} = 0.255$$

At 35°C = $\Delta G = -14.1$ kJ T = 308 K

$$\therefore \quad \Delta G = \Delta H + T\left[\frac{\partial}{\partial T}(\Delta G)\right]_P$$

$$-14.1 = \Delta H + 308 \times 0.255$$

or $$\Delta H = -14.1 - 78.54 = -92.64 \text{ kJ}.$$

Example 2. For the reaction $H_2(g) + O_2(g) \rightarrow H_2O(l)$, the enthalpy change and free energy change are –285.85 and 237.18 kJ respectively at 25°C. Calculate the value of free energy constant at 30°C.

Solution. We know that

$$\Delta G = \Delta H + T\left[\frac{\partial}{\partial T}(\Delta G)\right]_P$$

Given,

$$\Delta G = -237.18 \text{ kJ}$$

$$\Delta H = -285.85 \text{ kJ}$$

$$T = 273 + 25 = 298 \text{ K}$$

$$\therefore \quad -237.19 = -285.85 + 298\left[\frac{\partial}{\partial T}(\Delta G)\right]_P$$

or

$$\left[\frac{\partial}{\partial T}(\Delta G)\right]_P = \frac{-237.18 + 285.85}{298} = 0.1633$$

Assuming that $\left[\frac{\partial}{\partial T}(\Delta G)\right]_P$ remains constant over this range of temperature at 30°C, we can write

$$\Delta G = -285.85 + 303 \times 0.1633$$

$$= -285.85 + 49.48$$

$$= -236.73 \text{ kJ.}$$

Criteria of Spontaneity and Conditions of Equilibrium

(*i*) **Criteria of Spontaneity in terms of Entropy (S).** In a reversible process the entropy remains unaltered but in case of irreversible process there is an increase of entropy. Therefore,

(*a*) for a reversible process, $\Sigma dS = 0$ and

(*b*) for an irreversible $\Sigma dS > 0$

Let us consider a system with its surroundings (including reservoir). For an isothermal process in such a system the transfer of heat of work takes place between the system and the surroundings. Let $dS_{sys.}$ and $dS_{surr.}$ be the entropy changes in the system and surroundings respectively then,

(a) for reversible process $dS_{sys.} + dS_{surr} = 0$

(b) for irreversible process $dS_{sys} + dS_{surr} > 0$

or $$dS_{sys} + dS_{surr} \geq 0$$

If the change in surroundings is regarded as reversible and if *dq* heat is supplied by the surroundings reversibly, then

$$dS_{surr} = -\frac{dq}{T}$$

If dE is the increase in internal energy of the system and *d*W be the work done by the system, then

$$dS_{surr} = -\frac{dq}{T} = -\frac{dE + dW}{T}$$

Now $dS_{sys} + dS_{surr} \geq 0$ for reversible and irreversible process

or $dS_{sys} - \dfrac{dE + dW}{T} \geq 0$

or $TdS_{sys} - dE - dW \geq 0$ (1)

If the workk done is mechanical it can be written as

$TdS_{sys} - dE - PdV \geq 0$ [$\because dW = PdV$]

Now if E, V be kept constant, then

$$dS_{E,V} \geq 0$$

The inequality sign is for real changes (irreversible) and the equality sign is for reversible process.

Thus $dS_{E,V} > 0$ for irreversible process

$dS_{E,V} = 0$ for reversible process (equilibrium)

(ii) If S, V are kept constant for an isothermal change (T constant)

$$-dE \geq 0 \text{ or } dE \leq 0$$

the inequality sign for irreversible process and equality sign is for reversible process (equilibrium).

(iii) **In Terms of Change of Enthalpy**

$$H = E + PV$$

or $dH = dE + PdV + VdP$

or $(-dE - PdV) = -dH + VdP$

Putting this value is (i)

$$TdS - dH + VdP \geq 0$$

or $dH - VdP - TdS \leq 0$

At constant S and P,

$$dH \leq 0$$

$dH < 0$ for irreversible process

$dH = 0$ for reversible process (equilibrium)

(2) Criteria of Spontaneity in Terms of Change in Work Function (A)

We know $dA = dE - TdS - SdT$

or $$TdS - dE = -dA - SdT$$

Putting this value in (i), we get

$$-dA - SdT - PdP \geq 0$$

or $$dA \leq -SdT - PdV$$

The inequality sign is for irreversible process and the equality sign is for reversible process.

In an *isothermal process*, if the system yields no work i.e., Temperature and volume remain constant, then

$$dA \leq 0$$

(i) For an Irreversible Process

$$dA_{V,T} = 0$$

(ii) For Reversible Process (Equilibrium State)

$$dA_{V,T} = 0$$

(3) Criteria of Spontaneity in Terms of Free Energy

We know

$$dG = dE + PdV + VdP - SdT - TdS$$

$$\therefore \quad TdS - dE - PdV = -dG + VdP - SdT$$

Putting this value in (i), we get

$$-dG + VdP - SdT \geq 0$$

or $$dG - VdP + SdT < 0$$

The sign of inequality is for real process (irreversible) and the sign of equality is for reversible process (equilibrium).

At constant pressure in an isothermal process

$$dG \geq 0$$

so in real process there is a decrease in free energy $dG < 0$

Thus

(i) For irreversible process $dG_{P,T} < 0$

(ii) For reversible process (equilibrium) $dG_{P,T} = 0$

The criteria of equilibrium is thus

$$dS_{E,V} = 0 \quad dH_{S,P} = 0$$

$$dA_{V,T} = 0, \quad dES,V = 0$$

$$dG_{P,T} = 0,$$

These criteria may be summarised as under:

(i) At constant E,V. $dS \geq 0$ *i.e.* $(\partial S)_{E,V} \geq 0$

(ii) At constant S,V, $dE \leq 0$ *i.e.* $(\partial E)_{S,V} \leq 0$

(iii) At constant S,P, $dH \leq 0$ *i.e.* $(\partial H)_{S,P} \leq 0$

(iv) At constant T,V, $dA \leq 0$ *i.e.* $(\partial F)_{T,V} \leq 0$

(v) At constant T,P, $dG \leq 0$ *i.e.* $(\partial G)_{T,P} \leq 0$

In these sign of equality is used for a reversible process in equilibrium state and sign of inequality for an irreversible (spontaneous) process.

Limitations of Criteria

These criteria have some limitations. These criteria only tell us about the feasibility of a process and do not tell us any thing about the rate of reaction. The spontaneous or irreversible process may occur in a fraction of a second or it may take a million years. On thermodynamics considerations the combination of $H_{2(g)}$ and $O_{2(g)}$ to form water is feasible (spontaneous) process. These gases may be mixed at room temperature and kept for many years without reacting. Thus a process may be thermodynamically feasible but may not take place under given conditions as in above case the two gases will react on passing an electric spark through them.

So thermodynamics does not tell us about rate of a reaction or conditions, it only predicts whether a particular reaction is feasible or not.

The Standard free Energy (G°)

It is the free energy of a substance at 1 atmosphere pressure. Under this condition when pressure is specified G is a function of temperature only. Thus we have

$$G_{(T,P)} = G^\circ{}_{(T)} + nRT \ln P$$

or

$$\frac{G_{(T,P)}}{n} = \frac{G^\circ{}_{(T)}}{n} + RT \ln P$$

or

$$\mu_{(T,P)} = \mu^\circ{}_{(T)} + RT \ln P$$

where μ is the free energy per mole and is known as *chemical potential.*

While using these equations pressure must be expressed in atmospheres.

Free Energy Change and Chemical Affinity

The decrease in Gibbs-Free energy function (G) at constant temperature and pressure gives the useful work done in a chemical reaction.

Mathematically,

$$\Delta G_{P,T} = \Sigma G_{P,T} \text{(Products)} - \Sigma G_{P,T} \text{(Reactants)}$$
$$= -W$$

So the magnitude of decrease of fee energy ($-\Delta G$) in a chemical reaction is a quantitative measure of the spontaneity or tendency of a chemical reaction to occur or chemical affinity of the reaction.

The decrease in free energy in a reaction at constant temperature and pressure indicates the feasibility of the reaction.

Thus for a chemical reaction

$$A + B = C + D$$

(i) If ΔG is negative, the reaction is spontaneous from left to right, e.g.,

$$Pb_{(s)} + Br_{2(l)} \rightarrow Pb\,Br_2(s) \qquad \Delta G = -62 \text{ kcal}$$

This reaction is spontaneous in forward direction.

(ii) If ΔG is positive, the reaction is spontaneous from right to left, e.g.,

$$2Ag + \frac{1}{2}O_2 \rightarrow Ag_2O \qquad \Delta G = 4.95 \text{ kcal}$$

500°C, 1 atmo.

so reverse reaction is spontaneous.

(iii) If $\Delta G = 0$, it indicates the equilibrium state, e.g.,

$$H_2O_{(l)} = H_2O_{(g)} \qquad \Delta G = O$$

at 100°C, 1 atmo.

So this reaction is in equilibrium under these conditions.

NUMERICALS

Example 1. For the reaction $N_{2(g)} + 3H_{2(g)} = 2NH_{3(g)}$ free energy change at 25°C and 35°C are –3.98 and –3.37 kcal. Calculate the heat of reaction.

Solution. $\Delta G = \Delta H + T\left(\dfrac{\partial \Delta G}{\partial T}\right)_P$

Given

$$G_1 = -3.98 \text{ kcal} \qquad T_1 = 298°K$$

$$G_2 = -3.37 \text{ kcal} \qquad T_2 = 308°K$$

$$\left(\frac{\partial}{\partial T}\Delta G\right)_P = \frac{-3.37 - (3.98)}{308 - 298} = .061$$

At 35°C $\quad \Delta G = -3.37$ kcal

$$T = 308^\circ K.$$

$$\therefore \quad \Delta G = \Delta H + T\left(\frac{\partial \Delta G}{\partial T}\right)_P$$

$$3.37 = \Delta H + 308 \times .061$$

or $$\Delta H = -3.37 - 18.79$$

$$= -22.16 \text{ kcal.}$$

Example 2. For the reaction $H_{2(g)} + \frac{1}{2}O_{2(g)} = H_2O_{(l)}$ the values of enthalpy change and free energy change are –68.32 and –58.69 kcal respectively at 25°C. Calculate the value of free energy change at 30°C.

Solution. We know that

$$DG = DH + T\left(\frac{\partial \Delta G}{\partial T}\right)_P$$

$$\Delta G = -68.32 + 303 \times .039$$

$$= -68.32 + 11.81$$

$$= -56.52 \text{ kcal.}$$

Example 3. Calculate ΔG for conversion of 5 moles of liquid benzene at 80°C to vapour at the same temperature and a pressure of 500 mm. Vapours behave like an ideal gas (Normal b pt. of benzene is 80°C).

Solution. The change may take place in following two steps:

(i) Liquid benzene $\underset{760\text{ mm}}{\overset{80^\circ C}{\rightleftharpoons}}$ Benzene vapour

(ii) Benzene vapour → Benzene vapour at 500 mm
at 760 mm

The step (i) is reversible while step (ii) is irreversible.

Thus for irreversible step

$$\Delta G = nRT\ 1n\ \frac{P_2}{P_1}$$

$$= 5 \times 1.987 \times 353 \times 2.303\ (-0.1818)$$

$= -1468.3$ cal.

and ΔG for reversible step is zero.

Hence $\Delta G_{Total} = -1468.3$ cal.

Example 4. Calculate ΔS for

$$S_{(r)} \rightarrow S_{(m)}$$

given that a 298°K and 1 atmosphere pressure

$$\Delta H = 66 \text{ cal}, \qquad \Delta G = 18 \text{ cal}.$$

Solution. We know

$$\Delta G = \Delta H - T\Delta S \qquad \text{(definition)}$$

$$\therefore \qquad \Delta S = \frac{\Delta G - \Delta H}{T}$$

$$= \frac{66 - 18}{298} = .161 \text{ cal/degree.}$$

Example 5. Using the data given, predict whether the following reaction is feasible at 298 K? Find out the temperature at which the reaction will proceed in the opposite direction.

$$CO_{(g)} + H_2O_{(g)} \rightarrow CO_{2(g)} + H_{2(g)}$$

Given ΔH_f° (CO, g) $= -110.52$ kJ mol^{-1}

ΔH_f° (H_2O, g) $= -241.83$ kJ mol^{-1}

ΔH_f° (CO_2, g) $= -393.54$ kJ mol^{-1}

S° (CO) $= 197.91$ e.u.

S°(H_2O) $= 188.72$ e.u.

S(CO_2) $= 213.64$ e.u.

S°(H_2) $= 130.59$ e.u.

Solution. The reaction will be feasible at 298 K if at this temperature ΔG comes out to be negative.

Using the relation,

$$\Delta 6^\circ = \Delta H^\circ - T\Delta S^\circ$$

We can find the value of ΔG°.

Now, $\Delta H^\circ = -393.54 - (-110.52) - (-241.83)$

$$= -41.19 \text{ kJ.}$$

$$\Delta S^\circ = 213.64 + 130.59 - 197.91 - 188.72$$

$$= -42.4 \text{ J mol}^{-1} \text{ degree}^{-1}.$$

$$\Delta G^\circ = -41.19 - 298 \times (-42.4 \times 10^{-3})$$

$$= -41.19 + 12.64$$

$$= -28.55 \text{ kJ.}$$

Since ΔG for the given reaction is negative at 298 K, so the reaction is feasible at this temperature.

The reaction will proceed in the opposite direction if ΔG is positive. So we have to calculate the minimum temperature when ΔG becomes zero. At any temperature greater than this minimum temperature the reaction will proceed in opposite direction.

Thus, we have

$$\Delta G^\circ = \Delta H^\circ - T_{min}\, \Delta S$$

or $$O = \Delta H^\circ - T_{min}\, \Delta S^\circ$$

or $$T_{min} = \frac{-41.19 \times 10^3}{-42.4} = 9712.5 \text{ K}$$

At any temperature higher than 971.5 K, the reaction will go in opposite direction.

Example 6. Calculate ΔG when 1 mole of CO_2 at 1 atmosphere pressure and 298 K is heated to 400 K at constant pressure.

Given $S^\circ_{(CO_2)} = 213.64$ e.u., C_p $(CO_2) = 37.49$ J/°/mol

[Consider that C_p remains contant]

Solution. We know the reaction,

$$dG = -SdT$$

$$\therefore \qquad \Delta G = -\int_{298}^{400} SdT$$

Moreover entropy is also a function of temperature and at constant pressure

$$dS = \frac{C_v\, dT}{T}$$

or

$$S = C_p \ln T + \text{constant}$$

At 298 K,

$$S = 213.64 \text{ J/°/mol.}$$

Hence value of constant can be found

$$\text{Constant} = 213.64 - 37.49 \ln 298$$

$$= 0.056 \text{ e.u.}$$

$$S = C_p \ln T + 0.056$$

$$\text{Then} \quad \Delta G = -\int_{298}^{400} (0.056)dT - \int_{298}^{400} C_p \ln TdT$$

$$= -(0.056)(400\text{–}298) - \left[C_p \cdot T \ln T - C_p \cdot T\right]_{298}^{400}$$

$$= -(0.056)(102) - (37.49)(400 \ln 400$$

$$-298 \ln 298 - 400 + 298)$$

$$= -22.38 \text{ kJ.}$$

Ellingham diagrams

The diagram on next page is quite useful in metal industry as it is used to predict if a reaction is likely to proceed or not. In this diagram $\Delta G°$ is plotted against T for whole series of oxidations, all of which show a straight forward reaction of metal with oxygen. A nest of these lines (including those for C and H) on the same axes is called **Ellingham diagram.**

In this diagram a sudden shift of gradient must indicate a sudden shift of 'chaos factor' associated with a reaction.

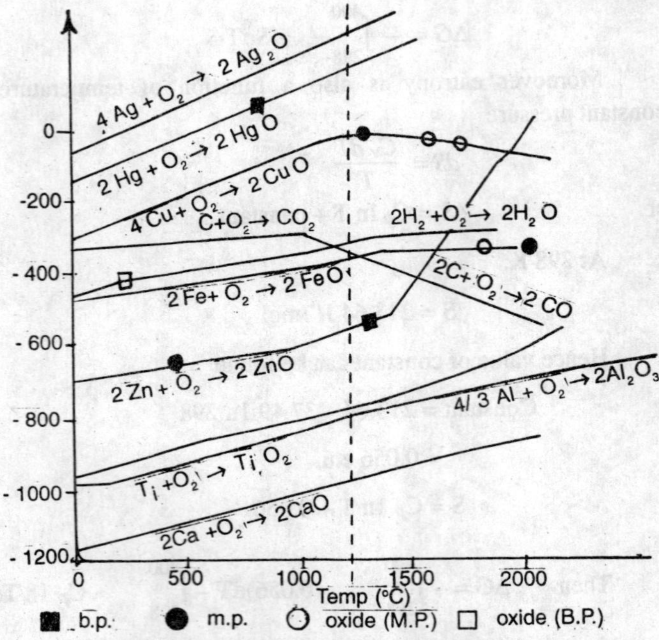

Fig. 9.1. Temp. of major switch over in carbon line.

To understand the importance of the diagram consider the 'carbon line' and 'zinc line' at 1200°C together. The reactions and their free energies are as follows:

(i) $2\,Zn + O2 \rightarrow 2\,ZnO$; $\Delta G = -280$ kJ mol^{-1}

(ii) $2C + O2 \rightarrow 2\,CO$; $\Delta G = -500$ kJ mol^{-1}

So taken individually both these reaction will take place but CO is stabler of the two oxides (has the more negative free energy). They can betreated like a pair of mathematical equations. Thus we can get

$2C + 2\,ZnO \rightarrow 2CO + 2\,Zn$; $\Delta G = -220$ kJ mol^{-1}.

Since for this reaction ΔG is negative, so carbon can reduce ZnO to Zn at 1200°C.

This negative free energy can be seen in the diagram from the fact that carbon line is beneath the zinc line.

Thus this diagram allow quick predictions of the outcomes when any two of the reactions compete for oxygen.

In all cases the reaction on the lower line can drive the upper reactions backward, with the upper oxide giving its oxygen to the lower 'metal'. The *carbon line* is of particular significance because in the form of coke, it is the most readily available reducing agent for smelting.

Following information can be obtained from Ellingham diagram.

(i) The temperature necessary for thermodynamic feasibility of a smelting reaction.

(ii) which reducing agent can perform the task.

(iii) What free energy will apply to any given reaction.

10

Thermodynamic Deductions from Free Energy Functions

Applications of Gibbs-Helmholtz Equations

These equations are of general applicability to physical and chemical changes. A few of them

(*i*) though derived for reversible process, are applicable to irreversible process as they involve ΔG, ΔH, ΔA and ΔE which are state functions and thus do not depend on the path of the process.

(*ii*) are useful in calculating the heats of reaction, i.e., values of ΔH or ΔE, if ΔG or ΔA at two different temperatures are known.

(*iii*) If ΔG value of a process is known at a given temperature, these equations can help us to find the value at another temperature provided the difference of temperature is so small that the rate of change of free energy with temperature remains practically constant over this temperature range.

(iv) These can be used to derive a relation between temperature, heat of reaction and E.M.F of a reversible cell. For such a cell, we have

$$-\Delta G = nEG' \qquad (1)$$

where ΔG is the decrease in free energy and nG′ coulombs of electricity are obtained by the cell in a reversible manner, E is the e.m.f. of the cell.

In the previous chapter we have obtained

$$-\Delta G = \Delta H - \Delta nG'T\left(\frac{\partial E}{\partial T}\right)_P \qquad (2)$$

so by combining (1) and (2), we get

$$-nEG' = \Delta H - nG'T\left(\frac{\partial E}{\partial T}\right)_P$$

where *n* and G′ are constant.

Simplifying this equation, we get

$$E = \frac{\Delta H}{nG'} + T\left(\frac{\partial E}{\partial T}\right)_P \qquad (3)$$

where G′ = 96500 coulombs.

Using this equation (3) we can calculate the heat of reaction from voltae measurements at different temperatures.

(v) **Calculations of Entropy and Enthalpy Change.** These calculations can be easily carried out for isothermal process by the use of Gibbs-Helmholtz equations.

(vi) **Entropy of a Reversible Ion.** The entropy of a reversible ion can be calculated if standard electrode potential is known at two different temperatures.

The Clapeyron-Clausius Equation

This equation is widely used in one component, two phase systems. Some important such systems are:

(*i*) *At the boiling point* of a substance, an equilibrium exists between liquid and vapour phases of the same substance, *e.g.*, at 100°C, $H_2O_{(l)} = H_2O_{(g)}$. Even at room temperature in a closed vessel there is an equilibrium between water and its vapour.

(*ii*) *At the melting point* of a substance there is an equilibrium between the solid and the liquid phases, *e.g.*, at 0°C, ice = water.

(iii) *At the sublimation temperature* of a solid there exists an equilibrium between the solid and vapour state.

(iv) *At transition temperature* there exists an equilibrium between the two crystalline forms of the substance.

Derivation of the Clapeyron-Clausius Equation

It can be derived in a simple way as follows:

Suppose in a system 1 mole of a substance exists in two phases A and B. Let the pressure and temperature of the system be P and T respectively. If G_A and G_B are the free energies of the substance in two phases respectively. So at equilibrium,

$$G_A = G_B \quad (1)$$

If the temperature of the system is raised from T to T + dT, the pressure of the system increases from P to P + dP and the free energy of the substance in two phases becomes $G_A + dG_A$ and $G_B + dG_B$ respectively. As the system is still in equilibrium,

$$\therefore \quad G_A + dG_A = G_B + dG_B \quad (2)$$

From (1) and (2), we get

$$dG_A = dG_B \quad (3)$$

Again, $G = H - TS$ (definition)

$= E + PV - TS$ $(H = E + PV)$

$\therefore$ $dG = dE + PdV + VdP - TdS - SdT$

$= dq + VdP - TdS - SdT$ $(dq = dE + PdV)$

$= VdP - SdT$ (4) $[dq = TdS]$

As the various equilibria considered under Clapeyron-Clausius equations are such that work is done only due to the volume change, Eq. (4) is applicable to the system under consideration.

$$\therefore \qquad dG_A = V_A\,dP - S_A\,dT \qquad (5)$$

and
$$dG_B = V_B\,dP - S_B\,dT \qquad (6)$$

where V_A, V_B are the volumes and S_A, S_B are the entropies of phases A and B respectively.

From equation (3), (5) and (6) we get

$$V_A\,dP - S_A\,dT = V_B\,dP - S_B\,dT$$

or
$$\frac{dP}{dT} = \frac{S_B - S_A}{V_B - V_A}$$

$$= \frac{\Delta S}{V_B - V_A} \qquad (7)$$

ΔS is the molar entropy and $(V_B - V_A)$ represent the change in volume when 1 mole of the substance changes from phase A to phase B at temperature T.

If q is the heat absorbed when 1 mole of substance changes from phase A to phase B at temperature T, then

$$\Delta S = \frac{q}{T} \qquad (8)$$

$\therefore$ From (7) and (8).

$$\frac{dP}{dT} = \frac{q}{T(V_B - V_A)} \qquad (9)$$

Equation (9) is the desired Clapeyron-clausius equation. Thus for the equilibrium of various types we get.

Type 1. Solid-Liquid Equilibrium

Solid = liquid

The above equation (9) can be put as

$$\frac{dP}{dT} = \frac{\Delta H_f}{T(V_L - V_S)} \qquad (10)$$

or

$$\frac{dT}{dP} = \frac{T(V_L - V_S)}{\Delta H_f} \qquad (11)$$

[inverting equation 10]

Such an equilibrium between solid-liquid phase exists at melting point of the substance. For

ice = water equilibrium

ΔH_f is the molar heat of fusion of ice, V_S, V_L are the molar volumes of solid and liquid water respectively and T is the melting point (0°C in this case) of the water. $\frac{dT}{dP}$ thus represents the rate of change of melting point (T) with the external pressure P.

Equation (11) can be used to find the change in m. pt. of the substance with change of pressure. The following two possibilities are there:

(i) $(V_L - V_S)$ is negative, substance contracts on melting, *e.g.*, ice. In such cases $\frac{dT}{dP}$ is also negative and thus melting point is lowered with increase in pressure.

(ii) $(V_L - V_S)$ is positive, substance expands on melting, e.g., sulphur expands on melting. In such cases $\frac{dT}{dP}$ is positive thus melting point is increased with increase in pressure.

Type II. Liquid-Vapour Equilibrium

liquid = vapour

In such cases the Clapeyron-Clausius equation takes the form

$$\frac{dT}{dP} = \frac{T(V_V - V_L)}{\Delta H_v}$$

where T is the boiling point of the liquid, the liquid-vapour equilibrium exists at boiling point. ΔH_v is thus the molar heat of vapourisation. V_V and V_L are the molar volumes of the vapour and liquid respectively.

$\frac{dT}{dP}$ represents the variation of boiling point with applied pressure.

Type III. Equilibrium between Two Solid Modifications

$S_R = S_m$

or $$\alpha = \beta$$

In such cases the equation takes the form

$$\frac{dP}{dT} = \frac{\Delta H_t}{T(V_\beta - V_\alpha)}$$

where T is the transition temperature at which this equilibrium between two solid modifications exist. The two forms and α and β and V_β, V_α are the molar volumes in phases β and α respectively. ΔH_t is the molar heat of transition.

Integrated form of Clapeyron-Clausius Equation

The equation as applied to vapourisation equilibrium is

$$\frac{dP}{dT} = \frac{L_v}{T(V_v - V_L)}$$

The molar volume in liquid (V_L) is very small as compared to the molar volume in gaseous phase (V_v) so V_L may be neglected in comparison to V_v.

The above equation then becomes

$$\frac{dP}{dT} = \frac{L_v}{TV} \quad (1)$$

where T is the boiling point of the liquid and V is the volume in the gaseous phase. If vapour behaves like and ideal gas, at the boiling point, then

$$PV = RT \quad \text{(Gas equation)}$$

or $$V = \frac{RT}{P} \quad (2)$$

Thus $$\frac{dP}{dT} = \frac{L_v}{T.\frac{RT}{P}} \quad \text{[From (1) and (2)]}$$

or $$\frac{dP}{dT} = \frac{L_v . P}{RT^2}$$

or

$$\frac{dP}{P} = \frac{L_v}{RT^2} dT \quad (3)$$

Integrating (3) and assuming L_v is independent of temperature, we get

$$\int \frac{dP}{P} = \int \frac{L_v}{RT^2} dT$$

$$1n\ P = \frac{-L_v}{RT} \qquad (4)$$

If P_1 and P_2 are vapour pressures at the temperature T_1 and T_2, then from eq. (4), we have

$$1n\ P_1 = -\frac{L_v}{RT_1} \qquad (5)$$

and

$$1n\ P_2 = -\frac{L_v}{RT_2} \qquad (6)$$

$$\therefore \quad 1n\ P_2 - 1n\ P_1 = \frac{L_v}{R}\left(\frac{1}{T_1} - \frac{1}{T_2}\right)$$

or

$$1n\frac{P_2}{P_1} = \frac{L_v}{R}\left(\frac{T_2 - T_1}{T_1 T_2}\right)$$

or

$$\log\frac{P_2}{P_1} = \frac{L_v}{2.303} R\left[\frac{T_2 - T_1}{T_1 T_2}\right] \qquad (7)$$

Equation (7) is the integrated form of Clapeyron-Clausius equation.

Applications of Clausius-Clapeyron Equation

(*i*) *Determination of the latent heats.* Latent heat of vaporisation of a liquid can be calculated by using this equation in the integrated form. If P_1 and P_2 are vapour pressures of the liquid at two temperatures T_1 and T_2 respectively then L_v can be calculated.

Similarly, knowing the vapour pressure of a solid phase in equilibrium with liquid phase at two different tempertures. Latent heat of fusion can be calculated using the integrated form of Clapeyron-Clausius equation.

(*ii*) *Effect of pressure on the boiling point.* If the boiling point of a liquid at one pressure and its latent heat of vapourisation are known, the boiling point of the liquid at any other pressure can be calculated using the integrated form of the equation.

(*iii*) *Effect of pressure on the melting point.* If the melting point of a solid at one pressure and its latent heat of fusion are known, the melting point of the solid at any other pressure can be calculated by making use of the integrated form of the equation.

(*iv*) *Effect of temperature on the vapour pressure.* Knowing the vapour pressure of a liquid at one temperature and its latent heat of vapourisation, the vapour pressure of the liquid at any other temperature can be calculated using the integrated form of the equation.

(*v*) The integrated form of the equation is used in correcting the points for barometric fluctuations.

(*vi*) It is also used in the thermodynamics of elevation of boiling points.

(*vii*) It is also used to understand the thermodynamics of depression of feezing points.

NUMERICAL PROBLEMS

Example 1. Determine the latent heat of fusion of ice from the following data:

$V_L = 1.00$ c.c /gm, $V_S = 1.091$ c.c./gm,

M.P. of ice = 0°C at 1 atmo.

M.P. of ice at 2 Atmosphere = – 0.0075°C.

1 cal = 41.2 c.c. (*Bombay 1968*)

Solution. We know that

$$\frac{dP}{dT} = \frac{L_f}{T(V_L - V_s)}$$

Given dP = 1 atm.

dT = –.0075°C or – .0075°A

$T = 273°A$

$V_L = 1.00$ c.c./g $\qquad L_f = ?$

$V_s = 1.091$ c.c./g

$$\therefore \quad \frac{1}{-.0075} = \frac{L_f}{273(1 - 1.091)}$$

$$L_f = \frac{-273 \times .091}{-.0075} = 3312 \text{ c.c. atm.}$$

$$= \frac{3312}{41.2} = 80.0 \text{ cal/g.}$$

Example 2. The dissociation pressures of $CaCO_3$ at 1000°C and 900°C are 2940 mm and 790 mm respectively. Calculate the heat of dissociation of $CaCO_3$ within the given range of temperature.

[*Gujrat 1965*]

Solution. We know

$$2.303 \log \frac{P_2}{P_1} = \frac{\Delta H}{R}\left(\frac{T_2 - T_1}{T_1 T_2}\right)$$

where ΔH is the heat of dissociation.

Here $\quad P_1 = 790$ mm $\qquad P_2 = 2940$ mm

$T_1 = 900°C \qquad T_2 = 1000°C$

$= 1173°A \qquad = 1273°C$

$R = 2$ cals

$$2.303 \log \frac{2940}{790} = \frac{\Delta H}{2}\left(\frac{1273 - 1173}{1273 \times 1173}\right)$$

or $\qquad \Delta H = 3890$ cals

Heat of dissociation of $CaCO_3$ is

$\Delta H = 38990$ cal.

Example 3. Vapour pressure of water at 95°C and 100°C are 634 mm and 760 mm respectively. Calculate the latent heat of vapourization of water per gram between 83°C and 100°C.

Solution. We know that

$$2.303 \log \frac{P_2}{P_1} = \frac{L_v}{R}\left[\frac{T_2 - T_1}{T_1 T_2}\right]$$

Here $P_1 = 634$ mm $\quad P_2 = 760$ mm

$T_1 = 368°A \quad T_2 = 373°A$

$R = 2$ cal $\quad L_v = ?$

$$2.303 \log \frac{760}{634} = \frac{L_v}{2}\left[\frac{373 - 368}{373 \times 368}\right]$$

$\therefore \quad L_v = 9885$ calories per mole

Molecular weight of water = 18

$\therefore$ Latent heat of vapourisation/gm $= \frac{9885}{18}$

$= 549.2$ cal.

Example 4. The vapour pressure of water at 100°C is 760 mm. Calculate the vapour pressure of water at 95°C. Given that the latent heat of vaporization of water in this temperature range is 548 calories.

Solution. We know that

$$2.303 \log \frac{P_2}{P_1} = \frac{L_v}{R}\left[\frac{T_2 - T_1}{T_1 T_2}\right]$$

Here $P_2 = ?$ $\quad P_1 = 760$ mm

$T_2 = 368°A \quad T_1 = 373°A$

$L_v = 548 \times 18 \quad R = 2$ cal

$$\therefore \quad 2.303 \log \frac{P_2}{760} = \frac{548 \times 18}{2}\left[\frac{373 - 368}{373 \times 368}\right]$$

$$P_2 = 634.3 \text{ mm.}$$

Example 5. Either boils at 33.5°C at one atmosphere pressure. At what temperature will it boil at pressure of 750 mm. Given that the latent heat of vapourization of ether is 88.4 cals/gm.

[*Bombay 1950*]

Solution. We know that

$$2.303 \log \frac{P_2}{P_1} = \frac{L_v}{R}\left[\frac{T_2 - T_1}{T_2 T_1}\right]$$

Here $P_2 = 750$ mm $\quad P_1 = 760$ mm

$T_2 = ?$ $\quad T_1 = 33.5°C$

$= 306.5°A$

$L_v = 88.4 \times 74$ cal $\quad R = 2$ cal

$$\therefore \quad 2.303 \log \frac{P_2}{P_1} = \frac{88.4 \times 74}{2}\left[\frac{T_2 - 303.6}{T_2 \times 303.6}\right]$$

or $\quad T_2 = 305.8°K$

or $\quad T_2 = 32.8°C.$

$\therefore$ Boiling point of ether at 750 mm is 32.8°C.

Thermodynamics of Elevation of Boiling Point

The relation between the elevation in boiling point and the molality of the solution can be derived with the help of Clapeyron-Clausius equation as under.

The Clapeyron-Clausius equation may be written as

$$\frac{dP}{dT} = \frac{l}{T \times \upsilon}$$

where υ is the volume of 1 gm of vapour. If vapour behaves as an ideal gas, then $\quad PV = RT \quad$ (Gas equation)

where V is the volume of 1 mole of gas.

This can be written as

$$PM\upsilon = RT$$

where M is the molecular weight and υ the volume of one gram of vapour.

$$\therefore \qquad \upsilon = \frac{RT}{PM}$$

Putting this value of υ in Clapeyron-Clausius equation, we get

$$\frac{dP}{dT} = \frac{l\,MP}{RT^2}$$

where Ml represents the latent heat of evaporation of one mole of solvent.

$$\therefore \qquad \frac{dP}{dT} = \frac{LP}{RT^2}$$

or

$$\frac{dP}{P} = \frac{LdT}{RT^2}$$

or

$$d \ln P = \frac{L}{RT^2} dT$$

Now consider the vapour pressure of the solution at temperature T_0 and T, where T_0 is the boiling point of the solvent and T is the boiling point of solution. When the vapour pressure of the pure solvent at T_0 is P_0 and of she solution at T is P (Fig. 10.1)

When the solvent and solution boil they must have the same vapour pressure.

For pure solvent P_0 is the vapour pressure at its boiling point, so the vapour pressure of solution at its boiling point must be also P_0. The relation between Temperature T_0 and the pressures P and P_0 is given by

$$\int_P^{P_0} d \ln P = \frac{L}{R} \int_T^{T_0} \frac{dT}{T^2}$$

Integrating this equation, we get

$$\ln \frac{P_0}{P} = \frac{L_v}{R} \left(\frac{1}{T_0} - \frac{1}{T} \right)$$

where L_v is the molar heat of vapourization of the solvent at its normal boiling point.

The above equation can be simplified to give

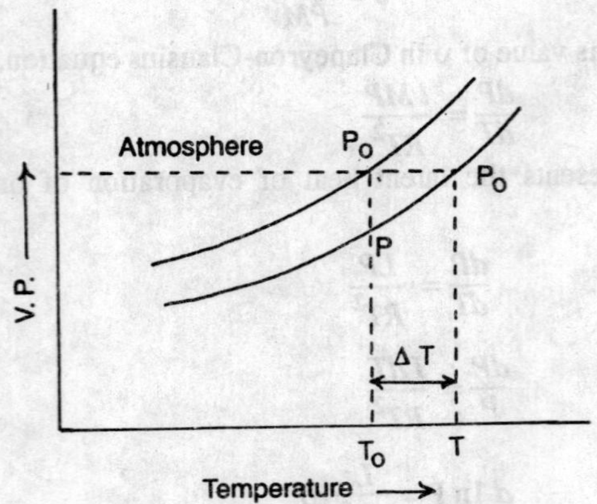

Fig. 10.1

$$\ln \frac{P_0}{P} = \frac{L_v}{R}\left[\frac{T - T_0}{T_0 \times T}\right]$$

$$= \frac{L_v}{R} \times \frac{\Delta T}{T_0 \times T}$$

Since T and T_0 differ by a very small quantity, for dilute solutions. In such a case T may be replaced by T_0. Then

$$\ln \frac{P_0}{P} = \frac{L_v}{R} \times \frac{\Delta T}{T_0^2}$$

or

$$\ln \frac{P_0}{P} = -\ln \frac{P}{P_0}$$

$$= -\ln\left(1 - \frac{P_0 - P}{P_0}\right)$$

$$= -\left(-\frac{P_0 - P}{P_0}\right)$$

[ln (1–*x*) = –*x* when *x* is small]

$$= \frac{P_0 - P}{P_0}$$

Hence $$\frac{P_0 - P}{P_0} = \frac{L_v}{R} \times \frac{\Delta T}{T_0^2}$$

According to Roults' law, the relative lowering of vapour pressure in a dilute solution is equal to the mole fraction of the solute in solution.

$$\therefore \qquad \frac{P_0 - P}{P_0} = \frac{n}{N}$$

where n, N are the number of moles of solute and solvent respectively. For dilute solutions total number of moles of solute and solvent may be taken as equal to number of moles of solvent.

Hence $$\frac{n}{N} = \frac{L_v}{R} \times \frac{\Delta T}{T_0^2}$$

or

$$\Delta T = \frac{RT_0^2}{L_v} \times \frac{n}{N}$$

But $$n = \frac{w}{m} \quad \text{and} \quad N = \frac{W}{M}$$

$$\therefore \qquad \Delta T = \frac{RT_0^2}{L_v} \times \frac{wM}{mW}$$

$$= \frac{RT_0^2 w}{l_v \times m \times W}$$

where l_v, is the heat of vapourisation per gram of the solvent

$$\left(l_v = \frac{L_v}{M}\right)$$

Above equation can be written as

$$\Delta T = \frac{RT_0^2 w \times 1000}{1000 \times l_v \times m \times W}$$

$$= \frac{K_b \times 1000 \times w}{m \times W}$$

where $$= \frac{RT_0^2}{1000 \times l_v} = K_b$$

K_b is a constant known as molal elevation constant.

Molal Elevation Constant or Molecular Elevation Constant (K_b). It is defined as the elevation in boiling point which would be produced by dissolving 1 mole of the solute in 1000 grams of the pure solvent.

Thermodynamics of Depression of Freezing Point

The diagram (Fig. 10.2) given shows the vapour pressure curves of the solvent and a dilute solution.

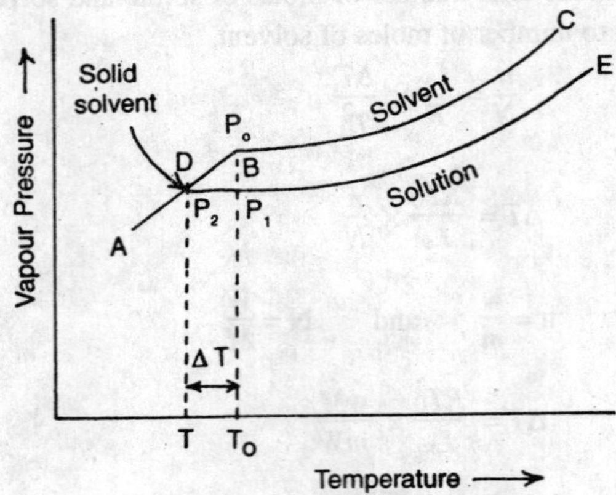

Fig. 10.2

The temperature T_0 corresponding to point B, where the curve AB of solid solvent and BC of liquid solvent, intersect is the normal freezing point of the pure solvent. The curve DE, vapour pressure curve of solution, intersects the curve of solid solvent at point D which corresponds to a lower temperature T. Thus T is the freezing point of the solution and $T_0 - T = \Delta T$ is the depression in freezing point.

From the curve we see that P_1 is the vapour pressure of the solution at temperature T_0 and P_2 is the vapour pressure of solution at temperature T.

Making use of the integrated form of Clapeyron-Clausius equation to the liquid = vapour equilibrium in solution. We have

$$\ln\frac{P_1}{P_2} = \frac{L_v}{R}\left[\frac{T - T_0}{T_0 \times T}\right] \tag{1}$$

where L_v is the molar heat of vapourisation at its normal freezing point P_0.

From the vapour pressure curve of the solid solvent, we find that P_0 is the vapour pressure of the solid solvent at temperature T_0 and P_2 at temperature T.

Again applying the integrated form of Clapeyron-Clausius equation to the solid = vapour equilibrium, we have

$$\ln\frac{P_1}{P_2} = \frac{L_s}{R}\left[\frac{T - T_0}{T_0 \times T}\right] \tag{2}$$

where L_s the molar heat of sublimation of the solid and P_0 is the vapour pressure of the solvent at its freezing point T_0.

Subtracting (1) from (2),

$$\ln\frac{P_1}{P_2} = \frac{L_s - L_v}{R}\left(\frac{T - T_0}{T \times T_0}\right)$$

$$= \frac{L_f}{R}\left(\frac{T - T_0}{T \times T_0}\right)$$

where $L_f = (L_s - L_v)$, the molar heat of fusion. Since $T_0 > T$ this equation may be written as

$$\ln = \frac{P_0}{P_1} = \frac{L_f}{R}\left(\frac{T_0 - T}{T_0 \times T}\right)$$

We know that T_0 and T are very close to each other in dilute solutions. This equation becomes

$$\ln\frac{P_0}{P_1} = \frac{L_f}{R} \times \frac{\Delta T}{T_0^2}$$

or

$$\ln \frac{P_0}{P_1} = -\ln \frac{P_1}{P_0}$$

$$= -\ln\left(1 - \frac{P_0 - P_1}{P_0}\right)$$

$$= -\left(-\frac{P_0 - P_1}{P_0}\right)$$

[$\ln(1-x) = -x$ when x is small]

$$= \frac{P_0 - P_1}{P_0}$$

By Roult's law,

$$\frac{P_0 - P_1}{P_0} = \frac{n}{N}$$

where n and N are the number of moles of solute and solvent respectively.

$$\therefore \quad \frac{n}{N} = \frac{L_f}{R} \times \frac{\Delta T}{T_0^2}$$

or

$$\Delta T = \frac{n}{N} \times \frac{RT_0^2}{L_f}$$

If w gm of solute of molecular weight m are dissolved in W gm of solvent of molecular weight M, then

$$\Delta T = \frac{w \times M}{m \times W} \times \frac{RT_0^2}{L_f}$$

$$= \frac{w\, RT_0^2}{m\, W} \frac{M}{L_f}$$

$$= \frac{w\, RT_0^2}{m \times W}$$

where $\left(l_f = \frac{M}{L_f}\right)$, l_f is the latent heat of fusion per gram of the solvent.

This equation can be written as

$$\Delta T = \frac{RT^2}{1000} \quad l_f \times \frac{1000w}{m \times W}$$

$$= \frac{K_f \times 1000w}{m \times W}$$

where K_f is known as molal depression constant or cryoscopic constant.

Molal Depression Constant (K_f). It is defined as the depression in freezing point which would be produced by dissolving one mole of the solute in 1000 gm of the pure solvent.

Relation Between Some Colligative Properties

(1) Between Osmotic Pressure and Elevation of Boiling Point

Consider a solution having osmotic pressure P and boiling point T+dT, where dT is the elevation in boiling point. The normal boiling point of solvent is thus T. If this solution is put in a vessel fitted with a semipermeable membrane and allowed to go through the following cycle:

(*i*) Some volume of solvent *dv* (weighing *dx* gm) is removed at a temperature T + dT (boiling point of solution). Work done in this process will be given by $w_1 = -P\,d\upsilon$.

(*ii*) Now Evaporate the solvent by heating it in an isothermal process. Since the process is isothermal so the heat absorbed in this process is the latent heat of vapourisation. Thus heat absorbed from the solution is Hdx, where H is the latent heat of evaporation per gram of solvent at (T + *d*T).

(*iii*) Cool this evaporated solvent and solution from (T + dT) to T. The heat evolved will be given by

Heat evolved = – (Total mass of solution × Sp. heat

× change in Temp.)

$= -H_A$.

(*iv*) Condense the isolated solvent vapours in solution at T. Heat given out in this process is given by

Heat given out = $-H_I$ dx

where H_1 is the latent heat of vapourisation per gram of the solvent at T.

(*v*) Heat the system adiabatically from T to T + *d*T to complete the cycle. Heat added is given by

Heat added = Total mass × Sp. heat × Change in Temp.

$= + H_A$.

Since the initial and final states are same, the process is cyclic so the algebraic sum of all work terms must be zero.

Hence $\quad -Pdv + Hdx - H_A - H_1dx + H_A = 0 \qquad (1)$

This is just like a Carnot cycle as the heat is absorbed at a higher temperature and a part of it is given out at a lower temperature. So putting this values in Carnot cycle equation,

$$\frac{Q_1 - Q_2}{Q_1} = \frac{T_1 - T_2}{T_1} \qquad \text{[Carnot theorem]}$$

We get $$\frac{Hdx - H_1dx}{Hdx} = \frac{(T + dT) - T}{T + dT}$$

or $$Hdx - H_1dx = Hdx \cdot \frac{dT}{T + dT} \qquad (2)$$

From (1) and (2), we have

$$Pdv = Hdx \cdot \frac{dT}{T + dT}$$

$$P = H \cdot \frac{dx}{dv} \cdot \frac{dT}{T + dT}$$

$$P = H \times \rho \frac{dT}{T + dT} \qquad \left[\rho = \frac{dx}{dv}\right]$$

$$P = \frac{H\rho\, dT}{T_b}$$

where T_b is the boiling point of solution and *dT* is the elevation of boiling *P* is the osmotic pressure.

(2) Between Osmotic Pressure and Depression of Freezing Point

Again consider a solution having osmotic pressure P and freezing point T–dT. The freezing point of pure solvent is T and thus dT is the depression in freezing point. This solution is taken in a vessel fitted with a frictionless, weightless piston and having the bottom made of a semipermeable membrane. This system completes the following cyclic process:

(*i*) A volume d_v (weighing dx gram) of the solvent at freezing point (T–dT) of the solution is removed. Work done in the process is given by $w_1 = -P\,dv$.

(*ii*) This removed solvent in isothermally frozen at temperature (T–dT). Heat given out in the process is $-Hdx$, where H is the latent heat of fusion per gram of the solvent.

(*iii*) Heat the solution and frozen solvent from (T–dT) to T adiabatically. Heat absorbed is given by

$$\text{Heat absorbed} = \text{Total mass} \times \text{Sp. heat} \times dT$$

$$= H_2$$

(*iv*) Allow the frozen solvent and solution to come in contact with each other through the semipermeable membrane. The Heat absorbed by the solvent in this process is given by H_1dx.

(*v*) Now cool the system adiabatically to (T–dT) to complete the cycle. Heat evolved is given by

$$\text{Heat evolved} = -\text{Total mass} \times \text{Sp. heat} \times \text{Change in temp.}$$

$$= -H_2.$$

Since the initial and final states are same, the process is cyclic, so the algebraic sum of all the work terms should be zero.

$$-Pd_v + H_1dx + H_2 - Hdx - H_2 = 0$$

[Heat absorbed and heat given out in step (iii) and (v) are equal in magnitude.]

H_1dx is the heat absorbed by the system at temperature T and Hdx

is the heat given out at temperature (T–dT) so it can be compared to a Carnot cycle.

$$\therefore \quad \frac{Q_1 - Q_2}{Q_1} = \frac{T_1 - T_2}{T_1}$$

or
$$\frac{H_1 dx - H dx}{H_1 dx} = \frac{T - (T - dT)}{F}$$

or
$$H_1 dx - H dx = H_1 dx \cdot \frac{dT}{T}$$

From these to equations,

$$P d\upsilon = H_1 dx \frac{dT}{T}$$

or
$$P = H_1 \frac{dx}{d\upsilon} \cdot \frac{dT}{T} = \frac{H_1 \rho \, dT}{T} \qquad \left[\rho = \frac{dx}{d\upsilon}\right]$$

Properties of Helmholtz Free Energy (A)

We know $\quad dA = -PdV - SdT$

i.e.,
$$\left(\frac{\partial A}{\partial V}\right)_T = -p$$

and
$$\left(\frac{\partial A}{\partial T}\right)_V = -S$$

From this equation we see that increase of temperature at constant volume increases the work function because S is always positive quantity. Greater the value of entropy, greater will be the decrease in A.

In the same way increase of volume will decrease the work function because p is also a positive quantity.

The above equations can also be written as

$$\left(\frac{\partial A}{\partial T}\right)_V = \frac{A - E}{T}$$

and
$$\left[\frac{\partial}{\partial T}(A/T)\right]_V = -\frac{E}{T^2}$$

These equations can also be applied to the chemical reactions and resulting equations are

$$\left[\frac{\partial}{\partial T}(\Delta A)\right]_V = -\Delta S$$

$$\left[\frac{\partial}{\partial T}(A/T)\right]_V = \frac{\Delta A - \Delta E}{T}$$

$$\left[\frac{\partial}{\partial T}\left\{\frac{\Delta A}{T}\right\}\right]_V = \frac{-\Delta E}{T^2}$$

It is possible to calculate the temperatures at which the chemical reactions will be spontaneous or non-spontaneous using these equations.

Properties of Gibbs' Free Energy (G)

We know $dG = VdP - SdT$

i.e.,
$$\left(\frac{\partial G}{\partial P}\right)_T = V$$

$$\left(\frac{\partial G}{\partial T}\right)_P = -S$$

From the equation it is clear that increase of temperature at constant pressure will decrease the free energy.

Increase of pressure at constant temperature will increase free energy. The larger the volume of the system, greater will be increase in G.

Similar to enthalpy and internal energy, the standard state for free energy function is most stable state at 1 atmosphere pressure at any temperature (Generally 298 K).

Example 1. What will be the free energy change for the following process:

$$H_2O\ (l, 100°C) \rightarrow H_2O\ (g, 100°C)$$

at 100°C.

Solution. We know

$$\Delta S = \frac{\Delta H_{vap}}{T_b}$$

and $$\Delta G = \Delta H_{vap} - T_b . \frac{\Delta H_{vap}}{T_b} = 0.$$

[Note that for any phase transition $\Delta G = 0$]

Law of Mass Action

For derivation of the law of chemical equilibrium, consider the equation

$$aA + bB + \ldots\ldots = cC + dD + \ldots\ldots$$

Let the concentrations of reactants in the beginning be C_A, C_B, ...respectively and the concentration of the products at the end be C_C, D_D....respectively. The Free energies of various substances at temperature T be denoted by μ_A, μ_B, μ_C, μ_D respectively.

Free energy of a multicomponent system at constant temperature and pressure is given by

$$(G)_{T,P} = n_1\mu_1 + n_2\mu_2 + n_3\mu_3 + \ldots\ldots\ldots$$

where n_1, n_2, n_3 etc. are the number of moles of various components and μ_1, μ_2, μ_3 etc. are their respective chemical potentials.

So, we have

Free eneryg of products $= c.\,\mu_C + d.\mu_D + \ldots$

Free energy of reactants $= a.\,\mu_A + b.\mu_B + \ldots$

$$\therefore \ \Delta G = (c.\mu_C + d.\,\mu_D + \ldots) - (a.\mu_A + b.\mu_B + \ldots) \quad (1)$$

But the chemical potential of a substance in any state is given by

$$\mu = \mu^{\circ} + RT \ln a \quad (2)$$

where μ° is the chemical potential in standard state of unit-activity and μ is its chemical potential in a state of activity a.

From (1) and (2), we get

$$\Delta G = [c(\mu^{\circ}C + RT\ 1n\ a_C) + d(\mu^{\circ}D + RT\ 1n\ a_D) + ...)]$$
$$- [a(\mu^{\circ}a + RT\ 1n\ a_A) + b(\mu^{\circ}B + RT\ 1n\ a_B) + ...]$$

Rearranging

$$\Delta G = [(c\mu^{\circ}{}_C + d\mu^{\circ}{}_D + ...) - (a\mu^{\circ}{}_A + b\mu^{\circ}{}_B + ...)]$$

$$+ RT\ 1n\ \frac{(a_C)^c \times (a_D)^d}{(a_A)^a \times (a_B)^b}$$

The first term in this equation represents the increase in free energy at unit activity of reactants and products, so it is equal to ΔG°.

Hence $\Delta G = \Delta G^{\circ} + RT\ 1n\ \frac{(a_C)^c \times (a_D)^d}{(a_A)^a \times (a_B)^b}$

or

$$\Delta G = \Delta G^{\circ} + RT\ 1n\ x$$

where $x = \frac{(a_C)^c \times (a_D)^d ....}{(a_A)^a \times (a_B)^b ....}$

If the reaction is in equilibrium, then the ΔG will be zero.

Hence $\Delta G^{\circ} = -RT\ 1n \left(\frac{(a_C)^c \times (a_D)^d ...}{(a_A)^a \times (a_B)^b ...} \right)_e$ (3)

Subscript *e* indicates that both reactants and products activities taken in this equation are those which exist at equilibrium.

Since ΔG° is standard free energy it must be constant at a given temperature and thus the term on right hand side of eq. (3) must be constant, so at constant temperature.

$$\left[\frac{(a_C)^c \times (a_D)^d ...}{(a_A)^a \times (a_B)^b ...} \right]_e = \text{constant} = K_a \quad (4)$$

where K_a is known as equilibrium constant.

If the system is ideal, its activities may be replaced by fugacities, pressure or by concentration. The equation may then be written as

$$\frac{p_C^c \times p_D^d \times \ldots}{p_A^a \times p_B^b \times \ldots} = K_p$$

and

$$\frac{[C]^b \times [D]^d \times \ldots}{[A]^a \times [B]^b \times \ldots} = K_c$$

From these equations, we get

$$\Delta G^\circ = -RT \ln K_a$$

$$= -RT \ln K_p \quad \text{(gases ideal)}$$

$$= -RT \ln K_c \quad \text{(solutions ideal)}$$

Second Method

The law of mass action can also be easily derived in the following way by Vant Hoff equilibrium Box.

Vant Hoff Equilibrium Box. Vant Hoff imagined a hypothetical device having certain characteristics for the purpose of derivation of this law. Thix box has the following characteristics:

(*i*) *Constant temperature.* It is always maintained at a constant temperature.

(*ii*) *Equilibrium.* At all times there exists an equilibrium between reactants and products in the box.

(*iii*) *Catalyst.* The box contains a catalyst to help bring about the equilibrium immediately whenever a reactants or a product is added to the box.

(*iv*) *Semi-Permeable Membrane.* It is provided with a large number of semi-permeable membranes. Each of these membranes is permeable to only one constituent of the reactions mixture consider the reaction.

$$aA + bB + ... \rightarrow cC + dD + ...$$

This reaction is taking place in the Vant Hoff equilibrium box. Let P_A, P_B, P_C, P_D be the partial pressures of A, B, C, D inside the box at constant temperature T. The various changes and cycles brought about in the box are:

(*i*) *Expansion of a moles of A.* '*a*' moles of A expand reversibly and isothermally from 1 atmosphere pressure to pressure p_A. The work done in this process is given by

$$w_1 = aRT \ 1n \frac{1}{P_A}$$

(*ii*) *Expansion of b moles of B.* The work done in expansion of 'b' moles of B from 1 atmosphere pressure to pressure p_B is given by

$$w_2 = bRT \ 1n \frac{1}{P_B}$$

(*iii*) *Introduction of a moles of A and b moles of* B at 1 atmosphere pressure into the equilibrium box and *withdrawal of c moles cf C and d moles of D* at 1 atmosphere pressure from the equilibrium box. Both these processes are carried out simultaneously so there is no change in contents of the equilibrium box.

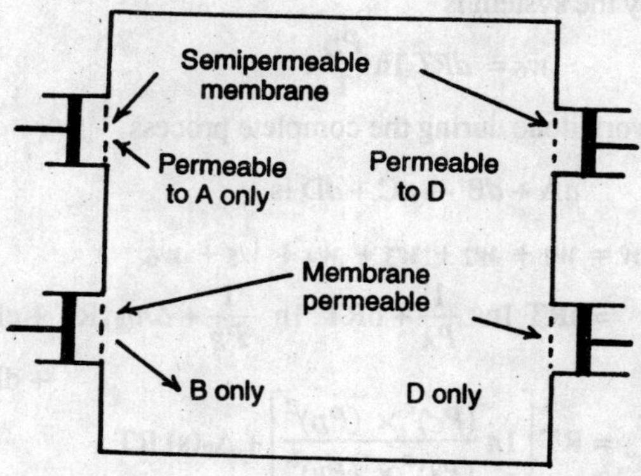

VANT HOFF BOX

Work done on the system to introduce 'A' moles of A and 'B' moles of B is given by

$$w_3 = -aRT - bRT. \qquad \left[\text{At 1 atmo. } \log \frac{1}{P} = 0\right]$$

work done by the system in withdrawing '*c*' moles of C and '*d*' moles of D.

$$w_4 = cRT + dDT$$

Net work done in this change is ($w_2 + w_4$)

or $$w_3 + w_4 = -aRT - bRT + cRT + dRT$$

$$= \text{RT}\,(c + d - a - b)$$

$$= \Delta n_{(g)}\,\text{RT}.$$

(*iv*) *Expansion of* 'c' *moles of C.* Allow this expansion to take place isothermally and reversibly from pressure p_C to 1 atmosphere pressure work done by the system is then

$$w_5 = cRT\ 1n\ \frac{P_C}{1}$$

(*v*) *Expansion of* 'd' *moles of D.* Allow this expansion to take place isothermally and reversibly from pressure p_D to 1 atmosphere. Work done by the system is

$$w_6 = dRT\ 1\text{n}\ \frac{P_D}{1}.$$

Total work done during the complete process

$$a\text{A} + d\text{B} \rightarrow c\text{C} + d\text{D is}$$

$$w = w_1 + w_2 + w_3 + w_4 + w_5 + w_6$$

$$= \text{aRT 1n}\ \frac{1}{P_A} + \text{bRT 1n}\ \frac{1}{P_B} + \Delta n_{(g)}\,\text{RT} + \text{cRT 1n P}_C$$

$$+ \text{dRT 1n P}_D$$

$$= \text{RT}\left[1n\,\frac{(P_C)^c \times (P_D)^d}{(P_A)^a \times (P_B)^b}\right] + \Delta_\text{n}(\text{s})\,\text{RT}$$

Since work done in an isothermal and reversible process is

definite quantity and $\Delta n_{(g)}$ RT is also constant at constant temperature.

Therefore,

$$RT \ 1n \ \frac{(P_C)^c \times (P_D)^d}{(P_A)^a \times (P_B)^b} \text{ is constant at constant}$$

temperature is represented by K_P

$$\frac{(P_C)^c \times (P_D)^d}{(P_A)^a \times (P_B)^b} = K_P.$$

This is the mathematical expression for the law of mass action (or chemical equilibrium).

Vant Hoff Isotherm. This can be deduced either.

(*i*) thermodynamically or

(*ii*) Vant Hoff equilibrium box method.

(i) Thermodynamically

Consider a general reaction

$$aA + bB \rightarrow cC + dD$$

where *a, b, c, d* are the moles of gaseous substances A, B, C, D respectively the free energy of the system is given by

$$G = H - TS \quad (1)$$

$$= E + PV - TS$$

differentiating, we get

$$dG = dE + PdV + VdP - SdT - TdS \quad (2)$$

or

$$dG = dq + VdP - TdS - SdT$$

$$= VdP - SdT$$

At constant temperature,

$$(dG)_T = VdP$$

Hence free energy change for 1 mole of a gas at constant temperature is given by

$$dG = VdP$$

$$= \frac{RT}{P}dP \qquad \text{[for ideal gas PV = RT]}$$

Integrating,

$$\int dG = \int \frac{RT}{P} dP$$

or
$$G = G^\circ + RT \ln P \qquad (3)$$

where G° is a constant of integration and is known as standard free energy.

If pressure is 1 atmosphere, then $G = G^\circ$

Thus *standard free energy* of a substance is defined as the free energy per mole of the substance, when the substance is under a pressure of 1 atmosphere.

Let G_A, G_B, G_C, G_D be the free energies per mole of gases A, B, C, D when their partial pressures are P_A, P_B, P_C, P_D respectively. Let G_A°, G_B°, G_C°, G_D° be their standard free energies respectively.

From Eq. (3), we get the free energies of '*a*' moles of A, '*b*' moles of B, '*c*' moles of C and '*d*' moles of D, as

$$aG_A = aG_A^\circ + aRT \ln P_A \qquad (4)$$

$$bG_B = bG_B^\circ + bRT \ln P_B \qquad (5)$$

$$cG_C = cG_C^\circ + cRT \ln P_C \qquad (6)$$

$$dG_D = dG_D^\circ + dRT \ln P_D \qquad (7)$$

Free energy change for the general reaction

$$aA + bB \rightarrow cC + dD \text{ is}$$

$$\Delta G = \Sigma G \text{ (Products)} \rightarrow \Sigma G \text{ (reactants)}$$

$$= (cG_C + dG_D) - (aG_A + bG_B)$$

$$= (cG_C^\circ + dG_D^\circ) - (aG_A^\circ + bG_B^\circ)$$

$$+ RT \ln \frac{(P_C)^c \times (P_D)^d}{(P_A)^a \times (P_B)^b}$$

$$= \Delta G^\circ + RT\ \ln \frac{(P_C)^c \times (P_D)^d}{(P_A)^a \times (P_B)^b} \quad (8)$$

At equilibrium free energy change is zero

$$\therefore \qquad \Delta G = 0.$$

From equation (8),

$$0 = \Delta G^\circ + RT\ \ln K_p, \left[K_p = \frac{(P_C)^c \times (P_D)^d}{(P_A)^a \times (P_B)^b} \right] \text{at equilibrium.}$$

or
$$\Delta G^\circ = -RT\ \ln K_p \quad (9)$$

From (8) and (9) free energy change for reaction is

$$\Delta G = -RT\ \ln K_p + RT\ \ln \frac{(P_C)^c \times (P_D)^d}{(P_A)^a \times (P_B)^b}$$

or

$$-\Delta G = RT\ \ln K_p - RT\ \ln \frac{(P_C)^c \times (P_D)^d}{(P_A)^a \times (P_B)^b} \quad (10)$$

Equation (10) is the Van't Hoff reaction isotherm.

It may also be written as

$$\Delta G = \Delta G^\circ + RT\ \ln x$$

where x is the reaction quotient of activities of the products and reactants.

Since the reaction is carried out at constant temperature this is used to tell the direction in which a reaction tends to proceed.

(ii) The Van't Hoff Equilibrium Box Method

We have already derived the following result

$$K_p = \frac{(P_C)^c \times (P_D)^d}{(P_A)^a \times (P_B)^b}$$

From the equation $-\Delta G = RT\ 1n\ K_P - RT\ \Sigma n\ 1n\ P$

$\therefore\quad \Sigma n\ 1n\ P = (P_C)^c + 1n\ (P_D)^d - 1n\ (P_A)^a - 1n\ (P_B)^b$.

If we express it in terms of C instead of P, the equation becomes

$$-\Delta G = RT\ 1n\ K_C - RT\ \Sigma n\ 1n\ C$$

The sigificance of $\Sigma n\ 1n\ C$ is the same as $\Sigma n\ 1n\ P$.

Van't Hoff Reaction Isochore

It is a mathematical equation that gives the effect of change of temperature on the equilibrium constant K_p.

It can be deriveed in the following manner.

(i) Thermodynamically

Consider the general equation,

$$aA + bB \rightarrow cC + dD \qquad (1)$$

The Van't Hoff isotherm for the reaction is

$$\Delta G = -RT\ 1n\ K_p\ 1n\ \frac{(P_C)^c \times (P_D)^d}{(P_A)^a \times (P_B)^b} \qquad (2)$$

When the partial pressure of each of the constituent is separately one atmosphere.

$$\Delta G = \Delta G^\circ$$

and $$1n\ \frac{(P_C)^c \times (P_D)^d}{(P_A)^a \times (P_B)^b} = 0$$

Equation (2) becomes

$$\Delta G^\circ = -RT\ 1n\ K_p$$

or $$1n\ K_p = -\ \frac{\Delta G^\circ}{RT} \qquad (3)$$

Also $\Delta G^\circ = \Delta H^\circ - T\Delta S^\circ$ (4)

$$\therefore \quad \ln K_p = -\frac{\Delta H^\circ - T\Delta S^\circ}{RT}$$

$$= \frac{\Delta H^\circ}{RT} + \frac{\Delta S^\circ}{R} \quad (5)$$

If K_1 and K_2 are equilibrium constants for the reaction at temperatures T_1 and T_2 respectively then we get

$$\ln K_1 = -\frac{\Delta H^\circ}{RT_1} + \frac{\Delta S^\circ}{R} \quad (6)$$

$$\ln K_2 = -\frac{\Delta H^\circ}{RT_2} + \frac{\Delta S^\circ}{R} \quad (7)$$

Substracting (6) from (7),

$$\ln K_2 - \ln \quad K_1 = \frac{\Delta H^\circ}{R}\left(\frac{1}{T_1} - \frac{1}{T_2}\right)$$

or

$$\ln \frac{K_2}{K_1} = \frac{\Delta H^\circ}{R}\left(\frac{1}{T_1} - \frac{1}{T_2}\right) \quad (8)$$

or

$$2.303 \log \frac{K_2}{K_1} = \frac{\Delta H^\circ}{R}\left(\frac{1}{T_1} - \frac{1}{T_2}\right)$$

or

$$\log \frac{K_2}{K_1} = \frac{\Delta H^\circ}{2.303R}\left(\frac{1}{T_1} - \frac{1}{T_2}\right) \quad (9)$$

Equations (8) and (9) are know as Van't Hoff isochore. DH° in these equations represent the heat of reaction at 1 atmosphere.

If the volume is kept constant, we get

$$\ln \frac{K_2}{K_1} = \frac{\Delta E}{R}\left[\frac{1}{T_1} - \frac{1}{T_2}\right] \quad (10)$$

or

$$\log \frac{K_2}{K_1} = \frac{\Delta E}{2.303R}\left[\frac{1}{T_1} - \frac{1}{T_2}\right] \quad (11)$$

Example 1. For the reaction

$$2H_2\,(g, 1\text{ atm.}) + O_2\,(g, 1\text{ atm}) \rightarrow 2H_2O\,(g, 1\text{ atm.})$$

ΔG_{298} is equal to –228.60 kJ.

Find out the value of ΔG_{298} for the reaction

$$2H_2\,(g, 0\,1\text{ atm.}) + O_2\,(g, 0.2\text{ atm.}) \rightarrow 2H_2O\,(g, 0.3\text{ atm.})$$

Solution. The ΔG_{298} for the first reaction is the standard state free energy because all the species are at 1 atmosphere.

Hence

$$\Delta G_{298} = \Delta G^{\circ}_{298} + RT\,\ln \frac{P^2_{H_2O}}{P^2_{H_2}\,.\,P_{O_2}}$$

$$= -228.60 + 8.314 \times 298\,\ln \frac{(0.3)^2}{(0.1)^2\,(0.2)^2}$$

$$= -219.17\text{ kJ.}$$

Example 2. Given the equilibrium constants for the reaction

$$2HI = H_2 + I_2$$

at 393.8°C and 490.8°C as 1.643×10^{-2} and 2.184×10^{-2} respectively. If there is no change in ΔH° what is its value?

Solution. Using the relation,

$$\ln K_p\,(T_2) - \ln K_p\,(T_1) = \frac{\Delta H^{\circ}}{R}\left(\frac{1}{T_1} - \frac{1}{T_2}\right)$$

We get

$$\ln \frac{2.184 \times 10^{-2}}{1.643 \times 10^{-2}} = \frac{\Delta H^{\circ}}{8.314}\left(\frac{1}{763.8} - \frac{1}{666.8}\right)$$

or

$$\Delta H^{\circ} = \left(\frac{8.314 \times 763.8 \times 666.8}{97}\right)\left(\ln \frac{2.184}{1.643}\right)$$

$$= 12.42\text{ kJ.}$$

Roult's Law

The vapour pressure of a solution is always lower than that of pure solvent. Many attempts were made to find a relationship between the concentration of solution and lowering of vapour pressure. Roult in 1892 stated a law known as Roult's law. It states:

"The Relative lowering of vapour pressure is equal to the mole fraction of the solute."

Relative lowering of vapour pressure $= \dfrac{p^\circ - p_s}{p^\circ}$ when p°, p_s are the vapour pressures of pure solvent and solution respectively.

If n_1 and n_2 are the number of moles of solute and solvent respectively. The mole fraction of solute is given by mole fraction of solute $= \dfrac{n_1}{n_1 + n_2}$

$$\frac{p^\circ - p_s}{p^\circ} = \frac{n_1}{n_1 + n_2}$$

or

$$\frac{p^\circ - p_s}{p^\circ} = \frac{n_1}{n_2}$$

[n_1 can be neglected in denominator as it is very small.]

This law can be derived by the use of Carnot cycle. The following steps are involved:

(i) Consider a solution which contains n_1 moles of solute and n_2 moles of solvent. Remove with the help of a piston a quantity of solvent that contains one moles of solute $\left(\dfrac{n_2}{n_1}\right)$. The work done is given by

$$w_1 = -PV = -RT \qquad \text{(ideal gas)}$$

(ii) Convert this removed solvent into vapour reversibly by expansion at pressure p°. Neglecting the volume of liquid, work done is given by

$$w_2 = \frac{n_2}{n_1} p_0 v_1$$

(iii) Allow the vapour to expand till the pressure changes from p° to p_s. The work done per mole of vapour is given by

$$\left(\frac{p^{\circ}+p_s}{2}\right)(v_2-v_1)$$

Total work done is given by

$$w_3=\frac{n_2}{n_1}\left(\frac{p^{\circ}-p_s}{2}\right)(v_2-v_1)$$

(iv) Bring the vapour at p_s in contact with solution and condense it reversibly to the initial state. Work done is given by

$$w_4=-\frac{n_2}{n_1}p_s v_2$$

Since the whole cycle is kept at constant temperature and it is a cyclic process so total work done is zero.

Hence

$$-RT+\frac{n_2}{n_2}p_0 v_1+\frac{n_2}{n_2}\left(\frac{p_0+p_s}{2}\right)(r_2-v_1)-\frac{n_2}{n_1}p_s v_2=0$$

$$0=-RT+\frac{n_2}{n_1}\left[p_0 v_1\left(\frac{P_0+p_s}{2}\right)(v_2-v_1)-p_s v_2\right]$$

or $$-RT+\frac{n_2}{n_1}\left[p_0 v_1+\frac{p_0 v_2}{2}-\frac{p_0 v_1}{2}+\frac{p_s v_2}{2}-\frac{p_s v_1}{2}-p_s v_2\right]=0$$

or $$-RT+\frac{n_2}{n_1}\left[p_0 v_1+\frac{p_0 v_2}{2}-\frac{p_0 v_1}{2}+\frac{p_s v_1}{2}-\frac{p_s v_2}{2}\right]=0$$

or $$-RT+\frac{n_2}{n_1}\left[p_0\frac{v_1+v_2}{2}-p_s\frac{v_1+v_2}{2}\right]=0$$

or $$-RT+\frac{n_2}{n_1}(p_0-p_s)\left(\frac{v_1+v_2}{2}\right)=0$$

or $$-RT+\frac{n_2}{n_1}(p_0-p_s)v=0$$

or $$-RT + \frac{n_2}{n_1}(p_0 - p_s)\frac{RT}{p_0} = 0$$

or $$-1 + \frac{n_2}{n_1}(p_0 - p_s)\frac{1}{p_0} = 0$$

or $$\frac{p_0 - p_s}{p_0} = \frac{n_1}{n_2}$$

which is Roult's law.

NUMERICALS

Example 1. Calculate the free energy change of reaction $CuSO_4.3\,H_2O_{(s)} + 2H_2O_{(l)} \rightarrow CuSO_4.5\,H_2O(s)$ at 25°C.

Given that at 25°C the dissociation pressure of $CuSO_4.5\,H_2O$ is 5.06 mm of Hg and the vapour pressure of water is 18.03 mm of mercury.

Solution. Decrease in free energy per mole is

$$\Delta G = RT\ 1n\frac{P_2}{P_1}$$

Since in this equation two moles of water are involved.

$\therefore$ $$\Delta G = 2RT\ 1n\ \frac{P_2}{P_1}$$

or $$\Delta G = 2 \times 2.303 \times 2 \times 298 \log\frac{18.03}{5.06}$$

$$= 1499 \text{ cal.}$$

Example 2. Calculate the free energy change in calories accompanying the expansion of 1 mole of water vapour in equilibrium with liquid water at 25°C to a pressure of 1 atmosphere at the same temperature assuming the vapour to behave ideally. Vapour pressure of water at 25°C is 23.76 mm.

Solution. $\Delta G = RT\ 1n\frac{P_2}{P_1}$ [for one mole]

$$= 2.303\ RT \log\frac{P_2}{P_1}$$

$$= 2.303 \times 2 \times 298 \log \frac{760}{23.76}$$

$$= 2054 \text{ cal.}$$

Example 3. Calculate the work done when 1 mole of a gas expands isothermally and reversibly from 10 litres to 100 litres at 25°C.

Solution. Work done = change in free energy

or $$w = -\Delta G$$

or $$-w = \Delta G$$

$$\Delta G = RT \, 1n \frac{V_1}{V_2}$$

$$= 2.303 \, RT \log \frac{V_1}{V_2}$$

$$w = -2.303 \times 2 \times 2 \times 298 \log \frac{10}{100}$$

$$= 1372.6 \text{ cal.}$$

Example 4. At 1000°K water vapour at 1 atmosphere has been found to be dissociated into hydrogen and oxygen to the extent of $3 \times 10^{-5}\%$. Calculate the free energy decrease of the system. (Assume ideal gas).

Solution. $$2H^2O = 2H_2 + O_2$$

$$a = 3 \times 10^{-5}\%$$

$$= 3 \times 10^{-7}$$

$$P_{steam} = 1 \text{ atm.}$$

$$P_{H2} = 3 \times 10^{-7} \text{ atm.}$$

$$P_{O_2} = \frac{1}{2} \times 3 \times 10^{-7} \text{ atm.}$$

$$K_p = \frac{[P_{H_2}]^2}{[P_{H_2}]^2[P_{O_2}]}$$

$$= \frac{1}{(3 \times 10^{-7})^3 \times (1.5 \times 10^{-7})}$$

$$= \frac{1}{1.35 \times 10^{-20}}$$

Change in free energy,

$$\Delta G = -RT \ln K_p$$

$$= -2 \times 1000 \times 2.303 \log \frac{1}{1.35 \times 10^{-20}}$$

or $$-\Delta G = 90760 \text{ cal.}$$

Example 5. At 2000°C, the equilibrium constant for the reaction $CO + \frac{1}{2} O_2 \rightarrow CO_2$ is 1.07×10^2. Calculate th change in free energy in thee formation of 1 mole of CO_2 from one mole of CO and $\frac{1}{2}$ mole of O_2 both being at 1 atmosphere pressure.

Solution. Free energy change is given by

$$\Delta G = -RT \ln K_p + RT \ln \frac{P_{CO_2}}{P_{CO} \times (P_{O_2})^{1/2}}$$

$$\therefore \quad \Delta G = -2 \times 2273 \ln 1.07 \times 102 + 2 \times 2273 \ln \frac{1}{1 \times (1)^{1/2}}$$

$$= -2 \times 2273 \times 2.303 \log 1.07 \times 10^2 + 0$$

$$= -21240 \text{ cal.}$$

Example 6. The equilibrium constant for the reaction

$$A + B \rightarrow C + D; \qquad \Delta H = 23.03 \text{ kcal}$$

is .0625 at 102°C. Calculate the equilibrium constant at 127°C.

Solution. We know

$$\log \frac{K_2}{K_1} = \frac{\Delta H}{2.303R} \left[\frac{T_2 - T_1}{T_2 T_1} \right]$$

Given $K_1 = .0625$ $\qquad \Delta H = 23.03$ kcal

$T_1 = 375°A$ $\qquad K2 = ?$

$T_2 = 400°A$

$$\therefore \qquad \log \frac{K_2}{.0625} = \frac{23030}{2.303 \times 2}\left[\frac{400 - 375}{400 \times 375}\right]$$

$$= .8333$$

Taking antilog, we get

$$\frac{K_2}{.0625} = .6808$$

$$\therefore \qquad K_2 = .04255.$$

11

Nernst Heat Theorem and Third Law of Thermodynamics

It has always been found desirable to have some relationship between ΔG and ΔH in a chemical relation and to calculate ΔG from ΔH values. Many attempts were made in this direction. Joule and Thomson proposed that ΔG and ΔH are identical. They found that ΔG and ΔH values are same in case of Daniell cell and so it supports their proposition. Soon after Bertholet suggested, "when heat is given out in a reaction, the free energy of the system decreases".

This relationship was found to be true in case of condensed systems at ordinary temperature but failed in a number of other cases.

Gibbs-Helmholtz Concept

The Gibbs-Helmholtz equation gave a quantitative relationship between ΔG and ΔH for the first time

$$\Delta G = \Delta H + T\left[\frac{\partial(\Delta G)}{\partial T}\right]_P$$

But this equation too was found to have its limitations. These limitation can be illustrated if we consider some special cases.

(i) When $\left[\frac{\partial(\Delta G)}{\partial T}\right]_P$ is zero

Then $\Delta G = \Delta H$

Daniel cell is a well-known example of this type where ΔG and ΔH are found to be same.

(ii) When $G \approx 0$

From Gibbs-Helmholtz equation in such a case, we get

$$\Delta H = -T\left[\frac{\partial(\Delta G)}{\partial T}\right]_P$$

In such cases the values of ΔH are found to be quite large. We came across such cases for fusion of solids at their melting point when $\Delta G \approx 0$ and the values of ΔH are quite high.

(iii) When $T = 0$

From Gibbs-Helmboltz equation, we get

$$\Delta G = \Delta H$$

It is not important case as it is not possible to carry out the reaction at Absolute zero.

Except in cases given, it is possible to calculate the change in enthalpy of a reaction knowing the change in free energy but it is not possible to calculate change in free energy from the enthalpy of the reaction.

Richard Concept. Richard (1902) found the E.M.F of cells at low

temperatures and observed that $\left[\frac{\partial(\Delta G)}{\partial T}\right]$ is quite low at lower temperatures or that ΔG and ΔH are nearly same at extremely low temp.

or $$\underset{T\to 0}{Lt}\ \Delta G = \Delta H$$

Nernst Heat Theorm. Nernst (1906) postulated That, "for a process in a condensed system, the value of $\left[\frac{\partial(\Delta G)}{\partial T}\right]_P$ approaches zero asymptotically as the absolute zero is approached".

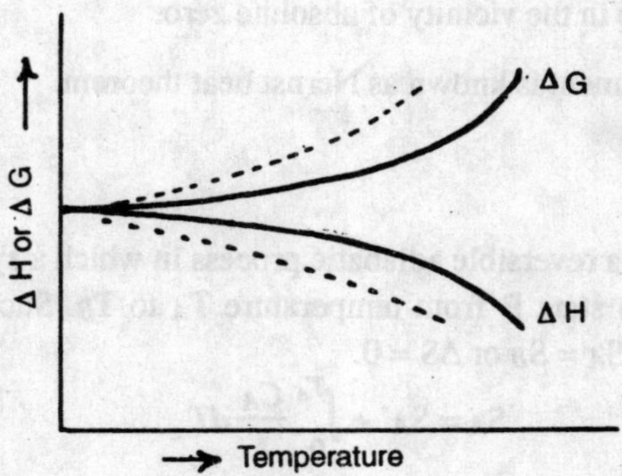

Fig. 11.1

It means ΔG and ΔH meet each other slightly above absolute zero and run coinciding with each other, to absolute zero. This behaviour is shown by dark lines.

The dotted line curve shows that the ΔG and ΔH become zero at absolute zero and not before that.

This can be written mathematically, as

$$\underset{T\to 0}{Lt}\ \frac{\partial}{\partial T}\ \Delta G = \underset{T\to 0}{Lt}\ (\Delta H) = 0$$

This equation is known as Nerst heat theorem.

To understand various consequences of this theorem. We consider the known equations.

$$\left[\frac{\partial}{\partial T}(\Delta G)\right]_P = -\Delta S$$

and

$$\left[\frac{\partial}{\partial T}(\Delta H)\right]_P = -\Delta C_P$$

Applying Nernst heat theorem to these equation, we get

$$\underset{T\to 0}{Lt}\ \Delta S = \underset{TÄ0}{Lt}\ \Delta C_P = 0$$

that means all processes must occur without entropy change and heat capacity change in the vicinity of absolute zero.

This statement is known as Nernst heat theorem.

Proof

Consider a reversible adiabatic process in which a system changes from state A to state B from temperature T_A to T_B. Such a process is isoentropic i.e. $S_A = S_B$ or $\Delta S = 0$.

But $$S_A = S_A^{\circ} + \int_0^{T_A} \frac{C_A}{T} dT$$

and $$S_B = S_B^{\circ} + \int_0^{T_B} \frac{C_B}{T} dT$$

$\therefore$ $$S_A^{\circ} + \int_0^{T_A} \frac{C_A}{T} dT = S_B^{\circ} + \int_0^{T_B} \frac{C_B}{T} dT \quad (1)$$

If during this isoentropic process the state B is at absolute zero i.e., $T_B = 0$. Then equation (1) becomes

$$\int_0^{T_0} \frac{C_A}{T} dT = S_B^{\circ} - S_A^{\circ} \quad (2)$$

Since in the equation (2) integral cannot be negative so it will hold good if $S_B^{\circ} > S_A^{\circ}$. This conclusion seems to be incorrect as it is not possible to attain absolute zero. Hence $S_B^{\circ} < S_A^{\circ}$.

Similarly considering the reverisble adiabatic process from state B to state A we can show $S_A^{\circ} < S_B^{\circ}$.

Hence $$S_A^{\circ} = S_B^{\circ}$$

That is the Nernst theorem.

Application. Nernst theorem has many applications.

1. Calculation of ΔG from ΔH

We know

$$\Delta C_P = T\left[\frac{\partial}{\partial T}(\Delta S)\right]_P$$

or $$d(\Delta S) \frac{\Delta C_P}{T} dT \quad (3)$$

This equation can be integrated for a change from T_1 to T_2 at constant pressure.

Integrating, we get

$$\Delta S = \int_{T_1}^{T_2} \frac{\Delta C_p}{T} dT + C \tag{4}$$

where C is an integration constant.

the value of $\frac{\Delta C_p}{T}$ is either finite or zero

$\therefore$ 'C' should be zero because in accordance with Nernst heat theorem,

$$\underset{T\to 0}{Lt} \ \Delta C_p = \underset{T\to 0}{Lt} \ \Delta S = 0 \tag{4a}$$

So we get

$$DS = \int_{T_1}^{T_2} \frac{\Delta C_p}{T} dT \tag{5}$$

Substituting this value of DS in Gibbs-Helmboltz equation, we get

$$\Delta G = \Delta H - T \int_{T_1}^{T_2} \frac{\Delta C_p}{T} dT$$

We can calculate ΔG by the use of this equation if we know the values of ΔH and ΔC_p.

The variation of heat capacity at constant pressure is given by

$$\Delta C_p = \alpha + \beta T + \gamma T^2 + \ldots . \tag{6}$$

If this expression is true at lower temperature only, when

$$T = 0, \Delta C_p = 0$$

Then $\underset{Lt\to 0}{\Delta C_p} = 0 = \alpha + 0 + 0$ or $\alpha = 0$

So we get $\Delta C_p = \beta T + \gamma T^2 + \ldots\ldots.$

Integrating, we get

$$\int_0^T d(\Delta H) = \int_0^T (\beta T + \gamma T^2 + \ldots) \, dT$$

or

$$\Delta H_T - \Delta H_0 = \frac{\beta T^2}{2} + \frac{\gamma T^3}{3} + \quad (7)$$

where ΔH_0 is the enthalpy at absolute zero.

This can also be written as

$$\Delta H_T = \Delta H_o + \frac{\beta T^2}{2} + \frac{\gamma T^2}{3} + \ldots \quad (8)$$

Substituting the value of ΔC_p in (5), we get

$$\Delta S = \int_0^T \frac{\beta T + \gamma T^2 + \ldots}{T} \ldots dT$$

$$= \int_0^T (\beta + \gamma T + \ldots)\, dT$$

$$= \beta T + \frac{\gamma T^2}{2} + \ldots$$

From Gibbs-Helmboltz equation at temperature T, we get

$$\Delta G = \Delta H_T - T\Delta S$$

or

$$\Delta G = \left(\Delta H_0 + \frac{\beta T^2}{2} + \ \ldots\right) - T\left(\beta T + \frac{\gamma T^2}{2} + \ldots\right)$$

or

$$\Delta G = \Delta H_0 - \frac{\beta T^2}{2} - \frac{\gamma T^2}{6} \quad (9)$$

using this equation we can calculate the values of ΔG from the thermal data.

The assumption used to derive this equation that the heat capacity is given by equation $\Delta C_p = \beta T + \gamma T^2 + \ldots$ is not true at low temperatures where it has been found that heat capacity is proportional to T^3 so this derivation is not correct.

2. Application to Gaseous Systems. The equilibrium constants of gaseous systems can be calculated by the use of Nernst heat theorem.

For solid = vapour the Clapeyron – Clausius equation gives

$$\frac{d(1n\, P^\circ)}{dT} = \frac{L_s}{RT^2} \quad (10)$$

where L_s is the latent heat of sublimation and P is the vapour pressure of solid.

The Kirchoff's equation gives

$$\Delta C_p = \frac{d(L_s)}{dT} \tag{11}$$

Integrating, we get

$$L_s = L_{s(0)} + \int_0^T \Delta C_p \, dT \tag{12}$$

$L_{s(0)}$ is the integration constant.

Substituting the value of L_s in (10), we get

$$\frac{d(\ln P^\circ)}{dT} = \frac{1}{RT^2}\left[L_{s(0)} + \int_0^T \Delta C_p dT\right] \tag{13}$$

Integrating we get

$$\ln P^\circ = -\frac{L_{s(0)}}{RT} - \frac{\int_0^T \Delta C_p \, dT}{RT} + \int_0^T \frac{\Delta C_p}{RT} dT + i \tag{14}$$

where 'i' is an integration constant

Equation (12) can be written as

$$\frac{L_s - L_{s(0)}}{RT} = \frac{\int_0^T \Delta C_p dT}{RT} \tag{15}$$

Substituting from (15) in (14), we get

$$\ln P^\circ = -\frac{L_s}{RT} + \int_0^T \frac{\Delta C_p}{RT} \mathrm{dT} + \mathrm{i} \tag{16}$$

We can calculate the vapour pressure and heat capacity values by making use of equation (16).

For a simple gaseous reaction

$A_{(g)} = B_{(g)}$, we have

$$\frac{d\ln P}{dT}=\frac{\Delta H}{RT^2}$$ [Van't Hoff Equation]

So integrating, we get

$$\ln K_P=-\frac{\Delta H}{RT}+\int_0^T\frac{\Delta C_P}{RT}\,dT+i$$

where '*i*' is the integration constant and is equal to algebraic sum to true chemical constants in this case.

$$\therefore \qquad \ln K_P=-\frac{\Delta H}{RT}+\int_0^T\frac{\Delta C_P}{RT}\,dT+\Sigma i$$

So value of K_P for gaseous reactions can be calculated if we know the values of ΔH and ΔC_P and that of integration constant.

3. Transition Temperature

$$S_{(\alpha)} = S_{(\beta)}$$

for this transition, we have $\Delta C_P=\beta T$. $[\gamma, \delta \ldots = 0]$

If we can measure the heat capacities of α, β forms at some known temperature then the value of constant β in ($\Delta C_P = \beta T$) can be calculated. β is found to be 2.30×10^{-5}. From thermal data ΔH_T at temp. T was found the ΔH_0 war calculated as 1.57.

$$\therefore \qquad \Delta H_0=\Delta H_T-\frac{\beta}{2}T^2$$

At the transition temperature, an equilibrium exists.

$$\therefore \qquad \Delta G=0$$

Hence $\Delta H_0-\frac{\beta}{2}T^2=1.57-1.15\times10^{-5}\,T^2=0$

or $$T=369.5^{\circ}K.$$

This calculated value is in agreement with the experimental value of this transition of sulphur from α to β form.

4. E.M.F of a Voltaic Cell

The E.M.F. of a cell can be calculated by making use of the thermal date, e.g.,

$Zn/ZnSO_4.7H_2O//Hg_2SO_4/Hg$

The following changes are involved

	Reaction	*Heat changes*
(i)	$Zn \rightarrow Zn^{2+}$	C_1
(ii)	$Zn^{2+} \rightarrow ZnSO_4$	C_2
(iii)	$ZnSO_4 + 7H_2O \rightarrow ZnSO_4\ 7H_2O$	C_3
(iv)	$Hg_{22+} \rightarrow Hg_2$	C_4

The net heat change $\Delta H_T = \Sigma C = -33360$ cal/gm Eq.

This value of ΔH is obtained experimentally.

Limitations of Nernst Theorem

(i) This theorem can not be applied to gases as no gas can exist at 0°A.

(ii) It can not be applied to liquids as well so it is true in case of solids only.

(iii) This theorem is not found to be true in case of substances which can exist in both crystalline and super cooled states. In accordance with this theorem in such cases there should not be any entropy change at 0°A but it has been experimentally found that entropy of super cooled liquid is more than that of crystalline state.

Third Law of Thermodynamics

Experiment shows that the fundamental feature of all cooling processes is that the lower the temperature achieved, the more diffcult it is to go lower, e.g., the colder a liquid is, the lower the vapour pressure and the harder it is to produce further cooling by pumping away the vapour. Generalizing from experience we may make the following statement.

By no finite series of processes is the absolute zero attainable.

This is known as principle of unattainability of absolute zero or *the third law of thermodynamics.*

Another statement of the third law, is based on the entropy change of a condensed system during *reversible, isothermal* process, and gives ΔS_T behaviour as T approaches zero e.g., the entropy change of a solid during a reversible isothermal compression may either be measured at different temperatures or be calculated from *second* TdS *equation,* leading to

$$\Delta S_T = S(T,P_1) - S(T,O) = -\int_0^{P_1} \left(\frac{\partial V}{\partial T}\right)_P dP$$

the value of ΔS_T decreases as T decreases. The following principle is thus accepted:

The entropy change associated with any isothermal reversible process of a condensed system approaches zero a the temperature approaches zero.

This statement is known as Nernst-Simon Statement of the third law of thermodynamics.

The original statement of Nernst was 'The temperature derivative of the change of Helmholtz function during an isothermal process approaches zero as the temperature approaches zero.

Many researchers, working on the entropy of pure crystalline substances at absolute zero, gave a fundamental principle known as third law of thermodynamics. There are a number of statements for this principle. A few of the workers in the field are T.W. Richards, Walter Nernst, Max Plank etc. Some of the statements of the third law are given below:

(i) Plank's statement. From the Nernst heat theorem we have already derived an equation, as

$$\underset{T\to 0}{Lt}\ \Delta C_p = 0 \qquad \text{[see equation 4(a)]}$$

and

$$\underset{T\to 0}{Lt}\ DS = 0 \qquad \text{[see equation 4(a)]}$$

From these equations, it can be stated that "All processes should occur without any change in entropy and heat capacity near the absolute zero temperature."

From this statement of Nernst heat theorem it is indicated that all substances in the solid state must have the same entropy at the absolute zero of temperature.

Combining these two facts stated above the third law of thermodynamics can be stated:

"*Entropy of a solid or liquid approaches zero at the absolute zero of temperature.*"

The mathematical expression of this statement can be written as

$$\underset{T \to 0}{Lt}\ S = 0$$

(ii) *Lewis and Randall's statement.* We know that the entropies of super cooled liquids and solutions is not zero at 0°A so the Plank's statement is true only for solids. Another statement was given by Lewis and Randall.

"Every substance has a finite positive entropy; but at 0°A the entropy may become zero and it so actually happens in case of perfectly crystalline solids."

Thus we can say that in perfect crystals at 0°A there is complete order as its entropy is zero. So in these perfect crystals there is not disorder, but in case of super cooled liquids or glassy solids some randomness always exists and in such cases entropy is never zero.

It can therofore be concluded that third law is true only for perfectly crystalline solids which possess zero entropy at 0°A.

(iii) Another statement of the third law given is as under

"It is impossible by any method, no matter how idealized, to reduce the temperature of any system to 0°A in a finite number of operations."

This means we can not cool a body to absolute zero. We have been able to approach temperature as low as 1×10^{-6} degrees absolute.

Application

(1) *The Third Law Entropy.* This law has been used to calculate the absolute entropies of solids, liquids and gases at different temperatures.

(i) *For solids.* Let a solid be heated from 0°A to a temperature T below its melting point.

Then solid (0°A, P) → solid (T, P)

Entropy change will be given by

$$\Delta S = S_T - S_0$$

$$= \int_0^T \frac{C_p}{T}\, dT$$

or

$$S_T = S_0 + \int_0^T \frac{C_p}{T}\, dT$$

the entropy value is minimum at 0°A and in case of perfectly crystalline solids it is zero at 0°A.

$$\therefore \qquad S_0 = 0$$

and

$$S_T = \int_0^T \frac{C_p}{T}\, dT$$

S_T is known as third law entropy, this is the absolute entropy of the solid at temperature T.

S_T can be evaluted if we know the heat capacity of the solid in the temperature range 0°A to T°A. It has been possible to carry out the measurements of heat capacities of solids upto the range of 10°A to 15°A. Below this temperature it is (C_p) determined from the Debye T-cubed law

This law is given as

$$C_p = \alpha T^3$$

where α is a constant and its value is determined from the C_p or C_v

values (these are same at such low temperature) at some low

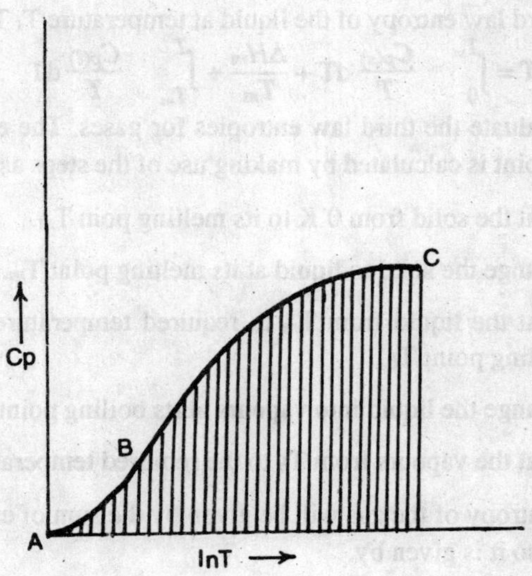

Fig. 11.2

temperature.

S_T the absolute entropy of solid at temperature T can be calculated by integrating the equation for S_T.

$$S_T = \int_0^T C_p \, d \ln T$$

or by plotting $\frac{C_p}{T}$ versus T or by plotting, C_p versus in T. The area of the curve gives the value of S_T between zero degree absolute and any desired temperature.

(ii) *For liquids and gases.* For liquids the third law entropy can be evaluated by the use of following steps:

(a) Heat the solid from 0°K to its melting point (T_m).

(b) Change solid to liquid at the melting point (T_m).

(c) Heat it from this temperature to the required temperature.

The sum of the entropy changes taking place in the above steps gives the third law entropy of the liquid at temperature T. Thus,

$$ST = \int_0^{T_m} \frac{C_{p(s)}}{T} dT + \frac{\Delta H_m}{T_m} + \int_{T_m}^{T} \frac{C_{p(l)}}{T} dT$$

To evaluate the third law entropies for gases. The entropy above its boiling point is calculated by making use of the steps as follows:

(i) Heat the solid from 0°K to its melting poin T_m.

(ii) Change the solid to liquid at its melting point T_m.

(iii) Heat the liquid from T_m to required temperature *i.e.*, upto its boiling point T_b.

(iv) Change the liquid into vapours at its boiling point T_b.

(v) Heat the vapours from T_b to the required temperature.

The entropy of the gas will be given by the sum of entropies of all these steps, so it is given by

$$S_T = \int_0^{T_m} \frac{C_{p(s)}}{T} dT + \frac{\Delta H_m}{T_m} + \int_{T_m}^{T_b} \frac{C_{p(l)}}{T} dT + \frac{\Delta H_{vap}}{T_b} + \int_{T_b}^{T} \frac{C_{p(g)}}{T} dT$$

If during this change the solid under study changes from one crystalline form to another, the entropy of transition should also be taken into consideration. Entropy of transition is given by

$$\Delta S_{Trans} = \frac{\Delta H_{Trans}}{T}$$

This should be added in the above expressions for solids, liquids and gases.

(iii) *Entropy change in chemical reactions.* The standard entropy of an element in its standard state is evaluated using the third law of thermodynamics. The standard state is generally taken at 1 atm. pressure and 25°C temperature.

To calculate the entropy change of the reaction

$$Fe_2O_{3(s)} + 3H_{2(g)} \rightarrow 2Fe_{(s)} + 3H_2O_{(l)}$$

We need to know the standard entropies of $Fe_2O_{3(s)}$, $H_{2(g)}$, $Fe_{(s)}$ and $H_2O_{(l)}$, knowing these the change in entropy for this reaction can be evaluated by

$$\Delta S^\circ = S^\circ_{\textit{products}} - S^\circ_{\textit{reactants}}$$

In this case making use of entropy tables we get

$$\Delta S^\circ = 2 \times 6.49 + 3 \times 16.716 - 21.5 - 3 \times 31.211$$

$$= -52.0 \text{ e.u.}$$

ΔS° is positive if more gaseous products are formed because the entropy of gases is much more in comparison to the entropies of the condensed phase.

Tests of Third Law

The third law can be tested in the following ways:

(i) If a substance X is allowed to change into its another allotropic modification Y at the transition temperature T then the Gibbs-Helmholtz equation gives the entropy change accompanying this transition.

$$\Delta G = \Delta H - T\Delta S \tag{1}$$

But the change in free energy at transition temperature is zero since it is a state of equilibrium.

$$\therefore \quad \Delta H - T\Delta S = 0 \tag{2}$$

or $$T\Delta S = \Delta H$$

or $$\Delta S = \frac{\Delta H}{T} \tag{3}$$

The absolute entropy for x as given by third law of thermodynamics is

$$S_x = \int_0^T \frac{C_x}{T}\, dT$$

and for y $$S_y = \int_0^T \frac{C_y}{T}\, dT$$

$$\therefore \qquad \Delta S = S_y - S_x = \int_0^T \frac{C_y}{T}dT - \int_0^T \frac{C_x}{T}dT \qquad (4)$$

The values from (3) and (4) are similar, as verified from the transition of grey tin to white tin at transition temperature in which case the values are found to be –1.87 e.u. and –1.94 e.u., as calculated from the Gibbs-Helmholtz equation and the third law respectively.

IInd Test. The values of absolute entropies for $Pb_{(s)}$ and $Br_{2(l)}$ have been found and from these the values of absolute entropies of $PbBr_{2(s)}$ are calculated. This value of ΔS° for formation of $PbBr_2$ is used to calculate ΔH° value by the application of the Gibbs-Helmholtz equation. The value so calculated has been found to be in close agreement with the experimental values for ΔH° so it is another evidence of third law.

Exceptions to the Third Law. Consider a process of mixing in which pure crystals of substance A are mixed with pure crystals of substance B. Then for this process of mixing the change in entropy is given by ΔS_{mix}

$$\text{Pure A} + \text{Pure B} \rightarrow \text{mixed crystals}$$

$$\Delta S_{mix} = S_{(mixed\ crystals)} - S_{(A)} - S_{(B)}$$

the entropy of pure crystals is taken as zero

$$\therefore \qquad S_A = 0, \qquad S_B = 0$$

Hence $\qquad \Delta S_{mix} = S_{(mixed\ crystals)}$

$$[-N\,k\,X_A \ln X_A + X_B \ln X_B]$$

where N is the total number of sites and is equal to $(N_A + N_B)$ where N_A and N_B are the number of atoms of A and B respectively. X_A and X_B are the mole fractions of A and B respectively.

We come across a number of crystals which do not follow the third law of thermodynamics such a crystals of CO_2 crystals of NO_2, crystalline ice, crystalline hydrogen etc. It can be explained on the basis of the fact that though these are chemically pure substances but do not have a perfectly ordered arrangement at absolute zero of temperature. It means that at sufficiently low temperatures there exists disordered states in solids. Some of these cases are discussed below:

(1) **CO.** There has been found to be a difference of 4.6 J k^{-1} $mole^{-1}$ in the spectroscopic value of entropy (160.1 J k^{-1} $mole^{-1}$) and calorimetric value of entropy (155.5 J k^{-1} $mole^{-1}$). This can be explained as follows:

In CO molecule a small electric dipole exists. If the crystals are arranged in the same or opposite direction, a small differene of energy is observed in the crystal. e.g.,

C—O C—O same direction

or O—C C—O opposite direction

Thus a molecule can be in one of the two directions in a crystal when molecular rotation in the crystal ceases. For N (Avagadro's number) molecules there are 2^N microstates and entropy can be calculated by using the equation:

$$S = k \ln W$$

$$= k \ln 2 = R \ln 2$$

$$= 5.77 \text{ J k}^{-1} \text{mole}^{-1}.$$

This is quite close to the difference observed in spectroscopic and calorimetric values in case of CO.

(2) **N_2O.** On similar considerations as above the difference in case of N_2O which is also a linear molecule can be explained. This molecule can possess orientations a NNO, NNO or ONN, NNO.

(3) **NO.** This molecule exists as a dimer in solid and has two isomers (i) and (ii)

Both isomers differ from each other in energies. For N (Avagadro's number) molecules of NO there are $2^{N/2}$ microstates. So the entropy at 0°K is given by

$$S = K \ln 2^{N/2}$$

$$= \frac{NK}{2} \ln 2 = 2.88 \text{ J k}^{-1} \text{ mole}^{-1}.$$

(4) **Water.** The difference in spectroscopic (188.5 J k^{-1} $mole^{-1}$) and calorimetric (185.3 J k^{-1} $mole^{-1}$) values of entropy in case of water can be explained on the basis of hydrogen bonding. A molecule of water can be oriented in the following six ways:

The chance of finding a hydrogen atom along two particular O—O direction from a given oxygen is $\frac{1}{4}$. So for N molecules there are $\left(\frac{3}{2}\right)^N$ microstates.

(I) (II) (III)

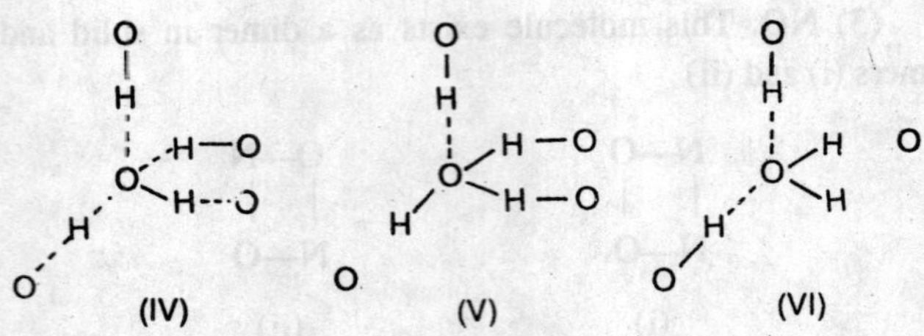

Fig. 11.3

Hence entropy is given by

$$S = k \ln \left(\frac{3}{2}\right)^N$$

$$= N k \ln \frac{3}{2}$$

$$= R \ln \frac{3}{2}$$

(5) **Hydrogen.** The difference in two values calculated spectroscopically (130.5 J k^{-1} $mole^{-1}$) and calorimetrically (124.5 JK^{-1} $mole^{-1}$) in case of hydrogen can be explained on the basis of the fact that ordinary hydrogen is a mixture of ortho- and para hydrogen. At ordinary temperatures these are present in the ratio of 3 : 1. At 0°K it will be a pure para form and this fact accounts for the difference.

Entropy and Degeneracy. A loss of energy takes place on cooling and at 0°A the system will be having the lowest energy. The system in this state is know as in *ground state.* If this minimum energy state is degenerate, then the constituent particles may be arranged to give more than one microstate at absolute zero. But equation S = k ln W indicates that if the system has just are microstate, then the thermodynamic probability W, is unity and so S = 0.

The third law can thus be equated to stating that the ground state of a system is non-degenerate.

Comparison of Third Law with other Laws

The third law is not based on experience but is important as its conclusions are in agreement with the experimental results. It does not introduce any new functions, it only places a limitation on the entropy value of a pure crystalline solid.

Standard State for the Entropy of Ideal Gas

$$ds = -\frac{R}{P} dP$$

gives the change in entropy with respect to pressure at constant temperature.

Integrating it, we get

$$\Delta S = \bar{S} - \bar{S}^{\circ}$$

$$= -R \ln \frac{P}{1}$$

where S° is the molar entropy, at 1 atmosphere pressure and at any temperature, in the standard state. So the above equation may be written as

$$\bar{S} = \bar{S}^{\circ} - R \ln P$$

In this equation P must be in atmospheres.

Example 1. How much will be the change in entropy from standard state at 298 K when the pressure is 600 mm.

Solution. We know

$$\Delta \bar{S} = -R \ln p$$

$$= -8.314 \times \frac{600}{760}$$

$$= +1.965 \text{ J mol}^{-2} \text{ deg.}^{-1}.$$

Example 2. If the entropy is standard state of O_2 is 205.00 e.u. What will be its entropy when the pressure is 600 mm.

Solution. $\Delta \bar{S} = \bar{S} - \bar{S}^{\circ} = 1.965$

$$\bar{S} = 205 + 1.965 \qquad \left[\Delta \bar{S} = -R \ln \frac{600}{760} \right.$$

$$= 206.965 \text{ e.u.} \qquad \left. = 1.965 \right]$$

12

Partial Molar Properties

Introduction to Partial Molar Quantities

We have derived various thermodynamics relations considering the system to be a closed one. In a closed system there is no change in mass and composition. We have also studied that the thermodynamic properties E, H, S, A and G are extensive properties. The values of extensive properties change with a change in the mass of the system. To apply various thermodynamic relations already derived, to open systems, where mass and composition may change, we have to consider the number of moles of various substances as variables like any other thermodynamic variable. G.N. Lewis called such a thermodynamic property as a *partial molar property.* e.g., Gibb free energy of a system, depends on the number of moles of each substance in the system (open system) in addition to temperature and pressure. So

$$G = f(T, P, n_1, n_2 \ldots)$$

where n_1, n_2...represents the number of moles of each component. For simplification let us consider a system of two components only. The total differential of G is given by

$$dG = \left(\frac{\partial G}{\partial T}\right)_{P, n_1, n_2} dT + \left(\frac{\partial G}{\partial P}\right)_{P, n_1, n_2} + \left(\frac{\partial G}{\partial n_1}\right)_{T, P, n_2} dn_1$$

$$\left(\frac{\partial G}{\partial n_2}\right)_{T, P, n_1} dn_2$$

The partial derivatives $\left(\frac{\partial G}{\partial n_1}\right)_{T, P, n_2}$ and $\left(\frac{\partial G}{\partial n_2}\right)_{T, P, n_1}$

are called partial molar free energies of components 1 and 2 respectively. In general, a bar over the symbol of any property denotes the partial molar property. The above two derivatives (partial) can thus be written as partial free energies of component n_1 and n_2 respectively

$$\bar{G}n_1 = \left(\frac{\partial G}{\partial n_1}\right)_{T, P, n_2}$$

and

$$\bar{G}n_2 = \left(\frac{\partial G}{\partial n_2}\right)_{T, P, n_1}$$

Partial molar property for ith component is thus represented by e.q. (1). For any thermodynamic property X.

$$\bar{X}_1 = \left(\frac{\partial x}{\partial n_1}\right)_{P, T, n_2, n_2 \ldots} \quad (1)$$

$$= \left(\frac{\partial X}{\partial n_i}\right)_{T, P, n_j} \quad J \neq i$$

Thus the partial molar quantities may be defined as

"The increase in extensive property (X) of a system on account of addition of one mole of its respective constituents, at constant temperature, pressure and moles of other constituents remaining unaltered, to such a large amount of system that this addition has a negligible effect on the composition of the system."

Physical Significance of Partial Molar Quantities

We have already derived the relations in which x_i represents the actual value of x per mole of the ith component in a system of known composition. This may or may not be equal to the corresponding molar value of x in the pure state. It is only in case of ideal systems that the partial molar value and molar value are the same as in such ideal systems there exists no interaction between the constituents. In non-ideal systems the molar value of X_i get modified to $\bar{X}_i$ due to interactions between various constituents of the system. The values of

$\bar{X}_i$ will not be same for all compositions. Thus the values of partial molar quantities not only depend on the nature of the particular substance but also on the nature and quantities of other substances present in the system.

Various partial molar properties are defined as

$$\bar{E}_i = \left(\frac{\partial E}{\partial n_i}\right)_{T,\,P,\,n_j} \qquad j \neq i$$

$$\bar{V}_i = \left(\frac{\partial V}{\partial n_i}\right)_{T,\,P,\,n_j} \qquad j \neq i$$

$$\bar{S}_i = \left(\frac{\partial S}{\partial n_i}\right)_{T,\,P,\,n_j} \qquad j \neq i$$

$$\bar{G}_i = \left(\frac{\partial G}{\partial n_i}\right)_{T,\,P,\,n_j} \qquad j \neq i$$

$\bar{G}_i$ is the most useful of the partial molar quantities and is called *chemical potential* represented by μ_i. Thus

$$\mu_i = \mu i = \left(\frac{\partial G}{\partial n_i}\right)_{T,\,P,\,n_j} \qquad j \neq i$$

(μi) *chemical potential* is an intensive property, since it is molar quantity.

The various thermodynamic relations already derived can be written in the form of partial molar properties as

$$\bar{H}_i = \bar{E}_i + R\bar{V}_i \qquad \bar{A}_i = \bar{E}_i - T\bar{S}_i$$

$$\bar{G}_i = \bar{H}_i - T\bar{S}_i \qquad \bar{G}_i = \mu \ \bar{H}_i - T\bar{S}_i$$

In case of single component systems the partial molar quantities are identinal with molar quantities, *i.e.*,

$$\bar{V}_i = \frac{V}{\Sigma n_i}, \qquad \bar{E}_i = \frac{E}{\Sigma n_i}$$

Like partial free energy all the partial molar quantities are intensive properties.

In case of a system containing mole numbers n_1, $n_2...n_m$ and having a total property m^t, it is found that increasing each mole number by the same factor α at constant T and P results in a new value αM^t of the property. Property showing this behaviour are *extensive property.* Extensive total properties M^t ($=n\overline{M}$) are related to the corresponding partial molar properties $\overline{M}_i$ by

$$nM = \Sigma n_i \overline{M}_i$$

Dividing this equation by the total number of moles n yields an expression for the molar properties M in terms of $\overline{M}_i$ and the mole fraction x_i

$$M = \Sigma xi \overline{M}_t$$

If we choose to deal with *masses* instead of *mole numbers* then n_i is replaced by m_i and n by m. In this case x_i is the weight fraction of ith component and $\overline{M}_i$ are then know as *partial specific properties.*

Determination of Partial Molar Properties

A number of methods can be used. Some of them are given below:

(*i*) *Direct Method.* From the definition of partial molar property, we have

$$\left(\overline{X}_i = \frac{\partial X}{\partial n_i}\right)_{T, P, n_j} \quad j \neq i$$

so to calculate the partial molar property a graph is plotted between the extensive property X at constant temperature and pressure, for various binary mixtures against the number of mole of one of them (n_2) keeping the number of moles of the other (n_1) component constant. When a tangent is drawn to such a curve the slope of the curve at any particular composition gives the value of $\overline{X}_2$. If we known X_2 then $\overline{X}_1$ can easily be calculated with the help of the reation,

$$X = n_1\overline{X}_1 + n_2\overline{X}_2$$

It is difficult to accurately determine the exact slope of the curve so this method is seldom used.

(ii) From apparent molar properties. If in a binary mixture having n_1 moles of A and n_2 moles of B, X_1 and X_2 denote the value of a particular property per mole of pure components 1 and 2 respectively, then the molar value (apparent) of the property for component 2 can be represented by ϕ_x, where

$$\phi_x = \frac{X - n_1 X_1}{n_2}$$

where X is the value of this particular property of the mixture.

or $$X - n_1 X_1 + n_2 \phi_x$$

Differentiating with respect to n_2 at constant temperature and pressure keeping n_1 constant so that $n_1 X_1$ is constant.

$$\left(\frac{\partial X}{\partial n_2}\right)_{n_1} = n_2 \left(\frac{\partial \phi_x}{\partial n_2}\right)_{n_1} + \phi_x$$

or $$X_2 = \left(\frac{\partial \phi_x}{\partial\, 1n\, n_2}\right)_{n_1} + \phi_x$$

where $$X_2 = \left(\frac{\partial x}{\partial n_2}\right)_{n_1}$$ [definition]

and $$\partial \ 1\text{n}\ n_2 = \frac{\partial n_1}{n_2}$$

$\therefore$ $$X_2 = m \left(\frac{\partial \phi_x}{\partial m}\right)_{n_1} + \phi_x$$

or $$X_2 = \left(\frac{\partial \partial_x}{\partial\, 1n\, m}\right)_{n_1} + \phi_x$$

If ϕ_x is determined for different values of n_2 keeping n_1 constant, partial molar property can be calculated from the slope of the curve obtained by plotting ϕ_x against n_2 or from the curve obtained by plotting fx versus ln n_2.

In place of graphical method analytical method can be used by expressing ϕ_x as a function of n_2.

(iii) *Method of intercepts*. This is a useful method as it gives simultaneously the values of partial molar properties of both the constituents of a binary mixture. The observed value of X of a system is given by

$$X = (n_1 + n_2)\,x$$

where n_1, n_2 are the moles of the two constituents and x is the mean value of a particular extensive property.

Differentiating this equation w.r.t. n_2 at constant n_1. Keeping the temperature and pressure constant, we get

$$X_2 = \left(\frac{\partial X}{\partial n_2}\right)_{n_1}$$

$$= x + (n_1 + n_2)\left(\frac{\partial x}{\partial n_2}\right)_{n_1}$$

the mole fraction of component 1 is

$$N_1 = \frac{n_1}{n_1 + n_2}$$

Differentiating this equation, keeping n_1 constant, we get

$$dN_1 = -\frac{n_1\, dn_2}{(n_1 + n_2)^2}$$

or
$$dN_1 = \frac{n_1}{n_1 + n_2} \cdot \frac{dn_2}{n_1 + n_2}$$

or
$$dN_1 = -N_1 \frac{dn_2}{n_1 + n_2}$$

or
$$\frac{n_1 + n_2}{dn_2} = -\frac{N_1}{dN_1}$$

multiplying by dX

$$(n_1 + n_2)\left(\frac{\partial x}{\partial n_2}\right)_{n_1} = -N_1\left(\frac{\partial X}{\partial N_1}\right)_{n_1}$$

or $$X_2 = x - N_1 \left(\frac{\partial X}{\partial N_1}\right)_{n_1}$$

A graph is drawn between the mean values of this property X for mixture of various composition and mole fraction N_1. If the partial molar property is to be determined at a point O (Fig. 12.1)

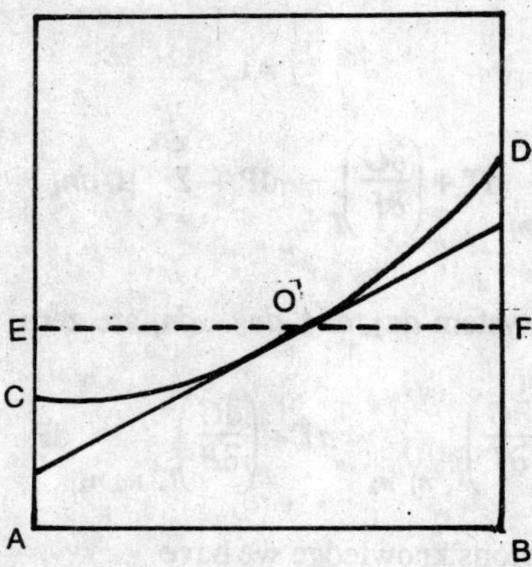

Fig. 12.1

(where partial molar property is required to be determined). Tangent CD at point O is drawn. A line || to base line is drawn. Then

$$\text{Slope CD} = \frac{dx}{dN_1}$$

and $$\text{CE} = N_1 \frac{dx}{dN_1} \text{ at O}$$

AE gives the value of x at this point O.

Chemical Potential (μ)

It is the most useful partial molar property. As already discussed the partial molar free energy $\overline{G}_i$ is known as chemical potential and is represented by a symbol μ_i for the *i*th component. So

$$\mu_i = \left(\frac{\partial G}{\partial n_i}\right)_{T, P, n_j} \quad j \neq i \tag{1}$$

For a system of *n* components, we can write

$$dG = \left(\frac{\partial G}{\partial T}\right)_{P\, n_j} dT + \left(\frac{\partial G}{\partial P}\right)_{T\, n_j} dP + \sum_{i=1}^{K} \left(\frac{\partial G}{\partial n_i}\right)_{T, P\, n_j} dn_i$$

$$j \neq i \tag{2}$$

or $$dG = \left(\frac{\partial G}{\partial T}\right)_{P\, n_j} dT + \left(\frac{\partial G}{\partial P}\right)_{T\, n_j} dP + \sum_{i=1}^{K} \mu_i\, dn_i \tag{3}$$

For a closed system $dn_1, dn_2, dn_3 \ldots dn_j$ are zero. Hence equation (3) becomes

$$dG = \left(\frac{\partial G}{\partial T}\right)_{T,\ n_1, n_2} dT + \left(\frac{\partial G}{\partial P}\right)_{T,\ n_1, n_2} dP \tag{4}$$

From our previous knowledge we have

$$dG = VdP - SdT \tag{5}$$

So from (4) and (5), we get

$$\left(\frac{\partial G}{\partial T}\right)_{P,\ n_1, n_2} = -S \tag{6}$$

and

$$\left(\frac{\partial G}{\partial P}\right)_{T,\ n_1, n_2} = V \tag{7}$$

From equation (3), (6) and (7), we get

$$dG = -SdT + VdP + \mu_1\, dn_1 + \mu_2\, dn_2 + \ldots$$

$$= -SdT + VdP + \Sigma_\iota \mu_i dn_i \quad (8)$$

Equation (8) gives a relation between change is free energy, temperature change, pressure change and the change in mole numbers.

Similar equations can be obtained for other functions such as dE, dH, dA etc., for multi-component open systems.

For Work Function (A)

We know from the definition of work function.

$$A = G - PV$$

So
$$dA = dG - PdV - VdP \quad (9)$$

From (8) and (9), we get

$$dA = (-SdT + VdP + \sum_{\iota=1}^{\kappa} \mu_1 dn_i) - PdV - VdP$$

$$= -SdT - PdV + \sum_{\iota=1}^{\kappa} \mu_i dn_i \quad (10)$$

For Enthalpy (H)

From definition of enthalpy

$$H = G + TS$$

$\therefore$
$$dH = dG + TdS + SdT \quad (11)$$

From (8) and (11), we get

$$dH = (-SdT + VdP + \sum_{\iota=1}^{\kappa} \mu_i dn_i) + TdS + SdT$$

$$= TdS + VdP + \sum_{\iota=1}^{\kappa} \mu_i dn_i \quad (13)$$

For Internal Energy (g)

$$E = H - PV$$

$$dE = dH - PdV - VdP \quad (13)$$

From (12) and (13),

$$dE = (TdS + VdP + \sum_{i=1}^{\kappa} \mu_i \, dn_i) - PdV - VdP$$

or

$$dE = TdS - VdP + \sum_{i=1}^{\kappa} \mu_i \, dn_i \qquad (14)$$

For closed systems in all these cases we get the equation

$$dG = -SdT + VdP$$

$$dA = -SdT - PdV$$

$$dH = TdS + VdP$$

$$dE = TdS - PdV$$

From equations (10), (12) and (14), we have

$$\mu_i = \left(\frac{\partial H}{\partial n_i}\right)_{S,\, P,\, n_1\, n_2 \ldots}$$

$$= \left(\frac{\partial E}{\partial n_i}\right)_{S,\, V,\, n_1\, n_2 \ldots}$$

$$= \left(\frac{\partial A}{\partial n_i}\right)_{T,\, V,\, n_1\, n_2 \ldots}$$

So chemical potential plays a different role though it is the same quantity.

Physical Significance of Chemical Potential

We have already seen that the increase in free energy per mole of the added substance is given by the following equation:

$$dG = -SdT + VdP + \Sigma\, bi\, \mu_i\, d\, n_i$$

where 'dn_i' is the small number of moles of component 'i' added to the system.

$$\mu_i = \left(\frac{\partial G}{\partial n_i}\right)_{T,\, P,\, n_i} \qquad (15)$$

This equation (15) gives us the significance of chemical potential.

"For any component 'i' in a multi component system, the value of μ_i *is given by the increase in free energy that takes place when an infinitesimal number of moles of that component is added to the system, per mole of the substance added."*

From this definition we find that the amount added is restricted to a very small quantity such that the composition of mixture and thus the value of μ_i does not change.

(ii) It is an extensive property and has the same value in all parts of the system in equilibrium condition.

(iii) It is also known as *escaping tendency.* More the chemical potential of a component in a system, more will be the escaping tendency of that component.

Escaping Tendency

To understand the escaping tendency let us consider a component 'i' present in two different regions (1) and (2) of a system and the values of chemical potential of this component in these two regions be represented by $\mu_{i(1)}$ and $\mu_{i(2)}$ respectively. If now dn moles of this component 'i' are transferred from region, (1) to region (2) then we have

$$dG_{(1)} = \mu_{i(1)}\ (d_n) \qquad (1)$$

and

$$dG_{(2)} = \mu_{i(2)}\ (d_n) \qquad (2)$$

Total change in free energy is

$$dG = dG_1 + dG_2$$

or

$$dG = -\mu_{i(1)} + \mu_{i(2)} \qquad (3)$$

If $\mu_{i(1)} > \mu_{i(2)}$ then the value of dG is negative and the transfer of 'i'th component from region (1) to region (2) will be spontaneous. This flow will continue till the value of $\mu_{i(1)} = \mu_{i(2)}$.

Thus we conclude that chemical potential of a component in a system is a direct measure of the flow or escaping tendency of that component.

Gibbs-Duhem Equation

We have already derived an equation for the chemical potential

$$dG = -SdT + VdP + \mu_1 dn_1 + \mu_2 dn_2 + \ldots \mu_1 dn_i$$

If temperature and pressure of the system are kept constant. Then we have

$$dG_{T,P} = \mu_1 dn_1 + \mu_2 dn_2 + \ldots \mu_i dn_i \quad (1)$$

Integrating (1), we get

$$G = \mu_1 n_1 + \mu_2 n_2 + \ldots + \mu_1 n_i \quad (2)$$

Exact differential of equation (2) is given by

$$dG = (\mu_1 dn_1 + \mu_2 dn_2 + \ldots) + (n_1 d\mu_1 + n_2 d\mu_2 + \ldots + n_1 d\mu_1) \quad (3)$$

From equation (1) and (3), we get

$$n_1 d\mu_1 + n_2 d\mu_2 + \ldots + n_i d\mu_i = 0 \quad (4)$$

Equation 2, 3 and 4 above are known as Gibbs-Duhem equations.

For a binary system, solution in equilibrium with its vapours at constant temperature and pressure we can derive another form of Gibbs-Duhem equation.

If such a system undergoes an infinitesimal small change in composition we have from equation (4).

$$n_1 d\mu_1 + n_2 d\mu_2 = 0 \quad (5)$$

n_1, n_2 are the number of moles of the two components of the binary system and μ_1, μ_2 are the chemical potentials respectively. If x_1 and x_2 are the mole fractions of the two components respectively then equation (5) becomes

$$x_1 \left(\frac{\partial \mu_1}{\partial x_1}\right)_{T,P} dx_1 + x_2\left(\frac{\partial \mu_2}{\partial x_2}\right)_{T,P} dx_2 = 0 \quad (6)$$

But $x_1 + x_2 = 1$ [for a binary system], differentiating, it we get

$$dx_1 + dx_2 = 0 \text{ or } dx_1 = -dx_2 \quad (7)$$

From (6) and (7), we get

$$\left(\frac{\partial \mu_1}{\partial \ln x_1}\right)_{T,\ P} - \left(\frac{\partial \mu_2}{\partial \ln x_2}\right)_{T,\ P} = 0 \quad (8)$$

This is another form of Gibbs-Duhem equation and is found useful in gas-liquid equilibria. Such type of equilibria is involved in the process of distillation.

Variation of chemical potential with temperature and pressure

We know that

$$\left(\frac{\partial G}{\partial T}\right)_{P,\, n_1,\, n_2} = -S$$

Differentiating *w.r.t* n_i, we get

$$\frac{\partial^2 G}{\partial T \partial n_i} = -\left(\frac{\partial S}{\partial n_i}\right)_{T,\, P,\, n_1,\, n_2\, ..} \quad (1)$$

But

$$\left(\frac{\partial G}{\partial ni}\right)_{T,\, P,\, n_1\ n_2\, ..} = \mu_i \quad \text{[definition]} \quad (2)$$

Differentiating (2) w.r.t. T, we get

$$\left(\frac{\partial^2 G}{\partial n_i \partial T}\right) = \left(\frac{\partial \mu_i}{\partial T}\right)_{T,\, n_1,\, n_2\, ..} \quad (3)$$

Since dG is an exact differential so comparing (1) and (3), we get

$$\frac{\partial^2 G}{\partial n_i \partial T} = \frac{\partial^2}{\partial T \partial n_i}$$ order of differentiation is immaterial in case of exact differential

so we get

$$\left(\frac{\partial \mu_i}{\partial T}\right)_{P, n_1, n_2,} = -\left(\frac{\partial S}{\partial n_i}\right)_{T, P, n_1, n_2} \quad (4)$$

or

$$\left(\frac{\partial \mu_i}{\partial T}\right)_{P, n_1, n_2} = -\bar{S}_i \quad (5)$$

where $\bar{S}_i$ is the partial molar entropy of component '*i*'.

Equation (5) thus gives us a change in chemical potential with a change in temperature. From this equation we can easily see that a decrease in temperature will result in an increase in chemical potential because the value of S is always positive.

In case of solids, liquids and gases the rate of change of chemical potential is found to be different as S in the following order in these cases.

$$S_{(s)} < S_{(l)} < S_{(g)}$$

The variation of chemical potential with respect to temperature can also be expressed in another manner. We know

$$G = H - TS \text{ (definition)}$$

Differentiating *w.r.t.* n_i at constant T, P and composition (except that of '*i*'th component), we get

$$\left(\frac{\partial G}{\partial n_i}\right)_{T, P, n_1 n_2 \ldots} = \left(\frac{\partial H}{\partial n_i}\right)_{T, P, n_1 n_2 \ldots} - T\left(\frac{\partial S}{\partial n_i}\right)_{T, P, n_1 n_2 \ldots} \quad (6)$$

or

$$\mu_i = \bar{H}_i + T\,\bar{S}_i \quad (7)$$

where Hi and $\bar{S}_i$ are the partial molar enthalpy and partial molar entropy of the 'i'th component.

From (5) and (7), we get

$$\mu_i = \bar{H}_i + T\left(\frac{\partial \mu_i}{\partial T}\right)_{P, n_1, n_2 \ldots} \quad (8)$$

It can also be written as

$$\left[\frac{\partial}{\partial T}(\mu_i/T)\right] = -\frac{\bar{H}_i}{T^2} \quad (9)$$

[Dividing Eq. 8 by T^2]

Equations (8) and (9) are equations similar to Gibbs-Helmholtz equation and express the effect of temperature on chemical potential.

Variation of Chemical Potential with Pressure

We have

$$\left(\frac{\partial G}{\partial n_i}\right)_{T, P, n_1 n_2 \ldots} = \mu_i \quad \text{[definition]}$$

Differentiating w.r.t. P, we get

$$\left(\frac{\partial G}{\partial n_i \; \partial P}\right)_{T, n_1 n_2 \ldots} = \left(\frac{\partial \mu_i}{\partial P}\right)_{T,, n_1 n_2 \ldots} \quad (10)$$

But we know

$$\left(\frac{\partial G}{\partial P}\right)_{T, n_1 n_2 \ldots} = V$$

Differentiating w.r.t. n_i, we get

$$\left(\frac{\partial G}{\partial P \, \partial n_i}\right) = \left(\frac{\partial V}{\partial n_i}\right)_{T, n_1 n_2 \ldots} \quad (11)$$

$$= \bar{V}_i \quad \text{[definition]}$$

From (10) and (11), we get

$$\left(\partial \mu_i\right)_{T, n, n_2} = \bar{V}_i \quad (12)$$

[Order of differentiation immaterial in case of exact differential]

$\bar{V}_i$ is the partial molar volume of 'i'th component.

$$V = (n_1 + n_2 \ldots)\frac{RT}{P} \quad (13)$$

Differentiating *w.r.t.* n_i, we get

$$\left(\frac{\partial V}{\partial n_i}\right)_{T, P, n_1, n_2} = \frac{RT}{P} \tag{14}$$

but

$$\bar{V}_i = \left(\frac{\partial V}{\partial ni}\right)_{T, P, n_1, n_2 \ldots} \qquad \text{[definition]}$$

or

$$V_i = \frac{RT}{P} \tag{15}$$

From (14) and (15),

$$\left(\frac{\partial \mu_i}{\partial P}\right)_{T, n_1, n_2} = \frac{RT}{P} \tag{16}$$

Equation (12) and (16) give the change of chemical potential with pressure.

Equation (12) may also be written as

$$d\mu_i = \bar{V}_i dP$$

Integrating this, we get

$$\int d\mu_i = \int \bar{V}_i \, dP \tag{17}$$

Equation (17) is useful in calculating the chemical potentials of solids, liquids and gases. This equation can be used if we know the pressure dependence of volume.

Chemical Potential of a Pure Solids and Pure Liquids

We have proved in the previous article that

$$d\mu i = \bar{V}_i \, dP$$

In case of pure solids and pure liquids the change in volumes negligible and $\bar{V}$ does not change much with pressure provided the temperature is kept constant. Integrating the above equation under such conditions, we get

$$\mu - \mu^\circ = \bar{V} \, (P - P^\circ)$$

where μ° is the chemical potential of solid or liquid at pressure P°. This pressure P° is generally taken as 1 atmosphere so μ° represent the chemical potential of 1 mole of pure solid or liquid at 1 atmosphere and is called standard chemical potential, so

$$\mu - \mu^\circ = \bar{V}(P - 1)$$

or $$\mu_i - \mu_i^\circ = \bar{V}\ (P_i - p_i^\circ) \qquad \text{[for 'i th component]}$$

where p_i and p_i° represent the partial pressure of '*i*'th species and pi° its pressure in pure state.

If the vapour pressure is more in pure state than the partial vapour pressure the value of $\mu_i - \mu_{i^\circ}$ will be negative, indicating that the mixing would be spontaneous, Otherwise the mixing may be partial.

Chemical Potential of an Ideal Gas

The chemical potential μ of an ideal gas a given temperature is a function of its pressure P expressed in atmosphere. It can be expressed as

$$\mu_{(T,P)} = \mu^\circ_{(T)} + RT\ 1n\ P \qquad (1)$$

$\mu_{(T)}^\circ$ is the standard chemical potential of the gas at 1 atmosphere pressure. Thus if the pressure in two regions of a gas phase is different there will be a spontaneous flow of gas from the region of higher pressure to a region of lower pressure till the pressure everywhere is same throughout the gaseous system.

The above relation (1) can also be derived in the following manner

$$V = (n_1 + n_2 + \ldots)\frac{RT}{P}$$

Differentiating w.r.t. n_i, we get

$$\left(\frac{\partial V}{\partial ni}\right)_{T,\,P,\,n_1,\,n_2\ldots} = \frac{RT}{P}$$

or

$$\bar{V}_i = \frac{RT}{P}$$

We have already calculated

$$\bar{V}_i = \left(\frac{\partial \mu_i}{\partial P}\right)_{T,\, n_1,\, n_2} \qquad \text{[see eq. (12) page 366]}$$

$$\therefore \qquad \left(\frac{\partial \mu_i}{\partial P}\right)_{T,\, n_1,\, n_2} = \frac{RT}{P}$$

Integrating this we get

$$\mu_i = \mu_i^{\circ} + RT \ln P$$

μ_i° is the integration constant and is equal to the standard chemical potential of the gas.

Chemical Potential of an Ideal Gas in a Mixture of Ideal Gases

Such an expression can be easily obtained if the pressure P in Eq. (1) of page 368 replaced by p_i the partial pressure of the gas in the gaseous mixture. Thus the required equation is

$$\mu_{(mix)(T,P)} = \mu^{\circ}{}_{i(T)} + RT \ln p_i$$

but $$p_i = x_i \times P$$

$$\therefore \qquad \mu_{(mix)(T,P)} = \mu_i^{\circ}{}_{(T)}\ RT \ln P + RT \ln x_i$$

or $$\mu_{i(mix)(T,P)} = \mu_{i\,pure}\,(T,P) + RT \ln x_i$$

$$[\because\ \mu_{i\,pure\,(T,P)} = \mu_i^{\circ} + RT \ln P]$$

Since x_i is always less than 1 so RT ln x_i is always negative and hence

$$\mu_{(i\,(mix)\,(T,P)} < \mu_{i(pure)\,(T,P)}$$

So in a mixture of ideal gases the chemical potential of an ideal gas is less than its chemical potential in pure state, provided the total pressure of the gaseous mixture and the pressure of the pure ideal gas are same.

It can be concluded that if a pure ideal gas and a mixture of ideal gases are separated from each other with the help of a permeable membrane which allows the pure to flow through it then the pure gas will flow to be mixture of ideal gases. This also explains the spontaneous mixing of gases.

Partial Molar Quantities of an Ideal Gas in a Mixture of Ideal Gases

Expressions can be derived for various partial molar quantities of any component 'i' in a mixture of ideal gases.

Starting from the equation

$$\mu_{i\,(mix)(T,P)} = \mu_i^{\circ}(T) + RT\ 1n\ P + RT\ 1n\ x_i \qquad (1)$$

Differentiating (1) w.r.t temperature keeping P and composition constant,

$$\left[\frac{\partial \mu_{i(mix)(T,P)}}{\partial T}\right]_{P,\,n_1,\,n_2\,\ldots} = \left(\frac{\partial \mu_i^{\circ}(T)}{\partial T}\right)_P + R\ 1n\ P + R\ 1n\ x_i$$

or $$-\bar{S}_t = -\bar{S}_i^{\circ} + R\ 1n\ P + R\ 1n\ x_i$$

or $$\bar{S}_{i(mix)} = \bar{S}_{i(pure)} - R\ 1n\ x_i$$

where $$S_{i(pure)} = S_i^{\circ} - R\ 1n\ P$$

If we differentiate (i) w.r.t. P at constant '1' and composition, we get

$$\left[\frac{\partial \mu_{i(mix)(T,P)}}{\partial T}\right]_{T,\,n_1,\,n_2\,\ldots} = \frac{RT}{P}$$

or $$\bar{V}_{i(mix)} = \frac{RT}{P} = \bar{V}_{i(pure)}$$

Thus the partial molar volume of an ideal gas is the same in the mixture as the partial molar volume in pure state.

For n moles of an ideal gas, we have

$$PV = nRT$$

or $$V = \frac{nRT}{P}$$

So, we get $$\bar{V}_i = \frac{V}{n}$$

where n is the total number of moles of all gases in the mixture of ideal gases.

Additivity Rules

The additivity rules are applicable to all partial molar quantities, i.e. Additivity rule is applicable to all extensive properties and can be derived using Eulers' theorem as applied to homogeneous functions of first degree w.r.t the number of moles of the substances present in the system at constant temperature and pressure. It should be clearly understood that Euler's theorem can be applied to the function Y in the form

$$Y = f(n_1, n_2 ...)$$

and not as $\quad Y = f(T, P, n_1, n_2 ...)$

because this function Y may not be a homogeneous function of first degree with respect to either T or P, e.g., G is not a homogeneous function of either T or P as shown in the following paragraph:

If G is a homogeneous function of first degree with respect to T and P, then applying Euler's theorem, we get

$$G = T\left(\frac{\partial G}{\partial T}\right)_P + P\left(\frac{\partial G}{\partial P}\right)_T$$

or $$G = -TS + PV \qquad (1)$$

But $$G = H - TS \qquad \text{(definition)}$$

$$= E + PV - TS \qquad [\because H = E + PV] \qquad (2)$$

Equation (1) and (2) are not same therefore G is not a homogeneous function of first degree with respect to T and P.

The additivity rules for various partial molar properties are summarized below:

(i) $\quad S = \Sigma_i n_i \bar{S}_i$

(ii) $\quad V = \Sigma_i n_i \bar{V}_i$

(iii) $H = \Sigma_i n_i \bar{H}_i$

(iv) $E = \Sigma_i n_i \bar{E}_i$

(v) $A = \Sigma_i n_i \bar{A}_i$

(vi) $G = \Sigma_i n_i \bar{G}_i$

or $G = \Sigma n_i, \mu_i$

Free Energy and Entropy of Mixing of Ideal Gases

The process of mixing in case of gases is a spontaneous and so there must be a decrease in free energy in such a process.

If we mix a number of gases at the same temperature and pressure the total volume of mixture will be equal to the sum of the individual volumes of the components we can calculate the change in free energy and entropy during this process in the following manner.

The initial conditions of the various components under a pressure P and temperature T can be shown a in Fig. 12.2(a) where each of the compartments has a pure gas. If now there partitions separating them are pulled out we get a mixture of these gases under the same pressure and temperature Fig. 12.2(b).

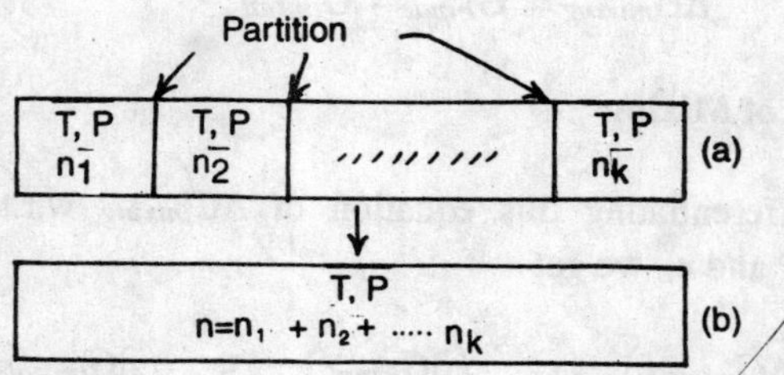

Fig. 12.2

The free energy and entropy of mixing are calculated as under. For pure gases the free energies are

$$G_1 = n_1 \mu_{1(pure)} \qquad G_2 = n_2 \mu_{2(pure)}$$

$$G_n = n_n \mu_{n(pure)}$$

So the total free energy of all the gases in the initial state is given by the sum of these individual free energy values of various components.

$$G_{initial} = n_1\mu_{1(pure)} + n_2\mu_{2(pure)} + ...$$

The free energy in final state is given by

$$G_{Final} = n_1\mu_1 + n_2\mu_2 + ...$$

Thus free energy of mixing

$$= G_{Final} - G_{initial}$$

$$\Delta G_{mixing} = n_1[\mu_1 - \mu_{1(pure)}] + n_2[\mu_2 - \mu_{2(pure)}] + ...$$

or $\quad \Delta G_{mixing} = RT\,(n_1 \ln x_1 + n_2 \ln x_2 + ...)$

$$[\mu_i - \mu_{i(pure)} = RT \ln x_i]$$

Again, $\quad n_i = x_i N$

$\therefore \quad \Delta G_{mixing} = NRT\,(x_1 \ln x_1 + x_2 \ln x_2 + ...)$

This expression gives us the free energy of mixing. For the formation of a mixture of ideal gases the free energy of mixing will be

$$\Delta G_{mixing} = NRT\,\Sigma_i\ x_i \ln x_i$$

or $\quad \Delta G_{mixing} = G_{Final} - G_{initial}$

Entropy of Mixing

Differentiating this equation of ΔG_{mixing} w.r.t temperature at contant P and n_i, we get

$$\left(\frac{\partial \Delta G_{mixing}}{\partial T}\right)_{P,\,n_1,\,n_2} = \left(\frac{\partial G_{Final}}{\partial T}\right)_{P,\,n_1,\,n_2} - \left(\frac{\partial G_{initial}}{\partial T}\right)_{P,\,n_1 ...}$$

$$= -\,(S_{Final} - S_{initial})$$

$$= -\,\Delta S_{mixing}$$

From the equation of ΔG_{mixing}

$$\Delta G_{mixing} = \text{NRT} \sum_i x_i \ln x_i$$

Differentiating w.r.t temperature, we get

$$\left[\frac{\partial(\Delta G_{mixing})}{\partial T}\right]_{P, n_1, n_2 \ldots} = \text{NR} \sum_i x_i \ln x_i$$

$$\therefore \quad -\Delta S_{mixing} = \text{NR} \sum_i x_i \ln x_i$$

or $$\Delta S_{mixing} = -NR \sum_i x_i \ln x_i$$

Thus the entropy of mixing is always positive because the minus sign will make this value positive as x_i is less than 1 and $\ln x_i$ will be negative. So on mixing there will be an increase in randomness.

A plot between $\frac{\Delta S_{mixing}}{NR}$ and x in case of a binary mixture can be drawn in accordance with the following equation

$$\Delta S_{mixing} = -n\text{R}\,[x \ln x + (1-x) \ln (1-x)]$$

The plot obtained is shown in Fig. 12.3

In the graph we find that the entropy of mixing has the maximum

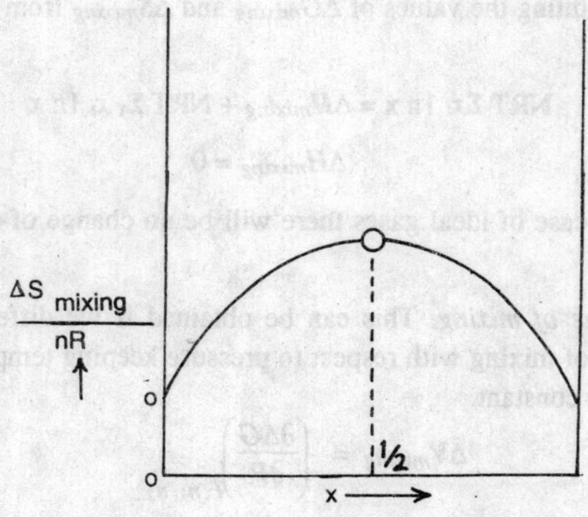

Fig. 12.3

value at x = $\frac{1}{2}$. From this the numerical value per mole of the mixture can be calculated.

$$\frac{\Delta S_{mixine}}{N} = -1.987\left[\frac{1}{2}\ln\frac{1}{2} + \frac{1}{2}\ln\frac{1}{2}\right]$$

So for a binary mixture, entropy of mixing depends on the composition of mixture and varies between 0 and 1.38 e.u.

Heat of Mixing and Volume of Mixing of Ideal Gases

Heat of mixing. It can be calculated as under, we know

$$\Delta G_{mixing} = \text{NRT} \sum_i x_i \ln x_i \qquad (1)$$

and

$$\Delta S_{mixing} = -\text{NR} \sum_i x_i \ln x_i \qquad (2)$$

From the definition of free energy

$$G = H - TS$$

or

$$\Delta G = \Delta H - T\Delta S$$

or

$$\Delta G_{mixing} = \Delta H_{mixing} - T\Delta S_{mixing}$$

Substituting the values of ΔG_{mixing} and ΔS_{mixing} from (1) and (2), we get

$$\text{NRT} \sum x_i \ln x = \Delta H_{mixing} + \text{NRT} \sum_i x_i \ln x_i$$

or

$$\Delta H_{mixing} = 0$$

So in case of ideal gases there will be no change of enthalpy on mixing.

Volume of mixing. This can be obtained if we differentiate the free energy of mixing with respect to pressure keeping temperature and composition constant.

$$\Delta V_{mixing} = \left(\frac{\partial \Delta G}{\partial P}\right)_{T, n_1, n_2 \ldots}$$

But

$$\Delta G_{mixing} = \text{NRT} \sum_i x_i \ln x_i$$

Since ΔG_{mixing} is not dependent on pressure so the derivative of ΔG_{mixing} with respect to pressure in zero.

or $$\Delta V_{mixing} = 0.$$

so there is no volume change when we prepare a mixture of ideal gases.

In describing these properties of solutions which can be collectively designated as *Property changes of mixing* and designated collectively as ΔM. They may be defined as the difference between an actual molar (or specific) property of a solution M and mole fraction (or mass fraction) average of the standard state (or specific) properties M_i° of the components of solution

$$\Delta M = M - \Sigma x_i M_i^\circ$$

Thus from this we can get the same relationships for various properties of mixing.

Volume change of mixing $\Delta V = V - \Sigma x_i M_i^\circ$ in case of ideal gases $\Delta V = 0$.

Enthalpy change of mixing $\Delta H = H - \Sigma x_i H_i^\circ$ and so on for other functions.

Chemical Potential of Ideal Solutions

For dilute solutions the chemical potential is given by

$$\mu_2 = \mu_2^\circ + RT \ln x_2 \tag{1}$$

where μ_2, x_2 and μ_2° are the chemical potential, mole fraction and a constant which depends on T, P and nature of solute and solvent respectively.

The chemical potential of a solvent in solution can be calculated by the use of Gibbs-Duhem equation which states

$$x_1\, d\mu_1 + x_2\, d\mu_2 = 0$$

or $$(1 - x_2)\, d\mu_1 + x_2\, d\mu_2 = 0 \qquad \text{[Binary solution]}$$

or $$d\mu_1 = \frac{x_2}{1 - x_2}\, d\mu_2$$

From equation (1), we get

$$d\mu_2 = RTd\ 1n\ x_1 + RT\frac{dx_2}{x_2} \qquad (2)$$

[T, μ_2° being constant]

From (1) and (2),

$$d\mu_i = -RT\frac{dx_2}{x_2} \times \frac{x_2}{1 - x_2}$$

$$= -RT\frac{dx_2}{1 - dx_2}$$

Integrating, we get

$$\mu_1 = \mu_1^\circ + RT\ 1n\ (1 - x_2)$$

μ_1° is a constant and is equal to chemical potential of solvent in solution.

We will now apply these equations to various laws to dilute solutions.

Applications of Chemical Potential

(1) **Nernst Distribution Law.** If to a system of two immiscible solvent at equilibrium, a solute is added then the solute distributes itself between the two solvents in such a way that the ratio of the amount of solute in two phases is constant at constant temperature.

Considering two immiscible liquids A and B at equilibrium and then adding a solute S to this system at equilibrium, we disturb the equilibrium and the solute gets distributed in the two phases when a new equilibrium is reached, we have

$$\frac{C_A}{C_B} = \text{constant} = K_D$$

where C_A, C_B are the concentrations of the solute S in two liquids A and B, K_D is the constant.

This result can be obtained on the basis of chemical potential as follows:

Chemical potential of solute (S) in liquid A is μ_A is given by

$$\mu_A = \mu_{A^\circ} + RT \ln a_A \qquad (1)$$

where α_A is the activity (concentrations) of solute (S) is liquid A.

Similarly

$$\mu_B = \mu_{B^\circ} + RT \ln a_B \qquad (2)$$

At equilibrium $\mu_A = \mu_B$ (T, P constant) From (1) and (2),

$$\mu_A^\circ + RT \ln \alpha_A = \mu_{B^\circ} + RT \ln a_B$$

or $$\ln \frac{a_A}{a_B} = \frac{\mu_{B^\circ} - \mu_{A^\circ}}{RT}$$

or $$\frac{a_A}{a_B} = e^{\frac{\mu_{B^\circ} - \mu_{A^\circ}}{RT}}$$

Since temperature has been kept constant μ_{A° and μ_{B° are constant

$$\therefore \qquad \frac{a_A}{a_B} = \text{constant} = k_D$$

μ_A° and μ_{B° are independent of concentration and so kD should not depend on concentration.

Activity = Activity coefficient × concentration = γC

or $$a_A = \gamma_A C_A$$

and $$a_B = \gamma_B C_B$$

so we get $$k_D = \frac{a_A}{a_B} = \frac{\gamma_A C_A}{\gamma_B C_B} = \frac{C_A}{C_B} \qquad \text{[For ideal solution } \gamma \to 1]$$

(2) *Henry's law*. This law is a statement for the solutions of gases in liquids. It states,

"When a gas is dissolved in any solvent at a given temperature, the concentration of the solution is directly proportional to the equilibrium pressure of the gas above the solution."

Mathematically,

$$p \propto x \text{ or } p = kx$$

where k is a constant. This constant is dependent of the temperature, nature of solute and nature of solvent.

This law is valid for very dilute solution at a sufficiently law vapour pressure so that vapour phase may be considered to be showing an ideal behaviour.

At high pressure, the vapour phase does not behave ideally and the Henry's law no more holds good in its simple form stated above. In such cases it is put as

$$f = k\,(T, P) \times x$$

where 'f' is the fugacity of solute, x is the mole fraction of solute.

Henry's law can be derived using the concept of chemical potential, as follows:

We know that chemical potential of dissolved gas in solution is given by

$$\mu_{2(l)} = \mu_{2(l)}^{\circ} + RT \ln x_2$$

x_2 is the mole fraction of the gas dissolved in the solution.

Chemical potential of gas above the solution is given by

$$\mu_{2(g)} = \mu_{2(g)}^{\circ} + RT \ln p_2$$

But at equilibrium

$$\mu_{2(g)} = \mu_{2(l)}$$

or $$\mu_{2(g)}^{\circ} + RT \ln p_2 = \mu_{2(l)}^{\circ} + RT \ln x_2$$

or $$\ln x_2 = \frac{\mu_2^{\circ}(g) - \mu_2^{\circ}(l)}{RT} + \ln p_2$$

$$= \ln k + p_2$$

$$= \ln k\, p_2$$

$\left(k = \dfrac{\mu_2^{\circ}(g) - \mu_2^{\circ}(l)}{RT}\right)$. It is a constant and its value is dependent on the temperature. At constant temperature it becomes

$x_2 = kp_2$ or $x_2 \propto p_2$

This is Henry's law.

Exact form of Henry's Law

This form is used at high pressure and can be derived using Gibbs-Duhem equation.

Consider a binary solution having n_1 moles of solvent and n_2 moles of solute.

In such a case the Gibbs-Duhem equation is

$$n_1\, d\mu_1 + n_2\, d\mu_2 = 0$$

Dividing by $(n_1 + n_2)$, we get

$$\frac{n_1}{n_1 + n_2} d\mu_2 + \frac{n_2}{n_1 + n_2} d\mu_2 = 0$$

or

$$x_1\, d\mu_1 + x_2\, d\mu_2 = 0 \tag{3}$$

x_1, x_2 are the mole fractions of solvent and solute respectively.

Since chemical potential is a function of T, P and mole fraction we can write.

$$\mu_i = f(T, P, x_i)$$

or $$d\mu_i = \left(\frac{\partial \mu_i}{\partial T}\right)_{P,\, x_i} dT + \left(\frac{\partial \mu_i}{\partial P}\right)_{T,\, x_i} dP + \left(\frac{\partial \mu_i}{\partial x_i}\right)_{T,\, P} dx_i$$

For an isobaric and isothermal process we have $dP = 0$, $dT = 0$ so the above equation becomes

$$(d\mu_i)_{T,\, P} = \left(\frac{\partial \mu_i}{\partial x_i}\right)_{T,\, P} dx_i$$

This can be put in the following form for the two components (solute and solvent) of a solution.

$$(d\mu_1)_{T,\, P} = \left(\frac{\partial \mu_1}{\partial x_1}\right)_{T,\, P} dx_1$$

and $$(d\mu_2)_{T,P} = \left(\frac{\partial \mu_2}{\partial x_2}\right)_{T,P} dx_2$$

Putting these values in equation (3), we get

$$x_1\left(\frac{\partial \mu_1}{\partial x_1}\right)_{T,P} dx_1 + x_2\left(\frac{\partial \mu_2}{\partial x_2}\right)_{T,P} dx_2 = 0 \tag{4}$$

But $x_1 + x_2 = 1$ [binary mixture]

Hence $dx_1 + dx_2 = 0$ [differentiating]

$\therefore$ $dx_1 = -dx_2$

Substituting in equation (4), we get

$$x_1\left(\frac{\partial \mu_1}{\partial x_1}\right)_{T,P} dx_1 - x_2\left(\frac{\partial \mu_2}{\partial x_2}\right)_{T,P} dx_1 = 0$$

or

$$x_1\left(\frac{\partial \mu_1}{\partial x_1}\right)_{T,P} - x_2\left(\frac{\partial \mu_2}{\partial x_2}\right)_{T,P} = 0$$

or

$$x_1\left(\frac{\partial \mu_1}{\partial x_1}\right)_{T,P} = x_2\left(\frac{\partial \mu_2}{\partial x_2}\right)_{T,P} \tag{5}$$

The chemical potential of solute and solvent is given by

$$\mu_1 = \mu_1^\circ + \text{RT } 1\text{n} f_1$$

and $$\mu_2 = \mu_2 + \text{RT } 1\text{n} f_2$$

Differentiating, we get

$$\left(\frac{\partial \mu_1}{\partial x_1}\right)_{T,P} = \text{RT}\left(\frac{\partial\, 1\text{n} f_1}{\partial x_1}\right)_{T,P} \tag{6}$$

and

$$\left(\frac{\partial \mu_2}{\partial x_2}\right)_{T,P} = \text{RT}\left(\frac{\partial\, 1\text{n} f_2}{\partial x_2}\right)_{T,P} \tag{7}$$

Substituting from (6) and (7) in (5), we get

$$\frac{x_1\, RTd\, 1\text{n} f_1}{dx_1} = \frac{x_2\, RTd\, 1\text{n} f_2}{dx_2}$$

or

$$x_1 \frac{d \ln f_1}{dx_1} = x_2 \frac{d \ln f_2}{dx_2} \qquad (8)$$

If the solvent molecules are supposed to obey Roult's law, then

$$f_1 = f_1^{\circ} x_1$$

or $\ln f_1 = \ln f_1^{\circ} + \ln x_1$ [Taking log of both sides]

Differentiating this equation, we get

$$\frac{d \ln f_1}{dx_1} = \frac{d \ln x_1}{dx}$$

$$= \frac{1}{x_1} \qquad \left[d \ln x = \frac{1}{x} dx \right]$$

Substituting this value in eq. (8), we get

$$x_1 \times \frac{1}{x_1} = x_2 \frac{d \ln f_2}{dx_2}$$

$$1 = x_2 \frac{d \ln f_2}{dx_2}$$

or $$d \ln f_2 = \frac{dx_2}{x_2} \text{d} \ln x_2$$

Integrating, we get

$$\ln f_2 = \ln x_2 + \ln k_2$$

or $$\ln f_2 = \ln k_2 x_2$$

or $$f_2 = k_2 x_2$$

This is exact form of Henry's law.

(3) *Solubility behaviour of ideal solutions.* In case of a saturated solution of a solid in a liquid if we add more of the solute, this solute remains undissolved and thus there exists an equilibrium between pure solute (solid) and the solute molecules in solution (liquid). So the chemical potential values of the solute in the two phases must be same at equilibrium.

Thus $$\mu_{2(s)} = \mu_{2(l)} \qquad \text{[T, P constant]}$$

For a small change it can be written as

$$d\mu_{2(s)} = d\mu_{2(l)} \qquad \text{(T, P constant)}$$

or $$d\mu_{2(s)}\,(T, P) = d\mu_{2(l)}\,(T, P, x_2) \qquad (9)$$

where x_2 is the mole fraction of dissolved solute

Thus $$d\mu_{2(s)} = \left(\frac{\partial \mu_{2(s)}}{\partial T}\right)_P dT + \left(\frac{\partial \mu_{2(s)}}{\partial P}\right)_T dP$$

At constant Pressure $dP = 0$

So $$d\mu_{2(s)} = \left(\frac{\partial \mu_{2(s)}}{\partial T}\right)_P dT$$

or $$d\mu_{2(s)} = -\left(\bar{S}_{2(s)}\right) dT$$

The equation (9) can be written as

$$d\mu_{2(l)} = \left(\frac{\partial \mu_{2(l)}}{\partial T}\right)_{P,\, x_2} dT + \left(\frac{\partial \mu_{2(l)}}{\partial P}\right)_{P,\, x_2} dP + \left(\frac{\partial \mu_{2(l)}}{\partial x_{2(l)}}\right)_{P,\, x_2}$$

At constant pressure $dP = 0$

$$\therefore \qquad d\mu_{2(l)} = \left(\frac{\partial \mu_{2(l)}}{\partial T}\right)_{P,\, x_2} dT + \left(\frac{\partial \mu_{2(l)}}{\partial x_{2(l)}}\right)_{T,\, x_2} dx_{2(l)}$$

$$= \bar{S}_{2(l)}\, dT + \left(\frac{\partial \mu_{2(l)}}{\partial x_{2(l)}}\right)_{T,\, x_2} d\mu_{2(l)}$$

So we get

$$-S_{2(s)}\, dT = -\bar{S}_{2(l)}\, dT + \left(\frac{\partial \mu_{2(l)}}{\partial x_{2(l)}}\right)_{T,\, P} dx_{2(l)}$$

or

$$\bar{S}_{2(l)}\, dT - S_{2(s)}\, dT = \left(\frac{\partial \mu_{2(l)}}{\partial x_{2(l)}}\right)_{T,\, x_2} dx_{2(l)}$$

In case of ideal solution, we have

$$\left(\frac{\partial \mu_{2(l)}}{\partial x_{2(l)}}\right)_{T,\, P} = RT\left(\frac{dx_{2(l)}}{x_{2(l)}}\right)$$

$$= RTd\ln x_{2(l)}$$

$$\Delta S_2\, dT = RTd\ln x_{2(l)}$$

where ΔS_2 is the difference in entropy of the solute intwo phases i.e., crystalline phase and solution phase.

At equilibrium

$$\Delta S_2\, dT = \frac{\Delta H_2}{T}$$

so we get

$$\frac{\Delta H_2}{T}\, dT = RTd\ln x_{2(l)}$$

or
$$\frac{\Delta H_2}{RT^2}\, dT = d\ln x_{2(l)}$$

where ΔH_2 is the enthalpy of a saturated solution, which may be equal to heat of solution and in this case (saturated solution) can be considered as heat of fusion of solid solute. The above equation then can be written as

$$\frac{\Delta H_{0fus}}{RT^2}\, dT = d\ln x_{2(l)}$$

If heat of fusion is taken as constant over this range of temperature this equation on integration gives

$$\int_{l}^{x_{2(l)}} d\ln x_{2(l)} = \frac{\Delta H_{2(fus)}}{R} \int_{T_0}^{T_1} \frac{dT}{T^2}$$

$$\ln x_{2(l)} = \frac{\Delta H_{2\,fus}}{R}\left[\frac{1}{T_0} - \frac{1}{T}\right]$$

where T_0 is the equilibrium temperature of solid solute and solution (saturated).

This is the expression for ideal solubility law.

The results given on next page can be directly derived from this equation.

(i) If we express the solubility in mole fraction, at a given temperature its value is same in all solvents with which it forms an ideal solution.

(ii) If two solutes having the same values of heats of fusion are dissolved in a given solvent, at constant temperature, the solute with lower melting point will be more soluble as compared to one having higher melting point.

(4) *Roult's Law.* It states that the relative lowering of vapour pressure $\left(\frac{p_0 - p_s}{p_0}\right)$ is equal to mole fraction of solute in solution.

Mathematically,

$$\frac{p_0 - p_s}{p_0} = x_2$$

where p_0, p_s are the vapour pressure of pure solvent and solution respectively and x_2 is the mole fraction of solute.

The chemical potential of solvent in case of dilute solution is given by

$$\mu_{1(l)} = \mu_1^\circ{}_{(l)} + \text{RT ln}(1 - x_2)$$

Chemical potential of solvent in vapour phase is given by

$$\mu_{1(g)} = \mu_1^\circ{}_{(g)} + \text{RT ln P}_s$$

where P_s is the vapour pressure of solvent vapours over solution.

At equilibrium,

$$\mu_{1(g)} = \mu_{1(l)}$$

$$\therefore \quad \mu_1^\circ{}_{(g)} + \text{RT ln P}_s = \mu_1^\circ{}_{(l)} + \text{RT ln}(1 - x_2)$$

when $\quad x = 0, p = p_0$

$$\therefore \quad \mu_1^\circ{}_{(g)} + \text{RT ln P}_0 = \mu_1^\circ{}_{(e)}$$

or $\quad \text{RT ln P}_s - \text{RT ln P}_0 = \text{RT ln}(1 - x_2)$

or $\quad \text{RT (ln Ps} - \text{ln P}_0) = \text{RT ln}(1 - x_2)$

or $\quad \ln \frac{P_s}{P_0} = \ln(1 - x_2)$

or $\quad \frac{P_s}{P_0} = (1 - x_2)$

or $$1-\frac{P_s}{P_0}=1-(1-x_2)$$

or $$\frac{P_0-P_s}{P_0}=x_2$$

This is Roult's law.

(5) *Law of Mass action.* For a general reaction

$$aA + bB + ... = cC + dD + ...$$

Free energies per mole at constant temperature T be denoted for various components as μ_A, μ_B, μ_C and μ_D etc. Total free energy values are given by

$$G = n_1\mu_1 + n_2\mu_2 + ...$$

or $$G = n_A\mu_A + n_B\mu_B + ...$$

where n_A,n_B...are the number of moles of various components.

So the Free energy of Products = $c\mu_C + d\mu_D + ...$

and of reactants = $a\mu_A + b\mu_B + ...$

$$\therefore \quad \Delta G = (c\mu_C + d\mu_D + ...) - (a\mu_A + b\mu_B + ...)$$

the chemical potential of any substance is given by

$$\mu = \mu^0 + RT \ln a$$

where μ^0 is the standard chemical potential of the substance and μ is its chemical potential of activity *a*.

Thus

$$\Delta G = [c(\mu^\circ_C + RT \ln a_C) + d(\mu_{D^\circ} + RT \ln a_D)]$$
$$- [a(\mu^\circ_A + RT \ln a_A) + b(\mu_{B^\circ} + RT \ln a_B)]$$

or $$\Delta G = [(c\mu^\circ_C + d\mu^\circ_D + ...) - (a\mu^\circ_A + b\mu_B^\circ + ...)]$$

$$+ RT \ln \frac{(a_C)^c \times (a_D)^d \times ...}{(a_A)^a \times (a_B)^b \times ...}$$

$$= \Delta G^0 + RT \ln k$$

where $$k = \frac{(a_C)^c \times (a_D)^d \times \ldots}{(a_A)^a \times (a_B^b) \times \ldots}$$

At equilibrium $\Delta G = 0$

$$\therefore \qquad \Delta G^0 = -RT \ln \frac{(a_C)^c \times (a_D)^d \times \ldots}{(a_A)^a \times (a_B)^b \times \ldots}$$

or $$\Delta G^0 = -RT \ln K_a$$

If all the reactants and products are in their standard states this equation becomes a constant.

or $$\frac{(a_C)^c \times (a_D)^d \times \ldots}{(a_A)^a \times (a_B)^b \times \ldots} = K_a$$

K_a is known as equilibrium constant.

This is law of mass action.

6. *Osmotic pressure and vapour pressure.* If a solution is separated from its pure solvent with the help of a semipermeable membrane, at constant temperature, the chemical potential of solution is less and there is a tendency of the spontaneous flow of solvent molecules into the solution. An equilibrium will be attained when osmotic pressure is equal to the hydrostatic pressure.

or $$\pi = p - p_o$$

where p_0 and p are the pressures of pure solvent and solution respectively. If μ^0 and μ are the chemical potential of pure solvent and solution, at equilibrium $\mu = \mu^0$

$$\therefore \qquad \mu^0 = \mu + \int_{P_0}^{P} \left(\frac{\partial \mu}{\partial P}\right)_T dP$$

$$= \mu + \int_{P_0}^{P} \bar{V}_1 \, dP$$

where $\bar{V}_1$ is the partial molar volume of solvent in solution.

If $\bar{V}_1$ is assumed to be constant this equation on integration give

$$\mu^0 = \mu + \bar{V}_1 (P - P_0)$$

the chemical potential of vapour of solvent over pure solvent is

$$\mu^0_{(vapour)} = \mu^0 + RT \ln P_0$$

where P_0 is the equilibrium vapour pressure of solvent in liquid and vapour phase. Chemical potential of vapours over solution is given by

$$\mu_{(vapour)} = \mu^0 + RT \ln P$$

where P is the partial vapour pressure of solvent is equilibrium with solution. At equilibrium,

$$\mu_{(vapour)} = \mu^0$$

and $$\mu_{(vapour)} = \mu$$

$$\therefore \quad \mu^0 + RT \ln P_0 = \mu^0 + RT \ln P + \bar{V}_1 (P - P_0)$$

or $$RT \ln P_0 - RT \ln P = \bar{V}_1 (P_0 - P_0)$$

or $$RT \ln \frac{P_0}{P} = \bar{V}_1 (P - P_0)$$

or $$RT \ln \frac{P_0}{P} = \bar{V}_1 \times \pi$$

This equation gives the relation between osmotic pressure and vapour pressure.

7. *Elevation of Boiling Point.* We have already seen that on addition of a non-volatile solute to a pure solvent the boiling point of the pure solvent is elevated.

In the process of boiling the vapour pressure of the substance is equal to the atmospheric vapour pressure at boiling temperature. In case of a solution of a non volatile solute at a given temp. and pressure an equilibrium will be reached between the vapour phase and liquid phase of the solvent. Thus at equilibrium $\mu_{1(l)} = \mu_{1(v)}$ where $\mu_{1(l)}, \mu_{1(v)}$ are the chemical potentials of the solvent in liquid phase and vapour phase respectively.

On addition of solute to such a solution, the equilibrium is disturbed and to again attain the equilibrium some of the solvent from vapour phase passes into liquid phase. Thus the solution shall have to be heated to a higher temperature, than the boiling point of pure solvent, so that the solvent molecules have a tendency to escape form the liquid phase. In this way the chemical potential of solvent will decrease in vapour phase and once again the chemical potential of solvent in liquid and vapour phase will be same when equilibrium will be reached.

This is illustrated in the following diagram.

At equilibrium we have $d\mu_{1(l)} = d\mu_{1(v)}$, chemical potential is a function of T, P and composition in case of solution but is a function of T, P in case of pure solvent.

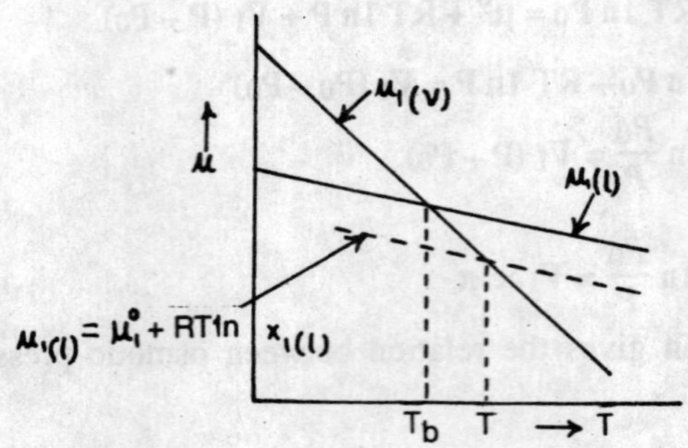

Fig. 12.4. Variation of chemical potential of vapour solvent and solution with temperature.

Thus

$$\mu_{1(l)} = (T, P, x_1)$$

$$\mu_{1(v)} = f(T, P)$$

or $$d\mu_{1(l)} = \left(\frac{\partial \mu_{1(l)}}{\partial T}\right)_{P, x_1} dT + \left(\frac{\partial \mu_{1(l)}}{\partial P}\right)_{T, x_1} dP$$

$$+ \left(\frac{\partial \mu_{1(l)}}{\partial x_1}\right)_{T, P} \qquad (10)$$

and $$d\mu_{1(v)} = \left(\frac{\partial \mu_{1(v)}}{\partial T}\right)_{P} dT + \left(\frac{\partial \mu_{1(v)}}{\partial P}\right)_{T} dP \qquad (11)$$

At constant pressure dP = 0 so these equations become

$$d\mu_{1(l)} = \left(\frac{\partial \mu_{1(l)}}{dT}\right)_{P,\, x_1} dT + \left(\frac{\partial \mu_{1(l)}}{\partial x_1}\right)_{T,\, P} dx_1 \qquad (12)$$

and

$$d\mu_{1(v)} = \left(\frac{\partial \mu_{1(l)}}{dT}\right)_{P} dT \qquad (13)$$

But $\left(\frac{\partial \mu_{1(l)}}{\partial T}\right)_{P,\, x_1} = -\overline{S}_{1(l)}$

Definitions

and $\left(\frac{\partial \mu_{1(v)}}{\partial T}\right)_{P} = -\overline{S}_{1(v)}$

Substituting these in above equations (12) and (13)

$$d\mu_{1(l)} = -\overline{S}_{1(l)}\, dT + \left(\frac{\partial \mu_{1(l)}}{\partial x_1}\right)_{T,\, P} dx_1 \qquad (14)$$

and

$$d\mu_{1(v)} = -\overline{S}_{1(v)}\, dT \qquad (15)$$

Since at equilibrium,

$$d\mu_{1(l)} = d\mu_{1(v)}$$

$$\therefore \quad -\overline{S}_{1(l)}\, dT + \left(\frac{\partial \mu_{1(vl)}}{\partial x_1}\right)_{T,\, P} dx_1 = -\overline{S}_{1(v)}\, dT$$

or

$$(\overline{S}_{1(v)} - \overline{S}_{1(l)})\, dT = -\left(\frac{\partial \mu_{1(v)}}{\partial x_1}\right)_{T,\, P} dx_1$$

or

$$\Delta \overline{S}_{vap}\, dT = -\left(\frac{\partial \mu_{1(l)}}{\partial x_1}\right)_{T,\, P} dx_1 \qquad (16)$$

$\Delta \overline{S}_{vap}$ is the entropy of vapourisation.

In case of ideal solution, we know

$$\mu_1 = \mu_1^{\circ} + RT \ln x_1$$

or

$$\left(\frac{\partial \mu_1}{\partial x_1}\right)_{T,\, P} = \frac{RT}{x_1}$$

at equilibrium $\Delta G = 0$

and $$\Delta S_{vap} = \frac{\Delta H_{vap}}{T} \tag{17}$$

So we get from (16) and (17),

$$\frac{\Delta H_{vap}}{T} dT = -RTd \ln x_1$$

or $$d \ln x1 = \frac{\Delta H_{vap}}{RT^2} dT \tag{18}$$

Since ΔH_{vap} does not depend on temperature, so integrating equation 18, we get

$$\int_1^{x_1} d \ln x_1 = \int_{T_b}^{T} \frac{\Delta H_{vap}}{RT^2} dT \tag{19}$$

where T_b and T are the boiling points of pure solvent and solution respectively

or $$\ln x_1 = -\frac{\Delta H_{vap}}{R}\left[-\frac{1}{T}\right]_{T_b}^{T}$$

$$\ln x_1 = -\frac{\Delta H_{vap}}{R}\left[\frac{1}{T} - \frac{1}{T_b}\right]$$

$$\ln x_1 = -\frac{\Delta H_{vap}}{R}\left[\frac{T_b - T}{T.T_b}\right]$$

$$= -\frac{\Delta H_{vap}}{R} \frac{\Delta T_b}{T_1 T_b}$$

For very dilute solutions $T \approx T_b$ and $\ln x_1 \approx -x_2$

$$\therefore \quad -x_2 = -\frac{\Delta H_{vap}}{R}, \frac{\Delta T_b}{T_b^2}$$

or $$\Delta T_b = \frac{RTb^2}{\Delta H_{vap}} x_2 \tag{20}$$

Equation (20) gives the relationship between elevation in boiling point (ΔT_b) and the amount of solute.

If n_1, n_2 are the number of moles of solute and solvent, then

$$x_2 = \frac{n_2}{n_1 + n_2}$$

for dilute solution n_2 is negligible as compared to n_1

Hence $$x_2 = \frac{n_2}{n_1}$$

Again $$n_2 = \frac{w_2}{M_2} \text{ and } n_1 = \frac{w_1}{M_1}$$

$$\therefore \quad x_2 = \frac{w_2}{M_2} \times \frac{M_1}{w_1}$$

$$= \frac{m.M_1}{1000} \tag{21}$$

where m is the molality of the solution. From (20) and (21), we get

$$\Delta T_b = \frac{RT_b^2 m M_1}{1000\, \Delta \bar{H}_{vap}} \tag{22}$$

$$= \frac{RT_b^2 m}{1000\, \lambda_{vap}} \tag{23}$$

where $\lambda_{vap} = \dfrac{\Delta \bar{H}_{vap}}{M_1}$ or latent heat of vapourisation.

Equation (22) can also be written as

$$\Delta T_b = k_b.m$$

where $$k_b = \frac{RT_b^2 M_1}{1000\, \Delta \bar{H}_{vap}}$$

$$= \frac{RT_b^2}{1000\, \lambda_{vap}}$$

k_b is a constant and is known as molal boiling point elevation constant.

These equations can be used to calculate Mol. wt. of solute if k_b, ΔT_b, w_1 and w_3 are known.

(8) *Gibbs adsorption equation.* The relationship that represents the extent of adsorption and the change of surface tension of a solvent when a solute is added to it is known as *Gibbs adsorption equation.* First derived by J. Williard Gibbs (1878) and then independently by J.J. Thomson (1888).

The equation can be derived by making use of chemical potential.

For a two component system

$$dG = -SdT + VdP + \mu_1\, dn_1 + \mu_2\, dn_2 + \gamma\, dA \quad (24)$$

where γ is the surface tension and dA is the increase in surface area. γdA compensates the increase in free energy due to increase in surface area.

Integrating (24), we get

$$G = \mu_1 n_1 + \mu_2 n_2 + \gamma A \quad (25)$$

where n_1, n_2 are the number of moles of solvent and solute respectively.

Differentiating (25), we get

$$dG = \mu_1 dn_1 + n_1 d\mu_1 + \mu_2 dn_2 + n_2 d\mu_2 + \gamma dA + A d\gamma \quad (26)$$

Comparing (24) and (26), we get

$$SdT - VdP + Adg + n_1 d\mu_1 + n_2\, d\mu_2 = 0 \quad (27)$$

At constant temperature and pressure, this gives

$$n_1\, d\mu_1 + n_2 d\mu_2 + A d\gamma = 0 \quad (28)$$

The system we are considering may be supposed to be made up of two portions:

(i) Surface phase, and (ii) Bulk phase.

(i) *Surface phase.* The above equation holds true only for this phase which is actually affected by surface force.

(ii) *Bulk phase.* The remaining portion of the system which is not affected by surface force is called the *bulk phase* and for this phase the Gibbs-Duhem equation is valid. Thus,

$$n_1^0\, d\mu_1 + n_2^0\, d\mu_2 = 0 \quad (29)$$

where n_1^0 and n_2^0 are the moles of solvent and solute in bulk phase respectively.

Multiplying (29) by $\frac{n_0}{n_1^0}$ and subtracting (28), we get

$$\text{Ad}\gamma + \left(n_2 - \frac{n_1 n_2^0}{n_1^0} \right) d\mu_2 = 0 \tag{30}$$

or

$$-\frac{d\gamma}{d\mu_2} = \frac{n_2 - \frac{n_1 n_2^0}{n_1^0}}{A} \tag{31}$$

Thus we find $\left(\dfrac{n_2 - \frac{n_1 n_2^0}{n_1^0}}{A} \right)$ is the excess concentration of solute per unit area. It is usually represented by T_2. Thus

$$T_2 = -\frac{d\gamma}{d\mu_2} \tag{32}$$

T_2 does not depend on the amount of surface phase it only depends on its nature. It is also known as surface concentration of solute per unit area of interface.

For solution, we know

$$\mu_2 = \mu_2^0 + RT\, 1n\, a_2$$

where a_2 is the activity of solute, differentiating this equation

$$d\mu_2 = RTd\, 1n\, a_2$$

assuming a_2 as constant,

We get

$$T_2 = -\frac{1}{RT} \frac{d\gamma}{d\, 1n\, a_2} \quad \text{[From equation] (32)}$$

$$= -\frac{a_2}{RT} \frac{d\gamma}{da_2} \left[1n\, a_2 = \frac{da_2}{a_2} \right] \tag{33}$$

Equation (33) is Gibbs adsorption equation.

In general,

$$T = -\frac{a}{RT} \cdot \frac{d\gamma}{da}$$

Replacing activity by concentration in case of dilute solutions.

$$T = -\frac{C}{RT} \frac{d\gamma}{dC}$$

(9) *Phase Rule.* The mathematical equation of phase rule is

$$F = C - P + 2$$

where F = degrees of freedom

C = Components

P = Phases

For a heterogeneous system of three phases (A, B, C) having three components (1, 2, 3) the chemical potentials in the three phases can be represented by

$$\mu_{1(A)}, \mu_{2(A)}, \mu_{3(A)}, \mu_{1(B)}, \mu_{2(B)}, \mu_{3(B)}, \mu_{1(C)}, \mu_{2(C)}, \mu_{3(C)}$$

Since the system under consideration is a closed system in equilibrium at definite temperature and pressure Gibbs-Duhen equation is applicable.

So, we have

$$\Sigma\mu\, dn = 0$$

If now a small quantity δn_1 is allowd to be transferred from phase A to phase B under equilibrium conditions, then

$$-\mu_{1(A)}\, dn_1 + \mu_{1(B)}\, dn_1 = 0$$

or

$$\mu_{1(A)} = \mu_{1(B)}$$

From here we conclude that for equilibrium between two phases the chemical potential of any component is the same in each phase. Applying it to other components, we get

$$\mu_{2(A)} = \mu_{2(B)}$$

$$\mu_{3(A)} = \mu_{3(B)}$$

Thus the equilibrium between three phases can be defined by only two independent equations.

This condition is true for P phases and C components, so we get

$$\mu_{1(A)} = \mu_{1(B)} = \mu_{1(C)} = \ldots \mu_{1(P)}$$

$$\mu_{2(A)} = \mu_{2(B)} = \mu_{2(C)} = \ldots\mu_{2(P)}$$

$$\mu_{c(A)} = \mu_{c(B)} = \mu_{c(C)} = \ldots\mu_{c(P)}$$

We thus have C(P–1) independent equations.

Thus degree of freedom F is

$$F = [P(C-1) + 2] + [C(P-1)]$$

$$= C - P + 2$$

This is the equation of phase rule.

NUMERICALS

Example 1. Calculate ΔH° and ΔG° at 1000°C for the synthesis of water gas

$$C_{(s)} + H_2O_{(g)} \rightarrow CO_{(g)} + H_{2(g)} \qquad \text{(i)}$$

Given $\Delta H_f^\circ\ (H_2O) = -241989$ J/Mol

$\Delta G_f^\circ\ (H_2O) = -241989$ J/Mol

$\Delta H_f^\circ\ (CO) = -110599$ J/Mol

$\Delta G_f^\circ (CO) = -137361$ J/Mol

and C_{pi}° (J/Mol K) $= a_t + b_t\,(T/K) + C_i\,(T/K)^2 + d_i\,(T/K)^{-2}$

Compound	a	10^3b	10^7C	10^{-5}d
$C_{(s)}$	17.166	4.271	0.0	–8.79
$H_2O_{(g)}$	30.379	9.621	11.85	0.0
$CO_{(g)}$	26.879	6.971	–8.21	0.0
$H_{2(g)}$	29.086	–0.831	20.14	0.0

Solution. The required values of ΔH^0 and ΔG^0 are as under.

The reaction for synthesis of water gas may be considered to be made up of

(i) $C_{(s)} + \frac{1}{2}O_{2(g)} \rightarrow CO_{(g)}$

(ii) $H_2O_{(g)} \rightarrow H_{2(g)} + \frac{1}{2}O_{2(g)}$

(i) is the formation of CO and (ii) is reverse of the formation of H_2O

So DH˚0 and G˚0 for synthesis of water gas will be given by

$$\Delta H^\circ_0 = \Delta H_f^0\,[CO(g)] - \Delta H_f^\circ\,[H_2O(g)]$$

$$= -110599 - (-241989)$$

$$= 131390 \text{ J/Mol}$$

$$\Delta G_{00} = \Delta G_{f0}\,[CO(g)] - \Delta G_f^0\,[H_2O(g)]$$

$$= -137361 - (-228746)$$

$$= 91385 \text{ J/Mol.}$$

Using the other data we can find the values of Δa, Δb, Δc and Δd.

$$\Delta a = (-1)(17.166) + (-1)(30.379) + (1)(26.879) + (1)(29.086)$$

$$= 8.42$$

Similarly,

$$\Delta b = -7.758 \times 10^{-3}$$

$$\Delta c = 8 \times 10^{-9}$$

$$\Delta d = 8.79 \times 10^{-5}$$

ΔH˚ and ΔG˚ can be calculated. For T = 1273 K and T_0 = 298˚K (25˚C the data given is for 25˚C).

The dimensionless temperature τ is

$$\tau = \frac{T}{T_0}$$

$$= \frac{1273}{298} = 4.272$$

$$\Delta H^\circ = 131390 + (298)\,(8.42)\,(4.272 - 1)$$
$$+ \frac{1}{2}(298)^2\,(-7.758 \times 10^{-3})\,[(4.272)^2 - 1]$$
$$+ \frac{1}{3}(298)^3\,(8 \times 10^{-9})\,[(4.272)^3 - 1]$$

$$= \frac{1}{298}(8.79 \times 10^5) \quad \left[\frac{4.272 - 1}{4.272}\right]$$

$$= 135922 \text{ J/Mol.}$$

$$\because \quad \Delta H^0 = \Delta H_0^0 + T_0(\Delta a)(\tau - 1) + \frac{T_0^2 \Delta b}{2}(\tau^3 - 1)$$

$$+ \frac{T_0^3 \Delta Cl}{3}(\tau^3 - 1) + \frac{\Delta d}{T_0}\left(\frac{\tau - 1}{\tau}\right)$$

Similarly ΔG° calculated using

$$\Delta G^0 = \Delta H_0^0 + (\Delta G_0^0 - \Delta H_0^0)_T + T_0(\Delta a)(\tau - 1 - \tau \ln \tau)$$

$$- \frac{T_0^2 \Delta b}{2}(\tau^2 - 2\tau + 1) - \frac{T_0^3 \Delta C}{6}(\tau^3 - 3\tau + 2)$$

$$- \frac{\Delta d}{2T_0}\left(\frac{\tau^2 - 2t + 1}{\tau}\right)$$

$$\therefore \quad \Delta G^\circ = 131390 + (91385 - 13390)(4.272)$$

$$+ (298)(8.42)(4.272 - 1 - 4.272 \ln 4.272)$$

$$- \frac{1}{2}(298)^2(-7.758 \times 10^{-3})[(4.272)^2 - 2(4.272) + 1]$$

$$- \frac{1}{6}(298)^3(8 \times 10^{-9})[(4.272)^3 - 3(4.272) + 2]$$

$$-\frac{1}{2}\left(\frac{1}{298}\right)(8.79 \times 10^5)\left[\frac{(4.272)^2 - 2(4.272) + 1}{4.272}\right]$$

$$= -46877 \text{ J/Mol.}$$

these values calculated for ΔG° and ΔH° at 1273°K are for the formation of water gas.

13

Activity, Activity Coefficient and Fugacity

Activity and Activity Coefficients

Activity of a Gas. The *fugacity* 'f' and the free energy per mole of a gas (G) are related to each other according to the following equation,

$$G = RT \ln f + B \qquad (1)$$

where B is a constant and it depends upon the temperature and the nature of the substance.

Since we can not evaluate the absolute value of 'G' so it is not possible to evaluate B. However, if all the measurements of free energy are made with reference to standard state, the energy so evaluated is know as G°. Let G° and f° be the standard free energy and standard state fugacity of the substance. So it becomes

$$G^\circ = RT \ln f^\circ + B \qquad (2)$$

From (1) and (2), we get

$$G - G^\circ = RT \ln \frac{f}{f}$$

or

$$G = G^\circ + RT \ln \frac{f}{f}$$

$$= G^\circ + RT \ln a \qquad (3)$$

where 'a' is called the activity

Thus $$\text{'a'} = \frac{f}{f^{\circ}}$$

Activity is thus defined as the ratio of fugacity of a substance in a particular state to the fugacity of the same substance in a standard state.

In the standard state,

$$G = G^{\circ}$$

In such a state equation (3) becomes

$$G^{\circ} = G^{\circ} + RT \ln a$$

Thus $$RT \ln a = 0$$

or $$a = 0$$

We thus conclude that the *standard state of a substance is that state when its activity (a) is equal to unity.*

Consider a mole of a substance having G_1 free energy in the initial state and G_2 be its free energy in the final state. Thus free change in given by

$$G_2 - G_1 = \Delta G = (G^{\circ} + RT \ln a_2) - (G^{\circ} + RT \ln a_1)$$

where a_1 and a_2 are the activities of the substance in initial and final states.

The above equation on simplification becomes

$$\Delta G = RT \ln \frac{a_2}{a_1} \quad (4)$$

The change in free energy was calculated

as $$\Delta G = RT \ln \frac{p_2}{p_1} \quad (5)$$

From equations (4) and (5) it is evident that pressure terms occurring in the equation for ideal gases have been replaced by activity terms in case of real gases.

So activity of a gas is a thermodynamic counterpart of pressure of a gas.

In case of solutions, on similar considerations, we can prove that activity is a thermodynamics counterpart of concentration of the solute.

Activity Coefficient of a Gas

For ideal gas $a = P$ but real gases $a \propto P$

or $$a = \gamma P$$

γ is known as activity coefficient,

Relation between fugacity and activity Coefficient of a gas

We have seen earlier that

$$G = G^{\circ} + RT \ln \frac{f}{f^{\circ}} \qquad \text{[equation 2]}$$

In the standard state of a gas $G^{\circ} = 1$ thus the above equation becomes

$$G = G^{\circ} + RT \ln f \qquad [\because \quad f^{\circ} = 1]$$

So in case of an ideal gas fugacity and pressure are equal.

At very low pressure all gases approach an ideal gas behaviour

Thus $f = P$ when $P \to 0$

i.e., $$\underset{P \to 0}{\text{Lt}} \frac{f}{P} = 1$$

This ratio $\frac{f}{P}$ is called activity coefficient (γ) of a gas

So $$\gamma = \frac{f}{P}$$

At very low pressure in case of real gases or in case of ideal gases $\gamma = 1$.

More the deviation of γ of a gas from unity more will be its non-ideality.

Activity and Activity Coefficients of Solution

Large deviations from ideal behaviour are brought about due to interionic attractions between ions, in case of solutions of electrolytes. In such cases the activity (a) and molar concentrations are not same. In case of such solutions activity is used in place of molar concentrations.

In solutions the activity is defined as the product of molar concentration of the solute and the factor known as activity coefficient

$$\alpha_A = \gamma_A X_A$$

or
$$\gamma_A = \frac{\alpha_A}{X_A}$$

γ_A is the activity coefficient.

For non-ideal solution $\gamma<1$

Choice of Standard State

The activity of a substance is always related to its activity in the standard state. If different standard states are chosen we will get different values for activity. Thus the nature of standard state has a thermodynamic importance. For two phases we select the standard states as under:

(i) For a gas the standard state at a constant given temperature is one when the fugacity (f) of the gas is equal to one.

or
$$a = \frac{f}{f^{\circ}}$$

$\therefore$ if $f^{\circ} = 1$, then $a = f$

so the activity of a gas is equal to its fugacity in the standard state.

For any gaseous component i, in the standard state $f_i = a_i = 1$.

(ii) The standard state is case of a liquid is that of the pure liquid at 1 atmosphere pressure, at any given constant temperature. So for liquids in standard state, we have

$$a = 1 \text{ and } f = f^{\circ}$$

for any liquid component i, in a solution the standard state is one of the pure state of this component such that $a_i = 1$.

(iii) In case of solids, it is the pure state of crystalline solid at 1 atmosphere pressure at any given constant temperature. So in this state activity, $a = 1$.

Determination of Activity

Various methods can be used for the determination of activity.

(1) Vapour Pressure Method

If the vapour pressure of solvent over a solution is p_s and over pure solvent is p_0. If p_s and p_0 are not very high than the vapours may be considered to behave like ideal gas and the vapour pressure are equal to fugacities of the solvent in solution and pure solvent respectively, i.e.,

$$\frac{p_s}{p_0} = \frac{f_1}{f_1^\circ}$$
$$= a$$

Thus the *activity of a solvent* in solution can be calculated if the values of p_s and p_0 are known.

Activity of the Solute. In case of volatile solute, by the some procedure as given above to measure the activity of a solvent, we get

Activity = Constant × Pressure

if a_2' is the activity, x_2' the mole fraction of solute and p_2' the partial mole fraction of the solute, then we get

$$a_2' = x_2'$$
$$= kp_2'$$

or $$k = \frac{x_2'}{p_2'} \qquad (1)$$

k is a proportionality constant.

If a_2, p_2 are the activity of solute and its vapour pressure respectively. Then

$$a_2 \propto p_2 \text{ or } a_2 = kp_2 \qquad (2)$$

From (1) and (2),

$$a_2 = \frac{x_2'}{p_2'} \times p_2 \quad (3)$$

Thus to find a_2 (activity of solute), we have to measure p_2', p_2.

But p_2 is very low as compared to p_2' so the results obtained by this method are not accurate.

Gibbs-Duhem equation gives better results. The Gibbs-Duhem equation is

$$n_1 d\mu_1 + n_2 d\mu_2 = 0 \quad (4)$$

where n_1, n_2 are the moles of two components respectively, $d\mu_1$ and $d\mu_2$ are the small changes in their chemical potential values respectively. If μ_1, μ_2 are the respective values for chemical potential and a_1, a_2 are their respective activities then, we have

$$\mu_1 = \mu_1^\circ + RT \ln a_1 \quad (5)$$

$$\mu_2 = \mu_2^\circ + RT \ln a_2 \quad (6)$$

Differentiating, we get

$$d\mu_1 = RT \ln a_1 \quad (7)$$

and

$$d\mu_2 = RT \ln a_2 \quad (8)$$

From (7) and (8), we get

$$n_1 RT \, d \ln a_1 + n_2 RT \, d \ln a_2 = 0$$

or

$$d \ln a_2 = -\frac{n_1}{n_2} d \ln a_1$$

Integrating, we get

$$\ln \frac{a_2}{a_1} = -\int_{n_1}^{n_1'} \frac{n_1}{n_2} \, d \ln a_1 \quad (9)$$

A graph between $\frac{n_1}{n_2}$ and $d \ln a_1$ is plotted then the area under the graph in the limits n_1, n_1' gives the value of integral of equation (9). The value of $\frac{a_2}{a_1}$ can then be evaluated, knowing the activity in one of the solutions (either a_1 or a_2) that in the other can be calculated.

(2) The Freezing Point Depression Method

This method gives very accurate results. For a solution having n_1 moles of solvent (water) and n_2 moles of solute. The following calculation are used.

If T_0 and T are the freezing points of pure solvent and solution respectively.

At the freezing point the fugacity of solvent (water) is same in both solid and liquid phases

$$f_{ice} = f_{water}$$

or $$f_{(s)} = f_{(e)} \text{ or } f_{(1)}$$

Since in each phase water is in pure state so the activity in each phase is equal to one

$$a_{ice} = a_{water} = 1$$

The freezing point of a solution is always lower than the freezing point of pure solvent thus $T < T_0$. The fugacity of water in solution is equal to that of ice.

$$f_{ice} = f_{water\ in\ solution}$$

But now the activity of water in solution is not equal to one thus the activity in its pure state and activity in solution are now different

$$a_{ice} \neq a_{water\ in\ solution}$$

(ice) (a1)

Then $$a_s = a_1$$

Considering the freezing process at any temperature T

$$H_2O_{(l)} \rightarrow H_2O_{(s)}$$

the chnage in thermodynamics potential is given by

$$\Delta G = RT\ 1n \frac{f_{ice}}{f_{water}}$$

$$= RT\ 1n\ a_s$$

or $$\Delta G = RT\ 1n\ a_1 \qquad [a_s = a_l]$$

or $$\frac{\Delta G}{T} = R\ 1n\ a_1$$

or $$\left[\frac{\partial}{\partial T}(\Delta G/T)\right]_P = R\left[\frac{\partial}{\partial T}(1n\ a_1)\right]_P \qquad (10)$$

In accordance with Gibbs-Helmholtz equation

$$\left[\frac{\partial}{\partial T}(\Delta G/T)\right]_P = -\frac{\Delta H}{T^2} \qquad (11)$$

From (10) and (11),

so $$R\left[\frac{\partial}{\partial T}(1n\ a_1)\right]_P = -\frac{\Delta H}{T^2} \qquad (12)$$

or $$\left[\frac{\partial}{\partial T}(1n\ a_1)\right]_P = -\frac{\Delta H}{RT^2}$$

$$= \frac{L}{RT^2} \qquad (13)$$

where L in the molar latent heat of fusion of ice [i.e. $-\Delta H = L$]

Applying Kirchoff's equation

$$\left(\frac{dL}{dT}\right)_P = \Delta C_p = (C_p)_l - (C_p)_s$$

or $$L = L_0 + \Delta C_p\,(T - T_0) \qquad (14)$$

or $$L = L_0 - \theta\Delta C_p \qquad (15)$$

Where θ is depression in freezing point and L_0 is the molar heat of fusion of solid at T_0.

From (13) and (15), we get

$$\frac{d\ 1n\ a_1}{dT} = \frac{L_0 - \theta\,\Delta C_p}{RT^2} \qquad (16)$$

But $$\theta = T - T_0$$

or $$-d\theta = dT$$

$$\therefore \quad \frac{1}{T^2} = \frac{1}{(T_0 - \theta_0)^2} \tag{17}$$

$$= \frac{1}{T_0^2}\left(1 - \frac{\theta_0}{T_0}\right)^{-2}$$

$$= \frac{1}{T_0^2}\left[1 + \frac{2\theta}{T_0} + \frac{3\theta^2}{T_0^2} + \ldots\right] \tag{18}$$

Making substitution from equation (17) into (16), we get

$$\frac{d \ln a_1}{-d\theta} = \frac{L_0 - \theta\Delta C_p}{R(T_0 - \theta)^2}$$

or

$$-d \ln a_1 = \frac{L_0 - \theta\Delta C_p}{R(T_0 - \theta)^2} d\theta$$

$$= \frac{L_0 - \theta\,\Delta C_p}{RT_0^2}\left(1 + \frac{2\theta}{T_0} + \frac{3\theta^2}{T_0^2} + \ldots\right) d\theta$$

or

$$-d \ln a_1 =$$

$$\left[\frac{L_0}{RT_0^2} + \frac{1}{RT_0^2}\left\{\left(\frac{2L_0}{T_0} - \Delta C_p\right)\theta + \left(\frac{3L_0}{T_0^2} - \frac{2\Delta C_p}{T_0}\right)\theta + \ldots\right\}\right] d\theta \tag{19}$$

Integrating equation (19) between limits $a_1 = 1$ to a_1 = activity of solution and θ between $\theta = 0$ to $\theta = \theta$, we get

$$-\ln_1 a = \frac{L_0\theta}{RT_0^2} + \frac{\theta^2}{RT_0^2}\left(\frac{L_0}{T_0} - \frac{\Delta C_p}{2}\right) + \ldots \tag{20}$$

In case of aqueous solutions

L_0 = 1438 cal/mole, R = 2 cal, ΔC_p = –9 cal and T_0 = 373 K, then we get

$$-2.303 \log a_1 = \frac{1438}{2 \times (373)^2}\theta + \left(\frac{1438}{373} - \frac{9}{2}\right)\frac{\theta^2}{2 \times (373)^2}$$

or $$\log a_1 = -4.2 \times 10^{-2}\theta - 2.2 \times 10^{-6}\theta^2 \quad (22)$$

Since θ can be determined experimentally activity of solvent (a_1) can be calculated using equation (21).

This method can also be used to find out the activity of solute. The relationship that exists between solute and solvent is

$$d\ln a_2 = -\frac{N_1}{N_2} d\ln a_1$$

$$= -\frac{n_1}{n_2} d\ln a_1 \quad (22)$$

If m moles of solute are dissolved per 1000 g of solvent, then

$$n_2 = \text{m and } n_1 = \frac{1000}{m_1}$$

$$\therefore \quad d\ln a_2 = \frac{1000}{mM_1} d\ln a_1 \quad (23)$$

From equation (23) and (19), we get

$$d\ln a_2 = \frac{1000}{mM_1}$$

$$\left[\frac{L_0}{RT_0^2} + \left(\frac{2L_0}{T_0} - \Delta C_p\right)\frac{\theta}{RT_0^2} + \ldots\right] d\theta \quad (24)$$

We know $k_f = \dfrac{RT_0^2}{1000 \times l}$

$$= \frac{RT_0^2 M_1}{1000 \times L_0} \quad [l \times M_1 = L_0] \quad (25)$$

Substituting in Eq. (24) from (25),

$$d\ln a_2 = \frac{1}{k_{(f)} \times m} d\theta + \frac{1}{k_{(f)} m}\left(\frac{2L_0}{T_0} - \Delta C_p\right)\theta d\theta + \ldots$$

$$= \frac{1}{k_f \times m} d\theta + \phi\frac{\theta}{m} d\theta \quad (26)$$

[higher powers neglected]

$$\phi = \frac{1}{k_f}\left(\frac{2L_0}{T_0} - \Delta C_p\right)$$

= Constant

We know $a_2 = (a\pm)^{\nu}$

or $\quad d \ln a_2 = \text{v}d \ln a\pm$ (27)

Substituting Eq. (27) in Eq. (26), we get

$$vd \ln a\pm = \frac{1}{kfm} d\theta + \phi \frac{\theta}{m} d\theta$$

or

$$d \ln a\pm = \frac{1}{v(kf)\, m} d\theta + \phi \frac{\theta}{m} d\theta \qquad (28)$$

Another function J– was introduced by Lewis and Randell to get a solution of equation (28). This function is defined as

$$J = 1 - \frac{\theta}{vkfm}$$

In case of very dilute solutions

$$m \to 0 \text{ as } \theta \to vkfm$$

$\therefore \quad$ Lt $m \to 0$; $J \to 0$

or
$$dj = \frac{\theta}{vkfm^2} dm - \frac{d\theta}{vkfm}$$

$$= (1 - \text{j}) \frac{dm}{m} - \frac{d\theta}{vkfm}$$

or
$$\frac{d\theta}{vkfm} = (1\text{–j}) \frac{dm}{m} - dj \qquad (29)$$

Substituting from Eq. (29) in Eq. (28)

$$d \ln a\pm = (1\text{–}j) \frac{dm}{m} - d + \phi \frac{\theta}{vm} \text{d}\theta$$

$$= (1\text{–}j)\, d \ln m - dj + f \frac{\theta}{vm} \text{d}\theta \qquad (30)$$

But $\text{a}\pm = (\text{m}\pm)(\gamma\pm)$

$$\therefore \quad d \ln g\pm + \text{d} \ln \text{m}\pm = (1\text{–j})\, \text{d} \ln m \text{–dj} + \phi \frac{\theta}{vm} \text{d}\theta$$

or
$$d \ln g\pm = -\text{j} \frac{dm}{m} - dj + \phi \frac{\theta}{vm} d\theta$$

for dilute solutions $m \to 0\ \gamma\pm \to 1$, $\ln \gamma\pm \to 0$

$$\therefore \quad \int_0^{\ln} \gamma\pm \ d \ln \gamma\pm = -\int_0^m \frac{dm}{m} - j + \int_0^m \phi \frac{\theta}{dm} d\theta$$

or

$$\ln \gamma\pm = -\int_0^m \frac{j}{m} dm - j + \int_0^m \frac{\phi}{\varpi} \frac{\theta}{m} d\theta \quad (31)$$

the right hand side integrals can be determined graphically by putting the graphs between (i) $\frac{j}{m}$ and m (ii) $\frac{\theta}{m}$ and θ.

So the activity coefficient of solute can be calculated.

Example 1. Find out the activity coefficient of H_2O at 273 K, Given $m_{(propylate)} = 0.13146$ mol kg^{-1}, T = –0.239°C, $\Delta H^\circ_{jus} = 6.01$ kJ mol^{-1}.

Solution. Using the equation

$$\ln a = \frac{\Delta H_{jus}}{R}\left[\frac{1}{T_0} - \frac{1}{T}\right]$$

We get

$$\log a = \frac{6.01 \times 10^2}{8.314}\left[\frac{1}{273} - \frac{1}{272.761}\right]$$

$$= \frac{(6.01 \times 10^3)(-0.239)}{8.314 \times 273 \times 272.761}$$

$$= -2.32 \times 10^{-3}$$

Taking antilog

or $\qquad a = 0.998$

We know

n (propylate) = 0.13146 mol

$$n(H_2O) = \frac{1000}{18}$$

$$= 55.55 \text{ mol}$$

Thus $$xH_2O = \frac{55.55}{55.55 + 0.13146}$$

$$= 0.997$$

Hence $\gamma_{H_2O} = \frac{0.998}{0.997}$
$= 1.0004$

(3) Solubility method. This method makes use of the fact that in case of saturated solution an equilibrium exists between the solid solute and solute in solution. The chemical potential of solid solute is constant therefore in saturated solution the chemical potential of solute in solution is also constant. Consider one molecule of a solute $H_{x+} B_{x-}$. This on dissociation gives x_+ cations and x_- anions as represented by the as follows equation

$$A_{x+} B_{x-} = x_+ A^+ + x_- B^- \qquad (33)$$

The chemical potential of an undissociated molecule is equal to the sume of the chemical potential of the ions. So we can take either of these values as the chemical potential value of the electrolyte in solution.

$$\therefore \qquad \mu = x_+ \mu_A + x_- \mu_B \qquad (33)$$

But $\mu_A^+ = \mu_A^{\circ +} + RT\ 2n\ a_A^+$ (35)

and $\mu_{\bar{B}} = \mu_{\bar{B}}^{\circ} + RT\ 1n\ a_{\bar{B}}$ (36)

From Eq. (33), (34) and (35), we get

$$\mu = x+ [\mu_A^{\circ} + RT\ 1n\ a_A^+] = x_- [\mu_{\bar{B}}^{\circ} + RT\ 1n\ a_{B^-}]$$

or $$x+\ 1n\ a_A^+ + x_-\ 1n\ a_{\bar{B}} = \frac{\mu - x_+ \mu_A^{\circ +1-1} \quad x_- \mu_{\bar{B}}^{\circ}}{RT} = K \qquad (36)$$

where k is a constant because, *mu*, $\mu_A^{\circ +}$, $\mu_{\bar{B}}^{\circ}$, x_+, x_-, R, T are all constants.

This equation may also be represented as

$$1n\ (a_A^+)^{x^+} \qquad + 1n\ (a_{\bar{B}})^{x_-} = k$$

or $$1n\ (a_A^+)^{x+} \qquad (a_{\bar{B}})^{x-} = k$$

or $$(a_A^+)^{x+} \qquad (a_{\bar{B}})^{x-} = k_s \qquad (37)$$

where k_s is called the activity solubility product.

But $a = fc$ (definition)

$$(f_+ . c_+)^{x+} (f_- C_-)^{x-} = k_s \quad (38)$$

If solubility is given in moles per litre as S.

Then $$C_+ = (x_+) S, C_- = (x_-) S \quad (39)$$

From (38) and (39), we get

$$(f_+ x_+ S)^{x+} (f_- x_- S)^{x_-} = k_s \quad (40)$$

$$f_+^{x_+} . (x_+)^{x_+} . S^{x_+} . f_-^{x_-} (x_-)^{x_-} . S^{x_-} = k_s \quad (41)$$

But $$x_+ . x_- = x$$

$\therefore$ $$f^{x+} \quad .S^x (x_+)^{x_+} (x_-)^{x_-} = k_s$$

or $$f^{x\pm} S^x = \frac{k_s}{(x_+)^{x_+} \; (x_-)^{x_-}}$$

or $$f^{\pm} S = \left[\frac{k_s}{(x_+)^{x+} (x_-^{x-1})}\right]^{1/x} \quad (42)$$

For two solutions having salts A and B then this equation can be written as

$$S_A (f\pm)_A = S_B (f\pm)_B = k' \quad (43)$$

We can use this equation to determine the mean activity coefficient of any sparingly soluble salt, by measuring the solubility in the presence of added salts at different ionic strengths, and extrapolating to infinite dilution. At infinite dilution $f\pm$ is unity the solubility values obtained at infinite dilution by extrapolation gives the value of constant k'. If we know one value of constant k' we can calculate mean activity coefficients by solubility measurements in any solution.

(4) *By the use of Gibbs-Duhem equation.* The Gibbs-Duhem equation is

$$n_1 \, d\mu_1 + n_2 \, d\mu_2 = 0 \quad (43a)$$

n_1, n_2 are the mole fractions of solute and solvent respectively in the solution. The relation between chemical potential (μ) and activity (a) is given by

$$\mu_1 = \mu_1^{\circ} + \text{RTd } 1\text{n } a_1 \quad (44)$$

$$\mu_2 = \mu_2^{\circ} = \text{RTd } 1\text{n } a_2 \quad (45)$$

From equations (43), (44) and (45) we get

$$n_1 d\,(\mu_1^{\circ} + \text{RTd } 1\text{n a}_1) + n_2 d\,(\mu_2^{\circ} + \text{RTd } 1\text{n } a_2) = 0$$

But $\quad d\,(\mu_1^{\circ}) = d(\mu_2^{\circ}) = \text{d(constant)} = 0$

Thus $\quad \text{RT}\,(n_1 d \text{ } 1\text{n } a_1 + n_2 d \text{ } 1\text{n } a_2) = 0$

or $$n_1 d \text{ } 1\text{n } a_1 + n_2 d \text{ } 1\text{n } a_2 = 0 \quad (46)$$

Since $\quad n_1 + n_2 = 1$

$\therefore \quad dn_1 + dn_2 = 0$

or $$n_1 \frac{dn_1}{n_1} + n_2 \frac{dn_2}{n_2} = 0$$

or $$n_1 d \text{ } 1\text{n } n_1 + n_2 d \text{ } 1\text{n } n_2 = 0 \quad (47)$$

From equations (46) and (47), we get

$$n_1\,(d \text{ } 1n \text{ } a_1 - d \text{ } 1n \text{ } n_1) + n_2\,(d \text{ } 1n \text{ } a_2 - d \text{ } 1n \text{ } n_2) = 0$$

or $$n_1 \, d \text{ } 1\text{n} \frac{a_1}{n_1} + n_2 \, d \text{ } 1\text{n} \frac{a_2}{n_2} = 0.$$

or $$n_2 \, d \text{ } 1\text{n} \left(\frac{a_1}{n_1}\right) = -n_2 \text{ d } 1\text{n} \frac{a_2}{n_2}$$

or $$d \text{ } 1\text{n} \left(\frac{a_1}{n_1}\right) = -\frac{n_2}{n_1} \text{ d } 1\text{n} \left(\frac{a_2}{n_2}\right)$$

Integrating between the limits

$$\int_{n_1'}^{n_1} \text{d } 1\text{n} \frac{a_1}{n_1} = -\frac{n_2}{n_1} \int_{n_2'}^{n_2} 1\text{n} \frac{a_2}{n_2}$$

$$\log \frac{a_1}{n_1} - \log \frac{a_1'}{n_2'} = -\int_{n_2'}^{n_2} \text{d} \log \frac{a_2}{n_2}$$

where a_1 and a_1' are the activities of solvent in two solutions in which mole fractions of the solvent are n_1 and n_1' respectively..By plotting a graph between log $\frac{a_2}{n_2}$ and $\frac{n_2}{n_1}$ the value of this integral can be calculated. This value is equal to the area under the curve between the limits and gives the difference between the two corresponding values of $\log \frac{a_1}{n_1}$. For very dilute solutions $n_1' = 1$ and $n_2' = 0$. Thus $a_1' = 1$.

Hence $\frac{a_1'}{n_1'} = 1$ and so $\log \frac{a_1'}{n_1'} = 0$.

The above equation in such a case becomes

$$\log \frac{a_1}{n_1} = -\int_0^{n_2} \frac{n_2}{n_1} d \log \frac{a_2}{n_2} \tag{48}$$

The value of this integral can be calculated graphically as above.

Activity of Solute from the Solvent

We know that

$$d \ln \frac{a_1}{n_1} = -\frac{n_2}{n_1} \text{d} \ln \frac{a_2}{n_2} \qquad \text{[Eq. 47]}$$

or

$$d \ln \frac{a_1}{n_2} = -\frac{n_2}{n_2} \text{d} \ln \frac{a_1}{n_1}$$

Integrating, we get

$$\log \frac{a_2}{n_2} - \log \frac{a_2'}{n_2'} = -\int_{n_2'}^{n_2} \frac{n_1}{n_2} \text{d} \log \frac{a_1}{n_1}$$

Value of this cannot be determined graphically so $\frac{a_2'}{n_2'}$ is determined as under:

$$\frac{a_2}{n_2} = \frac{a_2/n_2}{(a_2/n_2)*}$$

where $\left(\frac{a_2}{n_2}\right)^*$ is the $\left(\frac{a_2}{n_2}\right)$ value at infinite dilution.

$$\frac{a_2}{n_2} = \frac{(a_2/n_2)/(a_2'/n_2')}{(a_2'/n_2')\,(a_2/n_2)^*} = \frac{(a_2/n_2)\,(a_2'/n_2')}{(a_2/n_2)^*\,(a_2'/n_2')^*}$$

$$= \frac{\gamma_a/\gamma_2'}{(\gamma_2/\gamma_2')^*}$$

The value of (a_2/n_2) can be calculated if the activity coefficient values γ_2/γ_2' are extrapolated to infinte dilution.

(5) *From E.M.F. measurements.* In case of reversible galvanic cells the free energy change taking place for the chemical reaction occurring in the cell depends upon the activities of various species involved in the cell reaction. The activities thus can be measured by measuring the E.M.F of the cell. Consider any electrode concentration cell having two amalgam electrodes of different concentrations. These are dipped into a solution containing solute metal ions. e.g.,

$$Cd\text{—}Hg(a_1) \mid Cd\,SO_4 \mid Cd\text{—}Hg\,(a_2)$$

E.M.F of such a cell is given by

$$E = -\frac{RT}{F}\ln\frac{a_2}{a_1}$$

or
$$\ln a_2 = -\frac{EF}{RT} + \ln a_1$$

or
$$\ln a_2 - \ln n_3 = \frac{EF}{RT} + \ln a_1 - \ln n_2$$

or
$$\ln\frac{a_2}{n_2} = \left(-\frac{EF}{RT} - \ln n_2\right) + \ln a_1$$

or
$$\log\frac{a_2}{n_2} = \left(-\frac{EF}{2.303RT} - \log n_2\right) + \log a_1$$

If we construct a series of cells in which composition n_2 of the amalgam in right hand half cell is changed keeping the composition of the left hand half constant, measure the E.M.F of each such cell of the series and plot a graph between $\left(-\frac{EF}{2.303RT} - \log n_2\right)$ versus mole fraction n_2, we get a plot as shown in Fig. 13.1.

At infinite dilution when $n_2 = 0, \frac{a_2}{n_2} = 1$

or
$$\log\frac{a_2}{n_2} = 0$$

$$\therefore \quad \left(-\frac{EF}{2.303RT} - \log n_2\right) = \log a_1$$

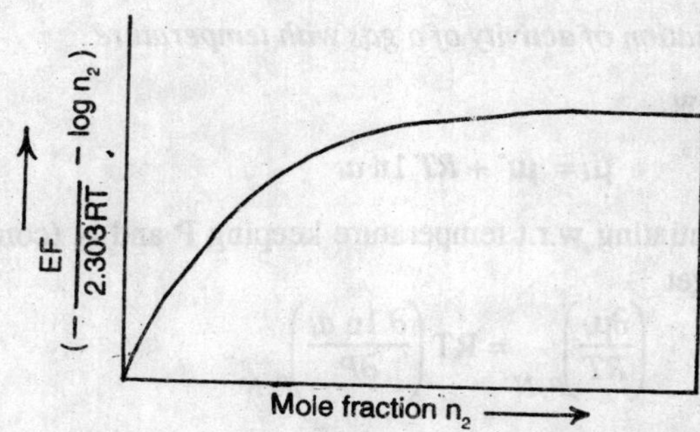

Fig. 13.1

so the intercept on the vertical axis for $n_2 = 0$ is equal to—log a_1.

In this way log (γ) can be calculated for any cell by adding log a_1. Knowing log $\frac{a_2}{n_2}$ etc. a_2 can be determined.

Variation of Activity and Activity Coefficient of a Gas

(a) *Variation of activity of a Gas with Pressure*

We know

$$\mu_i = \mu_i^\circ + RT \ln a_i$$

Differentiating with respect to pressure keeping temperature and composition (N) constant, we get

$$\left(\frac{\partial \mu_i}{RT}\right)_{T,N} = RT\left(\frac{\partial \ln a_i}{\partial P}\right)_{T,N}$$

But

$$\left(\frac{\partial \mu_i}{RT}\right)_{T,N} = \bar{V}_i$$

$$\therefore \qquad \bar{V}_i = RT\left(\frac{\partial \ln a_i}{\partial P}\right)_{T,N}$$

or

$$\frac{\bar{V}_i}{RT} = \left(\frac{\partial \ln a_i}{\partial P}\right)_{T,N}$$

$\bar{V}_i$ is the partial molar volume of the gas (i) in the mixture. This equation gives a relationship between pressure and activity of a gas.

(b) *Variation of activity of a gas with temperature*

We know

$$\mu_i = \mu_i^\circ + RT \ln a_i$$

Differentiating w.r.t temperature keeping P and N (composition) constant, we get

$$\left(\frac{\partial \mu_i}{RT}\right)_{P,N} = \mathrm{RT}\left(\frac{\partial \ln a_i}{\partial P}\right)_{P,N}$$

But $$\left(\frac{\partial \mu_i}{\partial T}\right)_{P,N} = -\bar{S}_i \qquad \text{(definition)}$$

$$\therefore \qquad -\bar{S}_i = \mathrm{RT}\left(\frac{\partial \ln a_i}{\partial P}\right)_{P,N}$$

or $$\left(\frac{\partial \ln a_i}{\partial P}\right)_{P,N} = -\frac{\bar{S}_i}{RT}$$

where $\bar{S}_i$ is partial molar entropy of gas (i) in the mixture

so $$\ln a_i = \frac{\mu_i}{RT} - \frac{\mu_i^\circ}{RT}$$

Differentiating this equation w.r.t temperature keeping P and N constant, we get

$$\left(\frac{\partial \ln a_i}{\partial P}\right)_{P,N} = \frac{1}{R}\left[\frac{\partial (\mu_i/T)}{\partial T}\right]_{P,N} - \frac{1}{R}\left[\frac{\partial (\mu_i^2/T)}{\partial T}\right]_{P,N}$$

or

$$\left(\frac{\partial \ln a_i}{\partial P}\right)_{P,N} = \frac{1}{R}\left(-\frac{\bar{H}_i}{T^2}\right) - \frac{1}{R}\left[\frac{-\bar{H}_i\,(ideal)}{T^2}\right]$$

where $$-\frac{\bar{H}_i}{T^2} = \left[\frac{\partial (\mu_i/T)}{\partial T}\right]_{P,N}$$

and $$-\frac{-\bar{H}_i\,(ideal)}{T^2} = \left[\frac{\partial (\mu_i^\circ/T)}{\partial T}\right]_{P,N}$$

or $$\left(\frac{\partial \ln a_i}{\partial T}\right)_{P,N} = \frac{\overline{H}_i\,(ideal) - \overline{H}_i}{RT^2}$$

These equations give variation of activity with temperature.

(c) *Variation of activity coefficients of a gas with pressure*

We know $$\gamma = \frac{a_i}{p_i}$$

or $$\ln \gamma = \ln a_i - \ln p_i$$

But $$p_i = N_i P$$

where N_i is the mole fraction of component (i) and P is the total pressure and pi is the partial pressure of component 'i' so from these two equations, we have

$$\ln \gamma = \ln a_i - \ln N_i P$$

$$= \ln a_i - \ln N_i - \ln P$$

differentiating w.r.t. P at constant temperature and composition (N)

$$\left(\frac{\partial \ln \gamma}{\partial P}\right)_{T,N} = \left(\frac{\partial \ln a_i}{\partial P}\right)_{T,N} - \left(\frac{\partial \ln P}{\partial P}\right)_{T,N}$$

Since N_i does not depend on pressure its differential is zero.

$$\therefore \quad \left(\frac{\partial \ln \gamma}{\partial P}\right)_{T,N} = \frac{\overline{V}_i}{RT} - \frac{1}{P}$$

$$= \frac{\overline{V}_i}{RT} - \frac{V_i}{RT}$$

$$= \frac{\overline{V}_i - V_i}{RT}$$

This equation gives the variation of activity coefficient with pressure.

FUGACITY, FUGACITY COEFFICIENT

Two functions of particular use in thermodynaimic treatment of solutions are the fugacity 'f' and fugacity coefficient 'ϕ'. In case of pure components, mixture and in solutions these can be defined as

$$dG_i = RTd\,(\ln f_i)\ (\text{Const } T)$$

$$\underset{P\to 0}{\text{Lt}}\left(\frac{f_i}{P}\right) = 1,\ \phi_i = \frac{f_i}{P} \qquad \text{Pure component}$$

$$dG = RTd\,(\ln f)\ (\text{Const } T)$$

$$\underset{P\to 0}{\text{Lt}}\ (f/p) = 1 \qquad \text{Mixture}$$

$$\phi = f/p$$

and $$d\bar{G}_i = RTd\,(\ln \hat{f}_i)\ (\text{Const } T)$$

$$\text{Lt}\ (\hat{f}_i/x_i P) = 1 \qquad \text{Component in solution.}$$

$$\hat{\phi}_i = \hat{f}_i/x_i P$$

All these equations are similar and can be summarised using the Gibbs function G, fugcity f, partial pressure P and fugacity coefficient ϕ, Thus

$$dG = RTd\,(\ln f)\ (\text{Const. } T) \tag{1}$$

$$\underset{P\to 0}{\text{Lt}}\ (f/p) = 1 \tag{2}$$

$$\phi = \frac{f}{P} \tag{3}$$

Combining (1), (2) and (3) for zero pressure limit of ϕ

$$\underset{P\to 0}{\text{Lim}}\ \phi = 1$$

The following table can be used to get these forms using appropriate variables:

	G	f	f	ψ
Pure component (i)	G_i	f_i	P	ψ_i
Mixture	G	f	P	ψ
Component (i) in solution	$\bar{G}_i$ or μ_i	$\hat{f}_i$	x_1 P	ψ_i

At constant 'I', we get

$$\Delta G = RT\ 1n\left(\frac{f_{final}}{f_{initial}}\right) \qquad (4)$$

$\Delta G = \Delta G_{final} - G_{initial}$ is quite useful as it allows for both pressure and composition changes in physical state. The units of 'f' are same as those of pressure.

Both 'f' and 'ϕ' are intensive properties but their extensive variables are $n\ 1n\ f$ and $n\ 1n\ \phi$ and not nf and $n\phi$. Because f_i and ϕ_i are not related to f and ϕ as partial molar properties the sign carat is used over f, ϕ instead of bar thus we write f, ϕ and not f, ϕ. But as in case of partial molar properties in case of these two properties also $f_i = f_i$ and $\phi = \phi_i$ for $x_i = 1$.

We have already derived.

$$\Delta G = \int_{P_1}^{P_2} n\ RT \frac{dP}{P} \qquad (5)$$

For 1 mole of an ideal gas we have

$$\Delta G = \int_{P_1}^{P_2} RT \frac{dP}{P} \qquad (6)$$

In case of real gases we cannot integrate eq. (6) as in such cases the gas equation PV = RT does not hold good and thus V is no longer equal to $\frac{RT}{P}$. Lewis introduced the fugacity function to remove this difficulty.

Consider an equilibrium system

$$Water_{(l)} = water_{(g)}$$

At this point the escaping tendency is same in all parts of the system (both liquid and gas).

This escaping tendency was called fugacity (f) by Lewis.

If (f) takes the place of P for real gases. It can be expressd as

$$\Delta G = \int_{f_1}^{f_2} nRT \frac{d}{P}$$

$$= n\mathrm{RT}\frac{f_2}{f_1} \quad (7)$$

so we get

fugacity α pressure

Another Definition

It (f) can also be defined as

$$dF = n\,\mathrm{RT}\,d\ \ln f \quad (8)$$

Integrating, we get

$$\mathrm{F} = n\ \mathrm{RT}\ \ln f + \mathrm{B} \quad (9)$$

where B is a constant of integration.

When $f = 1$, G = +G_0 then equation (9) becomes

$$G_0 = B + O \text{ or } G_0 = B$$

$$\therefore \quad G = G_0 + n\mathrm{RT}\ \ln f$$

where G_0 is the free energy of *n* moles of real gas when its fugacity happens to be unity.

Thus 'fugacity' is a 'fictitious pressure' which is used in order to retain for real gases simple form of equation which are applicable to ideal gases only.

Fugacity at Low Pressure

Fugacity ∝ Pressure

$$f \propto \mathrm{P}$$

This relation has already been derived for ideal gases. In case of real gases fugacity (f) is initially less than pressure (P) and becomes greater than P at higher pressure. This deviation from ideal gas behaviour can be correlated with the Van der Waals constants '*a*' and '*b*'. At low pressures the Van der Waals equation may be written as

$$\left(P + \frac{a}{V^2}\right)(\mathrm{V}) = \mathrm{RT} \quad \text{[forces of attraction are important]}$$

or $$Z \text{ (Compressibility factor)} = \frac{PV}{RT}$$

$$= 1 - \frac{a}{VRT}$$

So $Z < 1$

$\therefore$ f is less than P because of constant '*a*'.

At higher pressure, the Van der Waals equation may be written as

$$P(V - b) = RT \qquad \text{(excluded volume is important)}$$

or $$Z = \frac{PV}{RT} = 1 + \frac{bP}{RT}$$

So fugacity is greater than P. It is due to constant '*b*'.

So we find that neglecting '*b*' in Van der Waals equation makes '*f*' less than P and neglecting '*a*' makes '*f*' greater than P.

It can be shown easily that for a Van der Waals gas following relations hold good

(i) $$\ln f = \ln \frac{RT}{V-b} + \frac{b}{V-b} - \frac{2a}{RTV}$$

and (ii) $$\ln f = \ln P - \frac{aP}{(RT)^2} + \frac{bP}{RT} + \frac{ab\,P^2}{2\,(RT)^3}$$

The *compressibility factor* '*z*' used above can be defined as the ratio of V and V′ at the same T and P. Thus

$$Z = \frac{V}{V'}$$

But $$V' = \frac{RT}{P}$$

so $$Z = \frac{PV}{RT}$$

$Z = 1$ for ideal gas.

Z is a state function and is defined in terms of T, P, V. Various equations for Z as function of P, T and composition (or V, T and composition or P, V and composition) constitute *thermal equation of state.* Now an effort will be made to develop equations for calculations of 'f' or ϕ for a pure component or for a constant composition mixture from *thermal equations of state.*

In this equations subscript 'i' has been deleted just for simplicity However, these equations are valid for both cases.

We know

$$dG = RT\, d(\ln f) \text{ at constant T.} \tag{10}$$

Also for one mole of solution at constant temperature and composition, we know

$$dG\,(T, N) = VdP \tag{11}$$

From (10) and (11), we get

$$RT\, d(\ln f) = VdP \tag{12}$$

We also know

$$dG = RT\, d(\ln f) \text{ at Const. (T)}$$

From equation (3), we get

$$d(\ln f) = d(\ln f) + d(\ln P)$$

or

$$d(\ln f) = d(\ln f) + \frac{dP}{P} \tag{13}$$

substituting (13) in (12) and using definition of Z, we get

$$d(\ln \phi) = (Z-1)\frac{dP}{P} \text{ (const. T, N)} \tag{14}$$

Equation (14) can be integrated from zero pressure state [where $\phi = 1$] to a pressure P state when we get

$$\ln \phi = \int_0^P (Z-1)\frac{dP}{P} \text{ (const. T, N)} \tag{15}$$

Since $\phi = \phi P$

We get

$$\ln(f) = \ln P + \int_0^P (Z-1)\frac{dP}{P} \text{ (const T, N)} \tag{16}$$

Similar equations for which V, rather than P is variable can be easily derived. These are

$$1n\ \phi = (Z—1) — 1n\ Z \int_{\alpha}^{V} (Z—1)\frac{dV}{V} \text{ (const. T, N)} \qquad (17)$$

and $$1n\ \phi = 1n\frac{RT}{V} + (Z—) — \int_{\alpha}^{V} (Z—1)\frac{dV}{V} \text{ (const T, N)} \qquad (18)$$

In case of ideal gases Z = 1

$\therefore$ $$\phi_i = \phi = 1$$

and $$f_i = x_i = P$$

For ideal gas mixtures analogous equations are

$$\overset{A}{\varphi_i} = 1$$

and $$\overset{A}{f_i} = x_i\, P$$

where x_i is the mole fraction of component '*i*' in the gas phase.

Determination of Fugacity

Some of the methods used for the determination of fugacity of gases and liquids mixtures are described in following pages.

1. *Graphical Method.* As already proved in case of real gases at very low pressure $f = p$

But $$dG = VdP$$

$\therefore$ $$RTd\ 1n\, f = VdP \qquad \text{[for one mole]}$$

or $$\left(\frac{\partial\ 1n\, f}{\partial P}\right)_T = \frac{V}{RT} \qquad (19)$$

Introducing α where $\alpha = \frac{RT}{P} - V$, we get

$$RTd\ 1n\ f = \left(\frac{RT}{P} - \alpha\right)dP \quad \left[\because V = \frac{RT}{P} - \alpha\right] \qquad (20)$$

From (19) and (20), we get

$$RTd\ 1n\ f = \left(\frac{RT}{P} - \alpha\right)dP \qquad (21)$$

Integrating (21) we get

$$RT\int d\ln f = RT\int_0^P \frac{dP}{P} - \int_0^P \alpha\, dP$$

or

$$\int d\ln f = \int_0^P \frac{dP}{P} - \frac{1}{RT}\int_0^P \alpha dP$$

$$\ln f = \ln P - \frac{1}{RT}\int_0^P \alpha\, dP \qquad (22)$$

To evaluate the integral a graph is plotted between a/RT and P as shown in Fig. 13.2. The area under the curve gives the values of integral and thus 'f' can be calculated.

2. *Approximate calculation method.* At moderately low pressures, we can derive from Van der Wall's equation that PV for a real gas in a linear function of its pressure if temperatures is kept constant.

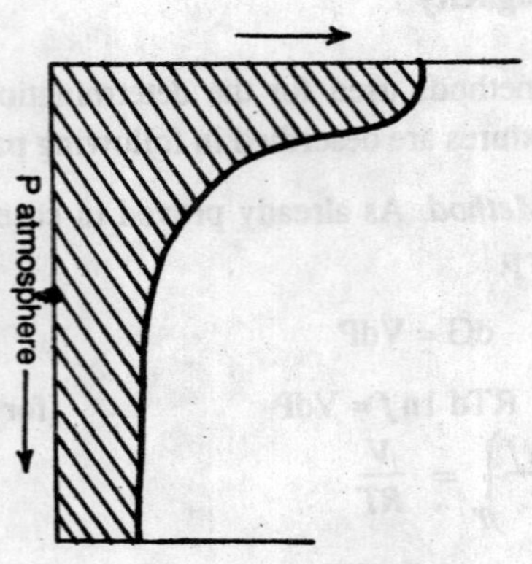

Fig. 13.2

Mathematically, we can write

$$PV = RT - \alpha P \qquad \rightarrow \left(\frac{\alpha}{R\tau} \times 10^4\right) atm^{-1}$$

or

$$\alpha = \frac{RT}{P} - V$$

a is constant over this moderate pressure range

Applying this condition to Eq. (22), we get

$$\ln f = \ln P - \frac{\alpha}{RT}\int_0^P dP \qquad [\because a \text{ is Constant}]$$

or $$\ln = \ln P - \frac{\alpha}{RT} P$$

or $$\ln \frac{f}{P} = -\frac{\alpha P}{RT} \qquad (23)$$

At moderate pressure $\frac{f}{P} = 1$

Hence $$\ln \frac{f}{P} = \frac{f}{P} - 1 \qquad [\log x \approx (x-1) \text{ when } x \to 1]$$

From equation (23), we have

$$\frac{f}{P} - 1 = \frac{aP}{RT} \qquad (24)$$

Substituting in Eq. 24 value of α from the mathematical from given above, we get

$$f = -\frac{P^2 V}{RT} \qquad (25)$$

From this equation fugacity of a gas can be calculated approximately if its pressure and molar volume are known. These results are quite accurate at low pressure and high temperature.

(3) *Generalized Method.* We have derived the equation,

$$a = \frac{RT}{P} - V = \frac{RT}{P}\left(1 - \frac{VP}{RT}\right)$$

The compressibility Z has been defined as $Z = \frac{PV}{RT}$

From the above two equations, we get

$$a = \frac{RT}{P}(1-Z) \qquad (26)$$

Putting this value of α in equation

$$\ln f = \ln P - \frac{1}{RT}\int_0^P \alpha \, dP \qquad [\text{Eq. 22}]$$

We get

$$\ln f = \ln P - \int_0^P \frac{1-Z}{P} dP \qquad (27)$$

Replacing pressure by reduced pressure, we get

$$\ln f = \ln P + \int_0^{\pi} \frac{Z-1}{\pi}\, d\pi \qquad \left[\pi = \frac{P}{P_c}\right]$$

or

$$\ln \frac{f}{P} = \int_0^{\pi} \frac{Z-1}{\pi}\, d\pi \tag{28}$$

We know that the compressibility factors of all gases are approximately equal at given reduced temperature and pressure. Thus a graph can be lotted by keeping $\frac{f}{P}$ as ordinate and π as abscissa, θ as the parameter [Fig. 13.3]. To draw Fig. 13.3 the values of Z are derived from Fig. 13.4 and the Integral in equation (28) is evaluated.

(4) From equation of State. If f and f^* are the fugacitees of a gas at pressure P and at low pressure P_0 respectively. Then from equation (9), we get

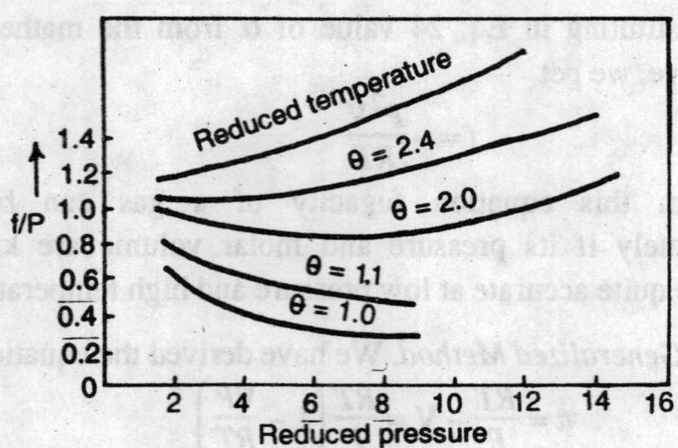

Fig. 13.3

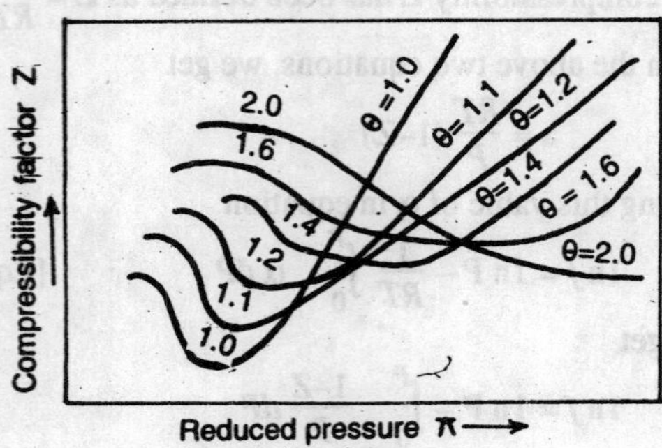

Fig. 13.4

$$RT\int_{f_0}^{f} d\ln f = \int_{P_0}^{P} VdP$$

or

$$\ln\frac{f}{f_0} = \frac{1}{RT}\int_{P_0}^{P} VdP \tag{29}$$

Now

$$\int_{P_0}^{P} VdV = [PV]_{V_0}^{V} - \int_{V_0}^{V} PdV \text{ [Integration by p.rH}$$

$$PV—P_0V_0 - \int_{V_0}^{V} PdV \tag{30}$$

If V_0 is volume at P_0 then

$$P_0V_0 = RT$$

Thus

$$\int_{P_0}^{P} VdP = PV - RT - \int_{V_0}^{V} PdV \tag{31}$$

or

$$\ln\frac{f}{f_0} = \frac{1}{RT}\left[PV - RT - \int_{V_0}^{V} PdV\right]$$

At very law pressure we have $\ln\frac{f}{f_0} = 1n\frac{f}{f_0}$

$$\therefore \quad \ln\frac{f}{P_0} = \frac{1}{RT}\left[PV - RT - \int_{V_0}^{V} PdV\right]$$

or

$$\ln f = \ln P_0 + \frac{1}{RT}\left[PV - RT - \int_{V_0}^{V} PdV\right] \tag{32}$$

The integral in this equation is calculated from equation of state like van der Waal's equation i.e.,

$$\left(P + \frac{a}{V^2}\right)(V-b) = RT$$

or

$$P = \frac{RT}{V-b} - \frac{a}{V^2}$$

Multiplying this equation by dV and integrating, we get

$$\int_{V_0}^{V} PdV = \int_{V_0}^{V} \frac{RT}{V-b} dV - \int_{V_0}^{V} \frac{a}{V2} dV$$

$$= RT \ 1n \frac{V-b}{V_0-b} + \frac{a}{V} - \frac{a}{V_0} \qquad (33)$$

Substituting from (33) in (32) we get

$$1n f = 1n \ P_0 + \frac{1}{RT}\left[PV - RT - RT \ 1n \frac{V-b}{V_0} - \frac{a}{V}\right] \qquad (34)$$

Again from the van der Waal's equation, we have

$$PV-RT = \left[\frac{RTV}{V-b} - \frac{a}{V}\right] - RT$$

or

$$PV-RT = \left[\frac{RTV}{V-b} - RT\right] - \frac{a}{V}$$

$$= \left[\frac{RTb}{V-b} - \frac{a}{V}\right] \qquad (35)$$

Substituting from Eq. 35 in Eq. 34

$$1n f = 1n \ P_0 + \frac{1}{RT}\left[\frac{RTb}{V-b} - \frac{a}{V} - RT \ 1n \frac{V-b}{V_0} - \frac{a}{V}\right]$$

$$= 1n \ P_0 + \frac{b}{V-b} - \frac{2a}{RTV} - 1n \frac{V-b}{V_0}$$

$$= 1n \ P_0 + \frac{b}{V-b} - \frac{2a}{RTV} - [1n \ (V-b) - 1n \ V_0]$$

$$= 1n \ P_0 + \frac{b}{V-b} - \frac{2a}{RTV} - \left[1n \ (V-b) - 1n \frac{RT}{P_0}\right].$$

$$[\therefore P_0V_0 = RT]$$

$$= \ln P_0 + \frac{b}{V-b} - \frac{2a}{RTV} - [\ln (V-b) - \ln RT] - \ln P_0$$

$$= \frac{b}{V-b} - \frac{2a}{RTV} + \ln \frac{RT}{V-b}$$

Fugacity can be calcalated from this equation if the values of '*a*' and '*b*' are known.

Change of Fugacity with Temperature

If G and G_0 are the molar free energy of gas in a given state and in the standard state respectively. The standard state here is the vapour state at low pressure. The change in free energy when one mole of this substance is converted will be

$$G_0 - G = RT \ln \frac{f_0}{f} \qquad (36)$$

Differentiating this equation w.r.t T at constant P, we get

$$\left(\frac{\partial G_0}{\partial T}\right)_P - \left(\frac{\partial G}{\partial T}\right)_P = R \ln \frac{f_0}{f} + RT\left(\frac{\partial \ln f_0}{\partial T}\right)_P$$

$$- TR\left(\frac{\partial \ln f}{\partial T}\right)_P \qquad (37)$$

At very low pressure f_0 becomes independent of temperature

$$\therefore \qquad \left(\frac{\partial}{\partial T} \ln f_0\right)_P = 0$$

From equation (36), we get

$$R \ln \frac{f_0}{f} = \frac{G - G_0}{T} \qquad (38)$$

Substituting from (38) in (37), we get

$$\left(\frac{\partial G_0}{\partial T}\right)_P - \left(\frac{\partial G}{\partial T}\right)_P = R \ln \frac{f_0}{f} - RT\left(\frac{\partial \ln f}{\partial T}\right)_P \qquad (39)$$

or $$R \ln \frac{f_0}{f} = \frac{G_0 - G}{T} \qquad (40)$$

From equation (39), and (40)

$$\left(\frac{\partial G_0}{\partial T}\right)_P - \left(\frac{\partial G}{\partial T}\right)_P = \frac{G_0 - H}{T} - \text{RT}\left(\frac{\partial}{\partial \text{T}} 1n f\right)_P$$

But

$$\left(\frac{\partial G}{\partial T}\right)_P = \frac{G - H}{T}$$

$$\therefore \quad \frac{G_0 - H_0}{T} - \frac{G - H}{T} = \frac{G_0 - G}{T} - \text{RT}\left(\frac{\partial}{\partial \text{T}} 1n f\right)_P$$

or

$$\left(\frac{\partial}{\partial \text{T}} 1n f\right)_P = \frac{H_0 - H}{RT^2} \quad (41)$$

where H_0–H gives the ideal heat of vapourisation.

Change of Fugacity with Pressure

We know

$$G = RT\, 1n f + C$$

where C is a temperature dependent constant.

Differentiating above equation w.r.t. pressure at constant temperature

$$\left(\frac{\partial G}{\partial P}\right)_T = \text{RT}\left[\frac{\partial\, 1n f}{\partial P}\right]_T$$

or

$$V = RT\left[\frac{\partial\, 1n f}{\partial P}\right]_T \qquad \left[V = \left(\frac{\partial G}{\partial P}\right)_T\right]$$

or

$$\left[\frac{\partial}{\partial \Pi} 1n f\right]_T = \frac{V}{RT} \quad (42)$$

Integrating this equation, we get

$$RT\, 1n \frac{f_2}{f_1} = \int_{P_1}^{P_2} V\, dP \quad (43)$$

Thus fugacity at any other pressure can be calculated if fugacity at some pressure is know.

Determination of Fugacity of a gas in a Mixture

Lewis—Rendull Rate. The same principle as used for determination of fugacity of pure gas is involved in this determination.

We have already derived

$$\left[\frac{\partial}{\partial T} \ln f\right]_T = \frac{V}{RT} \qquad \text{[equation (42)]}$$

or

$$RTd \ln f_i = V_i \, dP \qquad (44)$$

where V_i is the partial molar volume of component 'i' in the gaseous mixture.

Then $\alpha_i = \frac{RT}{P} - V_i$ [definition of α]

where P is the total pressure.

Eliminating V_i from the above two equations

$$d \ln f_i \, d \ln P = \frac{\alpha_i}{RT} dP \qquad (45)$$

If f_i, f_0 are the fugacities of component (i) at P and P_0 where P_0 is very low pressure. Integrating the equation (45) at constant temperature and composition, we get

$$\ln \frac{f_i}{f_0} = \ln \frac{P}{P_0} - \frac{1}{RT} \int_{P_0}^{P} \alpha_i \, dP \qquad (46)$$

But $(P_0)_i = N_i P_0$

or $\frac{P}{(P_0)_i} = \frac{P}{N_i P_0}$

where N_i is the mole fraction of (i) in the mixture

Taking log of both sides, we get

$$\ln \frac{P}{(P_0)_i} = \ln \frac{1}{N_i} + \ln \frac{P}{P_0}$$

or

$$\ln \frac{P}{(P_0)_i} = -\ln N_i + \ln \frac{P}{P_0}$$

or $$\ln \frac{P}{(P_0)} = \ln N_i + \ln \frac{P}{(P_0)_i}$$

Thus, we get

$$\ln \frac{f_i}{f_0} = \ln N_i$$

$$= \ln \frac{P}{(P_0)_i} - \frac{1}{RT} \int_{P_0}^{P} \alpha_i \, dP$$

or $$\ln f_i - \ln f_0 = \ln N_i + \ln P - \ln (P_0)_i - \frac{1}{RT} \int_{P_0}^{P} \alpha_i \, dP$$

or $$\ln f_i = \ln N_i + \ln P + \ln f_0 - \ln (P_0)_i - \frac{1}{RT} \int_{P_0}^{P} \alpha_i \, dP$$

$$= \ln N_i + \ln P + \ln \frac{f_0}{(P_0)_i} - \frac{1}{RT} \int_{P_0}^{P} \alpha_i \, dP \qquad (47)$$

At very low pressure,

$$\frac{f_0}{(P_0)_i} = 1$$

$$\therefore \qquad \ln \frac{f_0}{(P_0)_i} = 0$$

So equation (47) becomes

$$\ln f_i = \ln N_i + \ln P - \frac{1}{RT} \int_{P_0}^{P} \alpha_i \, dP \qquad (48)$$

If V_i and P are known then the values of α_i at various pressures can be calculated and the integral in equation (48) can be evaluated graphically. Thus fugacity f_i of any gaseous component can be determine in a mixture.

If the gases are mixed at constant temperature, $\Delta V = 0$. Then V_i of each gas is equal to the molar volume. In such a case (48) becomes

$$\ln f_i' = \ln P \frac{1}{RT} \int_{P_0}^{P} \alpha_i \, dP$$

or $$\ln f_i = \ln N_i + \ln f_i'$$

or $$\ln f_i = \ln N_i f_i'$$

or $$f_i = N_i f_i'$$

In such a mixture the fugacity fi of a component is equal to the products of its (N_i) mole fraction and its fugacity is pure state (f_i') at the total pressure of the mixture.

This statment is known as Lewis-Randell rule.

Duhem-Margules Equation

It is an equation which is applicable to liquid solutions only. It is applicable to all solutions (liquid) irrespective of the fact whether these solutions are ideal or not.

This equation gives a relationship between fugacity and mole fraction of a component in a liquid solution and can be derived from the Gibbs-Duhem equation easily, e.g.

According to Gibbs-Duhem equation, we have for a binary solution.

$$n_1 d\mu_1 + n_2 d\mu_2 = 0$$

or
$$\frac{n_1}{n_1+n_2} d\mu_1 + \frac{n_2}{n_1+n_2} d\mu_2 = 0$$

or
$$x_1 d\mu_1 + x_2 d\mu_2 = 0 \quad (49)$$

where x_1 and x_2 are the mole fractions of the components present in a binary liquid solution. Thus $x_1 + x_2 = 1$

But $d\mu = RTd\,(\ln f)$ (definition) (50)

from (49) and (50)

$$\therefore \quad x_1 \text{ RT} (\ln f_1) + x_2 \text{ RTd} (\ln f_2) = 0 \quad (51)$$

or
$$\frac{x_1}{dx_1} \text{d} (\ln f_1) + \frac{x_2}{dx_1} \text{d} (\ln f_2) = 0 \quad \text{[Dividing by } dx_1\text{]}$$

or
$$\frac{x_1}{dx_1} \text{d} (\ln f_1) = \frac{x_2}{dx_2} \text{d} (\ln f_2) \qquad dx_1 + dx_2 = 0$$

or $dx_1 = -dx_2$

or
$$\frac{d \ln f_1}{d \ln x_1} = \frac{d \ln f_2}{d \ln x_2} \quad (52)$$

extending the similar arguments it can be proved for a multi-component liquid solution having mole fractions x_1, x_2... of the various species present in it, that

$$\frac{d \ln f_1}{d \ln x_1} = \frac{d \ln f_2}{d \ln x_2} = \frac{d \ln f_3}{d \ln x_3}$$

Equation (53) is known as Duhem-Margules equation.

This can also be represented as

$$\frac{x_1}{f_1}\left(\frac{df_1}{dx_1}\right) = \frac{x_2}{f_2}\left(\frac{df_2}{dx_2}\right) \qquad (54)$$

If vapours show ideal behaviour then $f_i = p_i$ so making this substitution in (54), we get

$$\frac{x_1}{p_1}\left(\frac{dp_1}{dx_1}\right) = \frac{x_2}{p_2}\left(\frac{dp_2}{dx_2}\right) = \frac{x_3}{p_3}\left(\frac{dp_3}{dx_3}\right) = \ldots$$

For a binary system this equation becomes

$$\frac{x_1}{p_1}\left(\frac{dp_1}{dx_1}\right) = \frac{x_2}{p_2}\left(\frac{dp_2}{dx_2}\right) \qquad (55)$$

It can be calculated from these equations that any increase in mole fraction of a component of a liquid solution increases the partial pressure of that component. This will decrease the value of mole fraction of the other component and thus a corresponding decrease in the partial pressure of the other component.

Example. Find the fugacity of hydrogen at 100°C, at 300 atmosphere using approximate relation (i.e., $\ln f = \ln p - \frac{ap}{RT^2} + \frac{bp}{RT} + \frac{abp^2}{2(RT)^2}$. Given a = 0.2244 dm^6 atom. mol^{-2}, b = 0.0266 dm^3 mol^{-1}.

Solution. Putting the value of various terms in approximate equations, we get

$$\ln \gamma = -\frac{0.244 \times 300}{0.082} + \frac{0.0266 \times 300}{0.082 \times 373}$$

$$+\frac{0.244 \times 0.0266 \times 300}{2\,[0.082 \times 373]^2}$$

$$= -0.0781 + 0.2606 + 0.0101 = 0.2028$$

Taking antilog we get

Thus $\qquad \gamma = 1.225$

Now $\qquad f = p\gamma = 1.255 \times 300$

$$= 367.5$$

Application of Gibbs-Duhem-Margules Equations

This equation could be applied to total pressure curves.

Consider a binary system having a mole fraction x_1 of component 1 in liquid phase and a mole fraction y, of component 1n in vapour phase.

Then for such a system

$$p_1 = y_1 P \tag{56}$$

and

$$p_2 = y_2 P$$

$$= (1-y_1)\,P \tag{57}$$

Substituting from (56) and (57) into (55) we get

$$\frac{x_1}{y_1 P}\left(\frac{dp_1}{dx_1}\right) = \frac{x_2}{y_2 P}\left(\frac{dp_2}{dx_2}\right) \qquad x_1 = 1 - x_2, \quad dx_1 = -\,dx_2$$

$$\therefore \quad \frac{x_1}{y_1}\frac{dp_1}{dx_1} + \frac{x_2}{y_2}\left(\frac{dp_2}{dx_1}\right) = 0$$

or

$$\frac{dp_1}{dx_1} = -\frac{y_1 x_2}{x_1 y_2}\left(\frac{dp_2}{dx_1}\right)$$

or

$$\frac{dp_3}{dx_1} = -\frac{x_1 y_2}{x_2 y_1}\left(\frac{dp_1}{dx_1}\right) \tag{58}$$

But $$P = p_1 + p_2$$

$$\therefore \quad \frac{dp}{dx_1} = \frac{dp_1}{dx_1} + \frac{dp_3}{dx_1} \qquad \text{differentiating w.r.t. } x_1 \tag{59}$$

From (58) and (59),

$$\frac{dP}{dx_1} = \frac{dp_1}{dx_1} + \left[-\frac{x_1 y_2}{x_2 y_1}\left(\frac{dp_1}{dx_1}\right)\right]$$

$$= \frac{dp_1}{dx_1}\left[1 - \frac{x_1 y_2}{x_2 y_1}\right] \tag{60}$$

Equation (60) can be used in the fractional distillation. The change in pressure on changing the composition is given by this equation. It can be seen from the equation that addition of component 1 change the pressures in such a was that

$$\frac{x_1 y_2}{x_2 y_2} < 1$$

or in case of binary systems $(x_1 + x_2 = 1)$

and $$(y_2 + y_2 = 1)$$

$$\frac{x_1\,(1 - y_1)}{(1 - x_1) y_1} < 1$$

So $$x_1 < y_1 \qquad \text{or } y_1 > x_1$$

This is known as *Konovalov's first law.*

It states, "The vapour phase over a liquid mixture is richer in that component which is more volatile and its addition to the mixture increases the total vapour pressure".

On fractional distillation in case of such a mixture, the more volatile component (vapour pressure increasing or VPI component) will distill over first and can be collected leaving behind the less volatile component in the distillation flask.

This law will not be followed in case the mixture under any composition acquires a minimum or maximum boiling point. In such a state the mixture is known as an azeotropic mixture and the composition becomes same in both the liquid and vapour phase in such cases. So at the composition of minimum or maximum boiling point $\frac{dP}{dx_1} = 0$ and therefore

$$\frac{dp_1}{dx_1}\left[1 - \frac{x_1 y_2}{x_2 y_1}\right] = 0 \qquad \text{[from eq. 60]}$$

For all stable phase $\frac{dp_1}{dx_1}$ is positive thus $x_1 = y_1$ and $x_2 = y_2$

This is known as *Konovalov's second law.*

According to Konovalov's Rule the compositions of the vapour phase and the liquid phase, in a solution-vapour equilibrium are different. If the composition of vapour phase and liquid phase are same it is one of the following three cases:

(a) Vapour is in equilibrium with pure liquid component.

(b) Vapour is in equilibrium with a liquid solution showing minimum vapour pressure.

(c) vapour is in equilibrium with a liquid solution showing maximum vapour pressure.

Some binary azeotropic mixtures are listed below:

Table 13.1

Components with B. pt (°C)		*B. pt of azeotropic mixture (°C)*	Wt. % of A in the azeotropic mixture
A	*B*		
Minimum boiling point azeotropic mixture			
1. C_2H_5OH (78.3)	Water (100°)	78.17°	96
2. C_2H_5OH (78.3)	Benzene (80.2)	63.2	32.4
3. CCl_4 (76.75)	C_2H_5OH(78.3)	65.0	82.7
4. Water (100)	Pyridine (115.3)	92.5°	43.0
Maximum boiling point azeotropic mixture			
1. HNO_3 (83)	Water (100)	120.5	68.0
2. HCl (85°)	Water (100)	108.58	30.22
3. Methylacetate (5.71)	Chloroform (61.2)	64.8	23
4. Acetone (56)	Chloroform (61.2)	64.43	21.5

Numericals

Example 1. A mixture of chlorobenzene and water is distilled at an external pressure 740.2 mm. It is found to boil at a temperature of 90.3°C. Calculate the composition of the distillate (Vapour pressure of pure water at 90.3°C is 530.1 mm)

Solution. External pressure = 740.2m

Partial pressure of water, $p_A = 530.1$ mm

" " C_6H_5Cl, $p_B = 740.2 - 530.1$

$= 210.1$ mm

Mol. wt. of C_6H_5Cl, $M_B = 112.5$

" " H_2O, $M_A = 18$

we know

$$\frac{W_A}{W_B} = \frac{p_A M_A}{p_B M_B}$$

$$= \frac{530.1 \times 18}{210.1 \times 112.5}$$

$$= \frac{.404}{1}$$

So in 1.404 parts by weight of mixture we have 1 part by weight of chlorobenzene

$$\therefore \quad \% \, C_6H_5Cl = \frac{1}{1.404} \times 100$$

$$= 71.2\%$$

Example 2. A mixture of mitrobenzene (Mol. wt. 123) and water (Mol. wt. 18) boil at 90°C at an external pressure of 760 mm. Vapour pressure of pure water at 90°C is 733 mm. Calculate the weight composition of the distillate.

Solution. Proceeding as in example 1,

$$\frac{W_A}{W_B} = \frac{733 \times 18}{27 \times 123}$$

$$= \frac{397}{1}$$

So, Water : Nitrobeuze; 3.97 : 1.

Example 3. A cleaning solution is to be obtained by mixing equal masses of (*a*) acetone, and (*d*) dichloromethane both at 298°K. If the mixing is done adiabatically at a pressure of 101.325 kP$_a$ with the addition of negligible stirring work. What is the temperature of cleaning solution formed.

Given. Heat capacities of Pure '*a*' and '*d*'

$$Cp_{(a)} = 2173 \text{ J/kg K} \qquad Cp_{(d)} = 1193 \text{ J/kg K}$$

Heats of mixing of equal masses of (*a*) and (*d*) at 101.325 kP$_a$

T/K	ΔH/J/kg
293	12468
298	12380
303	12292

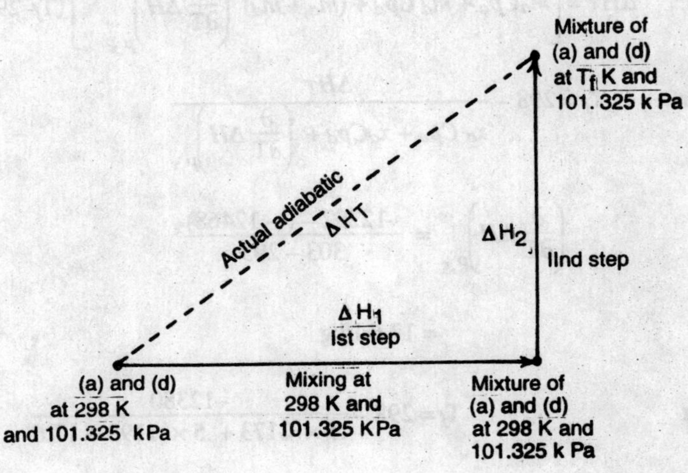

Fig. 13.5

Solution. The process may be considered as reversible in two steps as shown

We have $\Delta H_T = \Delta H_1 + \Delta H_2$

$$\Delta H_1 = \text{Heat of mixing at constant T, P}$$

$$= (m_a + m_d)\,\Delta H$$

where ΔH is the heat of mixing per unit mass of solution

$$\Delta H_2 = m_a + m_d \int_{298}^{T_f} C_p \, dT$$

$$= (m_a + m_d)(T_f - 298)\,C_p$$

where C_p is the specific heat capacity of the mixture and is independent of T for small temperature range under consideration

$$C_p = x_a C_{p_a} + x_d C_{p_d} + \Delta C_p$$

How $\Delta C_p = \left(\frac{\partial \Delta H}{\partial T}\right)_{P,x}$

Combining these, we get

$$\Delta H_T = \left[m_a C_{p_a} + m_d C_{p_d} + (m_a + m_d)\left(\frac{\partial}{\partial T}\Delta H\right)_{P,x} \right](T_f - 298)$$

or $$T_f = 298 - \frac{\Delta H_T}{x_a C_{p_a} + x_d C_{p_d} + \left(\frac{\partial}{\partial T}\Delta H\right)_{P,x}}$$

$$\left(\frac{\partial}{\partial T}\Delta H\right)_{P,x} = \frac{-12292 - (-12468)}{303 - 293}$$

$$= 17.6 \text{ J/kg}$$

or $$T_f = 298 - \frac{-12380}{.5 \times 2173 + .5 \times 1193 + 17.6}$$

$$= 305.3 \text{ K.}$$

Example. 4. A completely miscible binary system (1) benzene and (2) chloroform has VLE – K values at 343 K and 101.325 kPa as $K_1 = .719$ and $K_2 = 1.31$. Calculate the equilibrium compositions of the liquid and vapour phases and the fraction of the total system which is liquid for a total system composition of (a) 40% mole benzene (b) 50% mole benzene.

Solution. The sum of mole fractions is one in case of binary system

$$x_1 + x_2 = 1, y_1 + y_2 = 1$$

Again, we have

$$k_i = \frac{y_i}{x_i}$$

so $$k_1 x_1 + k_2 x_2 = 1$$

or $$x_1 = \frac{k_2 - 1}{k_2 - k_1}$$

$$y_1 = k_1 \frac{k_2 - 1}{k_2 - k_1}$$

For given T, P conditions there is only one equilibrium possible and hence there is only one set of *k*-values in two phase, two component system. These constant do not depend on over all composition as can be seen from above equation

$$\therefore \quad x_1 = \frac{1.31 - 1}{1.32 - .719}$$

$$= .525 \qquad x_2 = .475$$

and $$y_1 = (.719)(.525)$$

$$= .377 \qquad y^2 .= 623$$

For a total of 1 mole in this case we can write for L (fraction of the system in liquid state)

$$L = \frac{Z_1}{k_2 - 1} + \frac{Z_1 - k_1}{1 - k_1} \text{ where}$$

$$[Z_1 = x_1 L + y_1 (1 - L)]$$

for one mole

$$= \frac{.40}{1.31 - 1} + \frac{.40 - .719}{1 - .719}$$

$$= .155$$

and for second

$$L = \frac{.50}{1.31-1} + \frac{.50-.719}{1-.719}$$

$$= .834$$

Example 5. A solution of 1 mole of KOH in 4.6 moles of water has vapour pressure of 4.474 mm at 15°C. Vapour pressure of water at 15°C is 12.788 mm. Taking activity coefficient of pure water as one calculate: (i) the activity coefficient of water in solution: and (ii) the difference betwen the chemical potential of water in solution and in pure water.

Solution. At these low vapour pressure we can take the vapour pressure as equal to fugacity.

(i) $$a_{H_2O} = \frac{p}{p_0}$$

$$= \frac{4.474}{12.788}$$

$$= .3499$$

the mole fraction of water in solution

$$x_{H_2O} = \frac{.6}{1+4.6}$$

$$= .8214$$

Activity coefficient of water in solution is

$$\gamma_{H_2O} = \frac{a_{H_2O}}{x_{H_2O}}$$

$$= \frac{.3499}{.8214}$$

$$= .426$$

So γ_{H_2O} in solution = .426

(ii) The chemical potential of water in solution is

$$\mu = \mu^{\circ} + RT \ln a_{H_2O}$$

or $$\mu - \mu^\circ = 2 \times 288 \times 2.303 \log .3489$$

$$= -605 \text{ cal mole}^{-1}$$

This is the difference in chemical potential of water in solution (μ) and that of pure water (μ°).

14

Chemical Affinity

Introduction

The simple definition for this term was given by Albertus Magnus in 1250. He defined chemical affinity as under:

"*Chemical affinity is the force which accounts for the tendency of a reaction to occur.*"

Later attempts were made to arrange the elements in the order of their chemical affinity and table were made by 1718. *F.F. Geoffrey* gave '*Tables of affinity*' and the same was extended by Bergmann in 1775. This concept of chemical affinity could not explain the reversible reactions.

J.J. Thamson and Bertholet proposed that heat given out may be a measure of chemical affinity so they considered that only exothermic reactions are possible, so this concept could not be accepted as many endothermic reactions are spontaneous.

In the modern times it was proposed that maximum work obtainable during a reaction is a measure of chemical affinity. We know that any system can give maximum amount of work if carried out reversibly and isothermally. Thus *Vant Hoff* suggested:

"Affinity may be regarded as a measure of decrease of free energy accompanying a process".

So a chemical reaction will always proceed in the direction of a decrease of free energy. So the true criterion for the spontaneity of the process is the sign of ΔG is an accepted concept these days.

Criterion for Chemical Equilibrium

The decrease in free energy gives us an idea of the spontaneity of a chemical reaction, i.e.,

(i) If ΔG is negative, the process is Spontaneous.

(ii) If ΔG is positive, the process will not occur or it will occur in backward direction.

(iii) If ΔG is zero, the system is in equilibrium.

The condition for equilibrium with respect to variation in number of moles in a closed system is given by

$$(dG)_{T,P} = 0$$

For any heterogeneous system if the change in number of moles is due to a single chemical reaction, then we get

$$(dG)_{T,P} = \Delta G_{T,\,P} \,.\, d\varepsilon$$

and at equilibrium

$$\Delta G_{T,P}\, d\varepsilon = 0$$

To get general results for a system having r independent chemical reactions we can generalize it and obtain

$$(\Delta G_{T,P}) = 0 \; (J = 1, 2 ... r)$$

where $(\Delta G_{T,P})_j$ is the Gibbs function change of reaction for the jth independent chemical reaction.

Again $\qquad (\Delta G_{T,P})_j = \Sigma i V_{i,g}\, \mu_i \qquad$ [definition]

μ_i is a function of T, P and mole fraction of various species and thus constitutes γ implicit relationships that must hold among intensive variables at equilibrium. Thus the phase rule must also be satisfied and the additional r constraints be included. Thus we get

$$F = 2 - n + m - r$$

Thus the occurrence of a chemical reaction decreases the number of degrees of freedom of a PVT system at equilibrium. However, Duhem theorem remains unchanged.

To illustrate it let us determine the number of degrees of freedom at equilibrium of a chemically reactive system having $S_{(solid)}$ and $O_{2(g)}$, $SO_{2(g)}$, $SO_{3(g)}$.

Since in this system there are components ($m = 4$) and two phases (solid, gas) so $\pi = 2$, the number of independent reactions are 2 ($r = 2$) these are

$$S_{(s)} + O_{2(g)} \rightarrow SO_{2(g)}$$

and

$$S_{(s)} + \frac{3}{2} O_{2(g)} \rightarrow SO_{3(g)}$$

S(s) and $O_{2(g)}$ are assumed to be present at equilibrium.

In accordance with extended phase rule, we get

$$F = 2 - 2 + 4 - 2 = 2$$

Thus, this system has two degrees of freedom. Thus specification of any two intensive properties and the equilibrium mole fraction fixes the equilibrium state of the system.

To determine the equilibrium conditions for a gas mixture of known initial composition we apply Duhem's theorem which requires that two independently variable conditions be specified to make the problem determinate. So we eliminate component mole fractions in favour of initial compositions and the dimensionless reaction coordinate ξ to get the result as

$$y_i = \frac{y_{i0} + \gamma_i \xi}{1 + \upsilon \xi} \quad (i = 1, 2 \ldots m)$$

so a single equation relating the three variables T, P and ξ can be obtained by replacing y_i. However, the solution of problem is still complicated as ϕ_i depends on T, P and ξ.

In the simplest case of reacting mixture of ideal gas $\overset{A}{\phi_i} = 1$ and the T, P and composition dependent portion of the equation.

$$\overset{A}{\pi}\,(yi,\,\phi i)^{vi} = k \times \left(\frac{P/kPa}{101.325}\right)^{-v}$$

are distinct and separable and the solution for ξ, P or T is simple.

Example. Ethyl alcohol can be obtained by vapour phase hydration of ethylene as follows

$$C_2H_{4(g)} + H_2O_{(g)} \rightarrow C_2H_5OH_{(g)}$$

The feed to a reactor in which this reaction takes place is a gas mixture containineg 25 mol% ethylene, 75 mol % steam. What will be the composition of products if it occurs at 398 K and 101.3 *k*Pa. DG° at 398 k is 4530 J/mol.

Solution. Assuming ideal behaviour for gaseous phase. We have

$$\upsilon_1 = -1,\ \upsilon_2 = -1,\ \upsilon_3 = 1$$

So, $$\upsilon = \Sigma_{vi} = -1$$

$$\therefore \qquad y_1 = \frac{y_{10} + \upsilon}{1 + \upsilon}\,\frac{\xi}{\xi}\,\frac{.25 - 1\xi}{1 - \xi}$$

$$y_2 = \frac{y_{2_0} + \upsilon_2}{1 - \upsilon}\,\frac{\xi}{\xi}\,\frac{0.75 - 1\xi}{1 - \xi}$$

$$y_3 = \frac{y_{3_0} + \upsilon_3}{1 + \upsilon}\,\frac{\xi}{\xi}\,\frac{\xi}{1 - \xi}$$

so $$k = \exp\left(\frac{4530}{8.314 \times 398}\right) \qquad \left[k = \exp\left(\frac{-\Delta G^\circ}{RT}\right)\right]$$

$$= .254$$

From equation

$$\pi\ \ y_i^{yi} = \left(\frac{P/k\,Pa}{101.325}\right)^{-v} \times k, \text{ we get}$$

$$\left(\frac{.25 - \xi}{1 - \xi}\right)^{-1} \left(\frac{.75 - 1}{1 - \xi}\right)^{-1} \left(\frac{\xi}{1 - \xi}\right)^{+1} = (1)+1 \times .254$$

or

$$\frac{\xi\,(1 - \xi)}{(.25 - \xi)\,(.75 - \xi)} = .254$$

or $$\xi = .0395$$

So the equilibrium composition of the products is

$$y_1 = .219, y_2 = .740, y_3 = .041$$

By definition ξ (reaction coordinate) gives a measure of "yield" of reaction. Thus increase of x indicates a shift to right.

A large negative value of ΔG° will givee a large value of k and thus of k_y and a large equilibrium ξ wilel be obtained as compared to ξ for a reaction of identical geometry but smaller (negative) ΔG° value. Again a reaction with large positive value of ΔG° will yield small equilibrium ξ relative to that for a similar reaction with smaller positive value of ΔG°.

The sign and magnitude of ΔG° can thus indicate the expected yield of reaction. Large negative value of ΔG° imply high equilibrium conversion to the product species and large position value of ΔG° to small equilibrium conversion.

The effect of temperature and pressure and ξ values can be given by

$$\left(\frac{\partial \xi}{\partial T}\right)_P \left(\frac{d\xi}{dk_y}\right)\left(\frac{dk_y}{\partial T}\right)$$

But

$$\left(\frac{\partial k_y}{\partial T}\right)_P = \left(\frac{P/kPa}{101.325}\right)^{-\nu} \times \frac{dk}{dT}$$

or

$$\frac{dk}{dT} = \left[\exp\left(\frac{-\Delta G^\circ}{RT}\right)\right]\frac{d(-\Delta G^\circ/RT}{dT}$$

thus

$$\frac{dk}{dT} = \frac{k\,\Delta H^\circ}{RT^2}$$

or

$$\left(\frac{\partial \xi}{\partial T}\right)_P = \left[\frac{k_y}{RT^2} \times \frac{d\xi}{dk_y}\right]\Delta H^\circ$$

Again for

$$\left(\frac{\partial \xi}{\partial P}\right)_T = \left[\left(\frac{101.325}{P/kPa}\right)k_y\frac{d\xi}{dk_y}\right](-\upsilon)$$

The sign of $\left(\frac{\partial \xi}{\partial T}\right)_P$ and $\left(\frac{\partial \xi}{\partial P}\right)_T$ are determined solely by the sign of ΔH° and υ (=Σvi) respectively since the terms in square brackets are always positive. Following conclusions can thus be drawn:

(i) At a given temperature if ΔH° is positive the reaction is endothermic and in such a case increase in T at constant P increase the equilibrium ξ. If ΔH° is negative then the reaction is exothermic and an increase in T at constant P causes a decrease in equilibrium ξ.

(ii) If the total stoichiometric coefficient υ (=Σ_{vi}) is negative ah increase in pressure at constant temperature increase the equilibrium x. If υ is negative then an increase in *P* at constant 1—T decreases the equilibrium ξ.

The *sign of* ΔG can easily be calculated by the use of Gibbs-Helmholtz equation.

$$\Delta G = \Delta H - T\Delta S$$

(i) In exothermic reactions ΔH is negative and so we get a more negative value of ΔG. So such reactions will be spontaneous.

(ii) A positive value of ΔS will make ΔG negative and so such a process which is accompanied by increase in entropy will be spontaneous.

Generally these two tendencies are found to oppose each other and the direction of reaction is decided by the resultant of these two opposing tendencies. e.g.,

$$H_{2(g)} + \frac{1}{2}O_{2(g)} \rightarrow H_2O_{(l)} \qquad \Delta G = -56690 \text{ cal}$$

Negative value of ΔG shows thee spontaneity of the reaction and it is found to be so.

In case of Hydrogen peroxide the following two reactions are probable:

(i) $H_2O_2 \rightarrow H_2O + \frac{1}{2}O_2 \qquad \Delta G = -25090$ cal

(ii) $H_2O_2 \rightarrow H_2 + O_2 \qquad \Delta G = +31470$ cal.

the first reaction has a negative value of ΔG and actually H_2O_2

decomposes to give H_2O and O_2.

A negative value of ΔG only indicates the spontaneity of the reaction i.e., this reaction is thermodynamically feasible. It does not indicate that the reaction will take place under ordinary conditions. It only indicates that the reaction is possible if proper conditions are provided. In the above examples the reaction between H_2 and O_2 is thermodynamically feasible but hydrogen and oxygen can remain in contact with each other for a very long time without any reaction taking place. It is only in the presence of a catalyst that a reaction takes place.

Applications of Chemical Affinity

The most important application of the chemical affinity is to tell about the thermodynamic feasibility of the process under consideration. This fact is of great importance in industry. If a process is not thermodynamically feasible it would be futile to waste time and energy to carry out the process. If on thermodynamic consideration we find a process to be feasible it would be worthwhile to try and find and suitable conditions for that process to be carried out. If we are interested in manufacture of NH_4NO_2 by a new process say by the action of N_2 and water.

$$N_2 + 2H_2O \rightarrow NH_4NO_2$$

We calculate ΔG values for this reaction, if ΔG is negative only them this process will be feasible otherwise not and we can then predict if such a reaction has any chance of being carried out commercially or not.

Limitations

The thermodynamic feasibility of a process under certain conditions can only be predicted by calculations of ΔG. Nothing could be said about the conditions of reaction, rate of a reaction or the mechanism of a reaction by this value of ΔG.

Determination of Chemical Affinity

Some of the methods used for the determination of chemical affinity are given in following pages.

(1) *By the use of Gibbs-Helmholtz equation.* The equation is,

$$\Delta G = \Delta H + T\left[\frac{\partial \Delta G}{\partial T}\right]_P$$

In this ΔH can be determined calorimetrically, and $\left(\frac{\partial(\Delta G)}{T}\right)_P$ can be evaluated from the temperature coefficient of E.M.F. of the cell. Substituting these values in above equation ΔG can easily be calculated. It is a very accurate method.

(2) *E.M.F. of a cell* can be used to calculate the ΔG values of a chemical reaction.

We know $-\Delta G = nFE$

where *E* is the *E.M.F.* and *nF* is the number of coulombs of electricity required for the cell reaction to occur. Thus if we know the E.M.F. of a cell the above equation can be used to calculate ΔG values, e.g.,

If the E.M.F. of a $Zn/Zn^{2+}//Cu^{2+}/Cu$ cell is 1.1 V. then its chemical affinity can be calculated as under. We have

$$-\Delta G = nFE$$

$$= \frac{2 \times 1.1 \times 96500}{4.2} \text{ cal/Mol.}$$

$$= 50548 \text{ cal/Mol.}$$

So ΔG for this reaction is –50548 cal/Mol.

(3) *Fran Vant' Hoff isotherm.* This method can also be used. The Vant Hoff isotherm is

$$-\Delta G = RT \ln k_c - RT \Sigma n \ln c$$

If the initial concentration of reactants is unity and the final concentration of products is also unity the above equation becomes

$$-\Delta G^\circ = RT \ln k_c$$

Knowing the value of k_c we can calculate the chemical affinity for this reaction.

In case of gaseous reactions the above equation can be written as

$$-\Delta G^\circ = RT\ 1n\ k_p$$

and the chemical affinity in such a case can be calculated if we know the partial pressures of various reactants and products of the reaction.

(4) *By use of vapour pressure data.* For an isothermal process, we know

$$dG = VdP$$

Integrating it between the limits

$$\int_{G_1}^{G_2} dG = \int_{P_1}^{P_2} VdP$$

$$= \int_{P_1}^{P_2} \frac{nRT}{P} dP \qquad [PV = nRT]$$

$$\Delta G = nRT\ 1n \frac{P_2}{P_1}$$

$$= 2.303RT.\ n \log \frac{P_2}{P_1}$$

For one mole of a gas

$$\Delta G = 2.303\ RT \log \frac{P_2}{P_1}$$

Knowing the initial and final vapour pressures of various reactants and products ΔG can be calculated.

(5) *From solubility measurements.* This method can be used for less volatile substances.

We know

$$-DG = RT\ 1n \frac{S\alpha}{S\beta}$$

where $S\alpha$ = saturation solubility of one form

and $S\beta$ = saturation solubility of another form of substance in the same solvent and at the same temperature

(6) *Spectroscopic method.* For gaseous reactions, we have

$$-\Delta G = RT\ 1n\ k_p$$

The free energy change of any reaction for a particular reaction is also given by $-\Delta\left(\frac{G^0 - E^0}{T}\right)$ or $-D\left(\frac{G_0^0 - E_0^0}{T}\right)$

$$= -\frac{\Delta G_0^0}{T} + \frac{\Delta E_0^0}{T}$$

From these two equations, we get

$$-\frac{\Delta G^0}{T} = R\ 1n\ k_p$$

$$= -\Delta\left(\frac{G^0 - E_0^0}{T}\right) - \frac{\Delta E_0^0}{T}$$

ΔG_0^0 is the change in energy for the reaction at 0°k. The value of $\Delta E_0°$ can be calculated from spectroscopic heats of dissocition with the help of the following equation.

$$G_0 = E_0° + RT_2 \frac{\partial (1n\ Q^0)}{\partial T}$$

Thus by calculating E^0 spectroscopically the value of ΔG can be calculated.

(7) *By the third law of thermodynamics*

$$\Delta G^0 = \Delta H^0 - T\Delta S^0$$

Knowing the values of ΔH^0 calorimetrically and calculating ΔS^0 from absolute entropy tables we can get the chemical affinity value of a reaction.

15

The Specific Heat of Solids and Gases

Introduction

Specific heat of a substance is defined as the *amount of heat required to raise the temperature of 1 gm of the substance by* 1°C. In case of solids the specific heat increases with increase of temeperature and decreases with decrease of temperature. It approaches zero as the absolute zero temperature is approached. Many attempts by various scientists were made to explain this difference in specific heat of solids with a change in temperature. Some of the explanations given are discussed below:

(1) *Dulong and Petit's law. Lavoisier and Laplace* studied the specific heats of a large number of metals and determined these with great accuracy. They gave in tabular form the accurate values for the specific heats of a large number of metals. *Dulong and Petit* found that multiplication of the specific heat of a solid and its atomic weight always gave a value approximately equal to 6.4 and based on these observations they enunciated the law known as Dulong and Petits law.

"At room temperature, the product of the specific heat and the atomic weight of a solid is a constant for all elements and is approximately equal to 6.4."

Another weight × specific heat = 6.4 approx. This product of atomic weight and specific heat is called *atomic heat.*

Another statement for Dulong and Petits law can be:

"The atomic heats of solid elements at room temperature are approximately equal to 6.4."

Some exception to this law were also observed e.g.

(i) Atomic heats of some solid elements such as C, Si, B etc. are as low as 2 to 3 cal per degree.

(ii) A large variation is observed in atomic heat values with variation in temperature. Diamond and silicon the specific heats are found to change with temperature.

(iii) The law is not found valid at low temperatures.

So this law has a theoretical significance only.

(2) Another attempt to explain this was by *L. Boltzmann.* His attempt was based on Kinetic theory and is known as *classical theory.*

"An ideal solid is considered to be made up of atoms vibrating about their equilibrium position but not interacting with one another in any way. Thus it can be assumed that in such cases the atoms are oscillating about fixed positions and they posses Kinetic energy of translation. They also possess potential energy due to attraction. Thus total energy E is given by

$$\text{E} = \text{KE} + \text{P.E}$$

$$= \frac{p^2}{2m} + \text{V}_{(r)}$$

$$= \frac{P^2}{2m} + \frac{1}{2} m\, w_0^2 w^2 \qquad (1)$$

$V_{(r)} = \frac{1}{2} m\, w_0^2 x^2$, m = mass, w_0 = natural frequency of vibration of oscillator. Again

$$\text{E} = \text{N} \times E \qquad (2)$$

E is the mean energy of oscillator and N is the number of degrees of freedom. Applying statistics mechanics, we have

$$dN \propto e^{-E/kT}$$

or
$$dN = \text{Constant} \times e^{-E/kT}$$
$$= \text{constant} \times e^{-\left(\frac{p^2}{2m} + \frac{1}{2} m\, w_0^2 x^2\right)} \qquad (3)$$

The mean energy of each oscillator is

Thus
$$\bar{E} = \frac{\int_0^w E\,dN}{\int_0^w dN}$$

Substituting the values of E and dN from (1), (2) in (3)

We get

$$\bar{E} = \frac{\int_v \int_x \left(\frac{p^2}{2m} + \frac{1}{2} m w_0^2 x^2\right) e^{-p^2/2mkT} \cdot e^{-mw_0^2 x^2/2kT}\, dp_x dx}{\int_p \int_x e^{-p^2/2mkT}\, e^{-mw_0^2 x^2/2kT}\, dp_x dx}$$

or
$$\frac{\int_{-\infty}^{+\infty} \frac{p^2}{2m} e^{-p^2/2mkT}\, dp_x}{\int_{-\infty}^{+\infty} e^{-p^2/2mkT}\, dp_x} + \frac{\int_{-\infty}^{+\infty} \frac{1}{2} m w_0^2 x^2\, e^{-mw_0^2 x^2/2kT}\, dx}{\int_{-\infty}^{+\infty} e^{-mw_0^2 x^2/2kT}\, dx}$$

Using the standard integrals,

$$\int_0^\infty e^{-\alpha u^2}\, du = \frac{1}{2}\left(\frac{\pi}{\alpha}\right)^{1/2}$$

and

$$\int_0^\infty u^2 e^{-au^2}\, du = \frac{1}{4}\left(\frac{\pi}{\alpha^2}\right)^{1/2}$$

We have

$$\frac{\int_{-\infty}^{+\infty} \frac{p^2}{2m} e^{-p^2/2mkT}\, dp_x}{\int_{-\infty}^{+\infty} e^{-p^2/2mkT}\, dp_x} = \frac{2 \times \frac{1}{2m} \frac{1}{4} [\pi (2mkT)^3]^{1/2}}{2 \times \frac{1}{2} [\pi (2mkT)]^{1/2}} = \frac{k}{2}$$

and $\dfrac{\int_{-\infty}^{+\infty} \frac{1}{2} m w_0^2 x^2 \, e^{-m w_0^2 x^2/2kT} dx}{\int_{-\infty}^{+\infty} e^{-m w_0^2 x^2/2kT} \, dx}$

$$= \frac{2 \times \frac{1}{2} \times \frac{1}{4} [\pi (2mkT)^3]^{1/2}}{2 \times \frac{1}{2} [\pi (2kT)]^{1/2}} = \frac{kT}{2}$$

Substituting these values in equation (5), we get

$$E = \frac{kT}{2} + \frac{kT}{2} = kT \tag{6}$$

As on oscillator can vibrate about three mutually perpendicular directions, it means that the mean energy

$$E = 3\text{kT} \tag{7}$$

If there are N atoms in a mole, the total energy E is obtained by substituting eq. (7) in (2), we get

$$E = 3NkT = 3RT \tag{8}$$

Differentiating (7) with respect to temperature, we get

$$\left(\frac{dE}{dT}\right)_V = 3\text{R or } C_V = \text{ER} \left[\because \; C_V = \left(\frac{\partial}{\partial T}\right)_V \right]$$

or

$$C_V = 3 \times 1.98 \; [\because \text{R} = 1.98 \text{ cal. per gm. atom per A}^\circ]$$

or $\quad C_V = 5.94$ cal. per gm, atom per A°.

The above value of 5.94 is nearly equal to the value given by the *Dulong* and *Petit's* law. This is the derivation of the law of *Dulong* and *Petit* from kinetic theory.

(3) *Einstein theory of specific heats.* Classical theory based on the assumptions of kinetic theory failed to explain the decrease of specific heat with decrease of temperature. An attempt was made by Einstein who useed quantum mechanics in his calculations. So the energy of each linear oscillator is given by

$$E = nh\nu \qquad n = 0, 1, 2...$$

where the average energy of ane oscillator is totally different on quantum theory from the classical theory. Plank proposed that linear oscillator vibrates only with integral energy values 0, ε, 2ε... where ε = *hν*. Einstein used this concept and provided *kev* information for the dependence of heat capacity on temperature. the heat capacity at constant volume (C_v) is given by

$$C_v = \frac{\Delta E}{\Delta T}$$

and ΔE can be calculated using

$$\Delta E = N\Delta\,\bar{\varepsilon}$$

N = Number of atoms in a crystal and

$\Delta\bar{\varepsilon}$ = Change in average energy of the atom in a crystal by increase of ΔT temperature.

An equation was then derived to give $\bar{\varepsilon}$ dependence on temperature.

Einstein considered a mole of a monoatomic solid consisting of N mass points which can oscillate along all the three axes. Let ν be the frequency of vibrations. Then for oscillations in any one direction energy is given by

$$E_n = nh\nu, \qquad n = 0, 1, 2...$$

Though an atom in a solid is an oscillator in three dimensions but for simplification it is considered as one dimensional oscillator and after making calculations necessary corrections are applied.

For thermal equilibrium, the oscillators in a crystal are to be disturbed among various allowed energies in accordance with Boltzman distribution law such that the number of N_n of atoms with energy ε_n is related to number N_0 with energy $\varepsilon_0 = 0$ by the equation.

$$N_n = N_0\, e^{-\varepsilon_n/kT}$$

$$= N_0\, e^{-nh\nu/kTe}$$

where $\qquad N = N_0 + N_1 + N_2 + ...$

Substitution of Boltzmann factor gives

$$N = N_0 + N_0 e^{-h\nu/kT} + N_0 e^{-2h\nu/kT} + ...$$

$$= N_0 \sum_{n=0}^{\alpha} e^{-nh\nu/kT}$$

To calculate total energy of oscillation in are direction onee the atoms multiply energy E_n by the number of particles and add them, So

$$E = \varepsilon_0 N_0 + \varepsilon_1 N_1 + \varepsilon_2 N_2 + ...$$

$$= O N_0 + h\nu N_1 + 2h\nu N_2 + ...$$

$$= O + h\nu N_0 e^{-h\nu/kT} + 2h\nu N_0 e^{-2h\nu/kT} + ...$$

$$= N_0 \sum_{1}^{\infty} nh\nu e^{-h\nu/kT}$$

But

$$\bar{\varepsilon} = \frac{E}{N}$$

$$\therefore \qquad \bar{\varepsilon} = \frac{h\nu \Sigma n e^{-nh\nu/kT}}{\Sigma e^{-nh\eta/kT}}$$

ε = average energy of one dimensional oscillator.

To evaluate $\bar{\varepsilon}$ we complete the Σ. To do this we make the substitution.

$$\sum_{n=1}^{\alpha} -nh\nu/kT = \sum_{n=1}^{\alpha} Y^n = 1 + y + y^2 + y^3 ...$$

$$= \frac{1}{1-y}$$

Treating the series in numerator in the same way

$$\sum_{1}^{\alpha} ne^{-nh\nu/kT} = \sum_{1}^{\alpha} ny^n = y(1 + 2y + 3y^2 + ...)$$

$$= \frac{y}{(1-y)^2}$$

Thus $\qquad \bar{e} = \dfrac{h\nu y}{1-y}$

$$= \frac{h\nu\, e^{-h\nu/kT}}{e^{h\nu/kT}}$$

Multiplying numerator and denominator by $e^{h\nu/kT}$, we get

$$\bar{\varepsilon} = \frac{h\nu}{e^{h\nu/kT} - 1}$$

For three dimensional oscillator it will be three times.

$$\therefore \qquad \bar{\varepsilon} = \frac{3h\nu}{e^{h\nu/kT} - 1}$$

The heat capacity of a solid can thus be calculated at a high temperature and the denominator under these conditions is simplified by using the fundamental properties of exponential function.

$$e^{h\nu/kT} = 1 + \frac{h\nu}{kT} + \frac{1}{2}\left(\frac{h\nu}{kT}\right)^2 + \ldots$$

Since $\frac{h\nu}{kT}$ is very small so higher terms can be neglected.

Thus $\qquad e^{h\nu/kT}\ 1 + \frac{h\nu}{kT}$

and the equation of three dimensional oscillator then become

$$\bar{\varepsilon} = \frac{3h\nu}{e^{h\nu/kT} - 1} = \frac{3h\nu}{h\nu/kT} = 3\ \mathrm{kT}$$

so the total energy is given by

For 1 mole of atoms

$$E = N\bar{\varepsilon} = 3N\,kT$$

$$= 3RT \qquad [R = Nk]$$

$$\text{Heat capacity} = \frac{\Delta E}{\Delta T}$$

$$= 3R \qquad [\text{App. } 6\ \text{cal}/^{6}/\text{Mole}]$$

This theory can explain the specific heats of solids at high temperatures only where the results are found to be in close agreement with Dulong and Petit's law. If ε is sufficiently large a very high temperature is required to attain the classical value for atomic heat of solids. The heat capacities at low temperatures can not be explained by

Einstein's theory. However, if approximations are made even at low temperature it can be shown that specific heat of solids tends to approach zero as temperature approaches absolute zero.

This theory was thus successful in explaining the change of specific heats of solids with changes in temperatures and gave results to support Dulong and Petit's law at high temperatures. At low temperatures also it could explain the zero value of specific heat for solids at absolute zero of temperature.

Since it uses many approximations it has very serious limitations at low temperatures. The value of ν used in this theory can be calculated only empirically which can not be verified. Its assumption about the vibration of a particular atom with the same frequency is also not correct as it is under the field of other vibrating atoms.

(4) *Debye theory.* Debye recognised the differences, in the theoretical and experimental results obtained by Einstein's theory, at low temperatures and he considered the vibration of a crystal as a whole. He looked the problem as that continuous spectrum can result from modes of vibrations if these vibrations are treated to be elastic type vibrations. They could thus execute two types of waves

(i) Transverse vibrations with velocity $C_t = \sqrt{n/\rho}$

and (ii) Longitudinal vibrations with velocity

$$C_e = \sqrt{\left(k + \frac{4}{3} n\right)/\rho}$$

n in these denote the rigidity factor and k denotes the bulk modulus.

The number of vibrational modes that are possible, in an enclosed volume, V in second case are given by

$$\frac{4\pi\upsilon^2}{C_e^2}\, d\upsilon V$$

where V is the volume and the interval selected in V and $\nu + d\upsilon$.

In the same frequency interval the number of modes of vibrations in the first case would be

$$\frac{8\pi^2\,\upsilon^2\,d\upsilon}{C_t^3}\,\mathrm{V}$$

As in case of transverse vibrations are two waves at right angles to each other.

So for a solid capable of transmitting three independent vibrations in any given direction. Two transverse vibrations at right angles to each other and a longitudinal wave will be given by

$$\int_0^{\upsilon_m} 4\pi V \left(\frac{1}{c_t^3} + \frac{2}{cl^3}\right) \upsilon^2 \, d\upsilon$$

If 'N' atoms vibrate then it is a system of 3N oscillators. Hence

$$3N = \int_0^{\upsilon_m} 4p\, V \left(\frac{1}{C_t^3} + \frac{2}{C_l^3}\right) \upsilon^2 \, d\upsilon$$

or

$$\upsilon_m^3 = \frac{9N}{4\pi V \left[\frac{1}{C_t^3} + \frac{2}{cl^3}\right]}$$

In accordance with Plank, the energy of harmonic oscillator is given by

$$\frac{h\nu}{e^{h\nu/kt - 1}}$$

for 3N harmonic oscillators, we get

$$E = 3N \frac{h\nu^3}{e^{h\nu/kt - 1}}$$

or

$$E = 4\pi V \left(\frac{1}{c_t^3} + \frac{2}{cl^3}\right) \int_0^{\upsilon_m} \frac{h\nu^2}{e^{h\nu/kt} - 1} d\nu$$

or

$$E = \frac{9N}{\upsilon_m} \int_0^{\upsilon} \frac{h\upsilon^2 \, d\nu}{e^{h\nu/kt} - 1}$$

Differentiating w.r.t temperature, we get

$$Cv = \frac{9N}{\upsilon_m^3} \int_0^{\upsilon_m} \frac{e^{p\nu/kt}}{e^{h\nu/kt - 1}} \left(\frac{h\nu}{RT}\right)^2 \nu^2 \, d\nu$$

This equation on integration is called Debye's equation for the atomic heat of monoatomic solids.

On integrating the above eqeuation known as Debye heat capacity equation, we get

$$C_v = 3R\left[\frac{12}{x^3}\int_0^x \frac{\psi^3}{e^\psi - 1} - \frac{3x}{e^x - 1}\right]$$

[Put $\frac{h\nu}{kt} = y$, $\frac{h\nu_m}{k} = \theta$, $\frac{\theta}{t} = \psi_m = x$ to get the results of above integration].

The term in square bracket is a function of x known as Debye function represented by $D(x)$. Its value can be calculated for values of x ranging from 0 to a.

Limiting Cases

(1) In case of very high temperatures the above equation becomes

$$C_P = 3R\left[1 - \frac{1}{20}\left(\frac{\theta}{t}\right)^2 + \frac{1}{560}\left(\frac{\theta}{t}\right)^4 \ldots\right]$$

$\frac{\theta}{T}$ is very small and higher terms can thus be neglected.

then

$$C_v = 3R$$

this is in agreement with the classical value of C_v.

(2) In case oef very low temperatures T is very small so x and ψ become very large. $\frac{\theta}{t}$ tend to α. Hence we have

$$C_v = 3R \cdot \frac{12}{x^3}\int_0^3 \frac{\psi^3\, d\psi}{e^\psi - 1}$$

$$= 3R.\frac{12}{x^3}\cdot\frac{\pi^4}{15}$$

$$= 3R.\frac{12\pi^4}{15}\cdot\left(\frac{T}{\theta}\right)^3$$

$$= 77.94 \times 3R \times \left(\frac{T}{\theta}\right)^3$$

$$= 464.4 \left(\frac{T}{\theta}\right)^3 \text{ cal/°/gm atom.}$$

Thus at low temp. sp. heat varies as T^3. This is called *Debyes* T^3 law.

At T = 0°A, we get $C_v = 0$

Calculations of Characteristic Temperature

We know

$$\theta = \frac{h\upsilon_m}{k}$$

$$= \frac{h}{k}\left\{\frac{9N}{4\pi v\left(\frac{2}{c_t^3}+\frac{1}{c_t^3}\right)}\right\}^{1/2}$$

But

$$Ct = \left[\sqrt{\left(k+\frac{4}{3}\right)n}\right] \Big/ \rho \text{ and } c_l = \sqrt{n/\rho}$$

$$\therefore \qquad \theta = \frac{h}{k}\left(\frac{9N}{4\pi V}\right)^{\frac{1}{3}} \rho^{1/2}\left[\frac{2}{n^{3/2}}+\frac{1}{(k+4/3\, n)^{3/2}}\right]^{\frac{1}{3}}$$

thus θ can be calculated.

This theory predicts the heat capacity of a solid as a function of temperature and depends only on a characteristic frequency υ_m. In some cases such as lead the value reaches the classical heat capacity value at low temperatures while in others the classical value is reached only at high temperatures.

Weakness of the Theory

(1) It does not agree with theoretical foundations.

(2) It cannot be applied at low temperatures.

The theory was capable of explaining the change in specific heats of solids assuming that, on heating a solid elastic vibrations like continuous medium exists. Here the solid is taken as an isotropic body but actually the solids are definite crystals arranged in some definite form.

This equation according to *slater* is incorrect due to

(i) Velocity of elastic vibrations depends upon the wave length of vibration and so the change of velocity should also be taken into consideration.

(ii) In solids composed of N atoms we have to find independent modes of vibrations of atoms in the solid. *Born* assumed boundry conditions to get a solution and he showed that there will be 3N modes of vibrations. This number is the same as the number of degrees of freedom of solid which is different from Debye even in case of an isotropic medium.

5. *Born theory.* It is a modification of Debye theory. In case of Debye theory we assumed υ_m as the maximum frequency in case of both longitudinal and vibrational modes. In this theory it is proposed to cut the spectrum to get a common minimum wave length for both the longitudinal and transverse waves so for common minimum wave length.

$$\lambda_{\min} = \left(\frac{4\pi v}{3N}\right)^{2/3}$$

for longitudinal modes

$$\upsilon_l = c_l \left(\frac{3N}{4\pi V}\right)^{1/3}$$

and for vibrational modes

$$\upsilon_t = c_t \left(\frac{3N}{4\pi V}\right)^{1/3}$$

The total number of vibrational modes are 3N.

Thirring made interesting calculations for the specific heat at high temperature without finding an explicit form for spectrum and without the use of any more approximations. He expanded the mean energy of an oscillator in powers of $\frac{h\nu}{KT}$ to get

$$E = KT\left[1 - \sum_{n=1}^{\alpha} (-1)^n \frac{B_n}{2n\ !}\left[\frac{h\nu}{KT}\right]^{2n}\right\}$$

Bn here stands for Bernoulli numbers. The heat capacity of linear oscillator is given by

$$C_v = \frac{dE}{dT} = k\left[1 + \sum_{n=1}^{\alpha} (-1)^n \frac{B(2n-1)}{2n\ !}\left[\frac{h\nu}{KT}\right]^{2n}\right]$$

This theory Born and V. Karman's depends on the calculations of frequency spectrum as a dynamic crystal lattice.

These theories can be applied to isotropic body having all the atoms same. The atoms are arranged in lattice and not continuous in the solids has been recognised by these theories.

6. *Blackmann's lattice theory of specific heats*. All the theories discussed so far give an idea of specific heats of solids at low temperatures. The Debye theory does not treat crystal as continuous. Vast differences can exist in the crystals of elastic solid and lattice crystals.

Blackmans showed that there could be a very large difference and he took into account interactions between closed neighbours of the lattice and also next nearest neighbour particles.

This theory assumes that specific heats at high and low temperatures depend on different combination of forces between particles. In high temperatures region it is controlled by normal vibrations of the crystals where as in low temperatures region it is controlled by low frequency end of the spectrum.

Compassion with Debye Theory

Debye theory fails as law temperatures and now it is believed that vibrational spectrum is quite different from that assumed by Debye.

Specific Heats of Gases

The specific heat of a substance depends on the experimental conditions. In casee of solids and liquids practically there is no change in volume on heating and thus the work of expansion can be neglected. But in case of gas an appreciable expansion takes place on heating and so there is an appreciable amount of work done as work of expansion (PdV) so it becomes quite important to tell in case of gases whether the specific heat refers to the heating at constant volume or at constant pressure. Depending on experimental conditions there are two values of specific heats of gases.

(i) *Specific heat of a gas at constant volume.* It is the amount of heat required to raise the temperature of one gram of the gas by 1°C at constant volume and is represented by C_v.

(ii) *Specific heat at constant pressure.* It is defined as the amount of heat required to raise the temperature of 1 gm of the gas by 1°C keeping the pressure constant and is represented by C_p.

These two differ from each other by the amount of heat used up in doing the work of expansion in the second case.

So $\quad C_p = C_v +$ work of expansion

NUMERICALS

Example 1. Calculate Einstein's frequency for copper for which εE = 230K. Given $h = 6.6 \times 10^{-27}$ erg-sec. Boltzmann constant K = 1.37×10^{-16} ergs/°K. (*Agra 1976*)

Solution. Einstein's Frequency is given by

$$\frac{hV_E}{k} = \varepsilon E$$

Thus

$$V_E = \frac{\varepsilon_E \, . \, k}{h}$$

$$= \frac{230 \times 1.37 \times 10^{-16}}{6.6 \times 10^{-27}} \text{ sec}^{-1}$$

$$= 4.794 \times 10^{12} \text{ Sec}^{-1}$$

Example 2. Calculate Einstein's frequency for a case for which $\varepsilon E = 236$ K.

Given $k = 1.4 \times 10^{-16}$ ergs/°K, $h = 6.6 \times 10^{-27}$ erg sec.

Solution. We know

$$V_E = \frac{\varepsilon_E . k}{h}$$

$$= \frac{236 \times 1.4 \times 10^{-16}}{6.6 \times 10^{-27}}$$

$$= 5.0 \times 10^{12} \text{ sec}^{-1}$$

Example 3. For copper V = 4.8×10^{-12} cycles/sec (Hertz). Calculate the characteristic temperature ε_E. Given h = 6.6×10^{-27} ergs-sec, $k = 1.38 \times 10^{-16}$ ergs/°K (Agra 1970)

Solution. The characteristics temperature is given by

$$\varepsilon_E = \frac{h\upsilon}{k}$$

$$\therefore \qquad \varepsilon_E = \frac{6.6 \times 10^{-27} \times 4.8 \times 10^{12}}{1.38 \times 10^{-16}}$$

$$= 229.5\text{°K}.$$

Heat Capacities of Solids and Liquids

Heat capacities for each substance are to be found experimentally but in case of solids and liquids these are found as functions of temperatures by equations of the form

$$C_p = \alpha + bT + CT_2$$

or

$$C_p = \alpha + bT + CT^{-2}$$

where *a, b, c* which are constants must be specified for the substance. Heat capacities generally increase with increase of temperature. The effect of pressure in case solids and liquids is negligible.

The difference in heat capacities can be expressed in terms of volume expansivity and isothermal compressibility by substitution for the derivatives

$$\left(\frac{\partial V}{\partial T}\right)_P = V\beta \text{ and } \left(\frac{\partial P}{\partial T}\right)_V = \frac{\beta}{K}$$

Thus

$$C_P - C_v \frac{TV\beta^2}{K}$$

This difference is significant except at very low pressure.

β = Volume expansivity, K = isothermal compressibility

$$\beta = \frac{1}{V}\left(\frac{\partial V}{\partial T}\right)_P \text{ and } K = -\frac{1}{V}\left(\frac{\partial V}{\partial T}\right)_T$$

Example 1. For metallic copper at 300K the following values are known

$$C_P = 24.5 \text{ J/mol K}$$

$$\beta = 50.4 \times 10^{-6}\ \text{K}^{-1}$$

$$\text{K} = 7.78 \times 10^{-12}\ Pa^{-1}$$

$$\upsilon = 7.06 \text{ cm}^3\text{/mol}$$

Find C_v.

Solution. Making substitution in the equation

$$C_P - C_v = \frac{TV\beta^2}{K}$$

We get

$$C_P = C_v = \frac{300 \times 7.06 \times 10^{-6} \times (50.4 \times 10^{-6})^2}{7.78 \times 10^{-12}}$$

$$= .692 \text{ J/mol K}$$

$C_v = 24.5 - .692$

$= 23.808$ J/mol K.

Example 2. The heat capacity of quartz (SiO_2) at 100 kP$_a$ between 298 and 84 K is given by the equation

$$C_p \text{ (J/mol K)} = 46.98 + 34.33 \times 10^{-3} \times (T/K)$$

$$-11.3 \times 105 \times \left(\frac{T}{k}\right)^{-2}$$

If 1000 kg of Quartz is heated from 300 K to 700 K at 1000 PK$_a$, How much heat is required?

Solution. At constant pressure,

$$\Delta H = Q$$

Again, $dH = C_p dT$

Thus at constant pressure,

$$Q = \int_{T_1}^{T_2} C_p \, dT$$

Making substitutions in this equation

$$Q = 46.98 \times (T_2 - T_1) + \frac{34.33 \times 10^{-3}}{2}(T_2^2 - T_1^2)$$

$$-11.3 \times 105 \times \left(\frac{1}{T_1} - \frac{1}{T_2}\right)$$

Solving for given values of T_2 and T_1

$$Q = 23506 \text{ J/mol}$$

Molecular mass SiO_2 is .052 kg/mol

$$\therefore \quad Q = \frac{23506}{.052}$$

$$= 452040 \text{ J/kg}$$

For 1000 kg we requre heat

$$Q = 1000 \times 452 \times 10^3$$

$$= 452 \times 10^6 \text{ J}$$

= 452 MJ.

Example 3. If dH = C_pdT + V(1– βT) dP

Determine the enthalpy change of mercury for a change of state from .1013 MP_a and 373 K to 101.3 MP_a and 273K.

Following data is given for mercury:

T/K	$P \times 10^{-5}$/Pa	$V\times10^6$ (m^3/mol)	$\beta \times 10^6$K	C_p/(J/molK)
273	1.013	14.72	181	28.01
273	1013.0	14.67	174	28.01
373	1.013	—	—	27.51

Solution. Enthalpy is a state function. So a suitable path can be chosen, then

$$\Delta H = H_2 - H_1 = \int_{T_1}^{T_2} C_p dT + \int_{P_1}^{P_2} V(1 - \beta T)\, dP$$

The most convenient path with the given data is

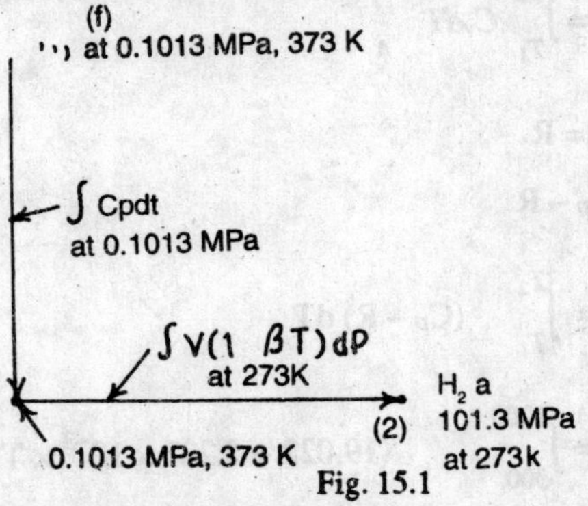

(f)
'') at 0.1013 MPa, 373 K

∫ Cpdt
at 0.1013 MPa

∫ V(1 βT) dP
at 273K

H_2 a
(2) 101.3 MPa
0.1013 MPa, 373 K
at 273k

Fig. 15.1

$$\Delta H = C_p^{ave} (T_2 - T_1) + V^{ave} (1 - \beta^{ave} T)(P_2 - P_1)$$

for P = .1013 for T = 273K

$$C_p^{ave} = \left(\frac{28.01 + 14.67}{2}\right) = 27.76 \text{ J/mal K}$$

$$V^{ave} = \frac{14.72 + 14.67}{2} \times 10^{-6} = 14.695 \times 10^{-6}\ m^3/mol$$

$$\beta^{ave} = \frac{181+174}{2} \times 10^{-6} = 177.5 \times 10^{-6}\, K^{-1}$$

$$\Delta H = 27.76 \times (-100) + 14.695$$

$$\times 10^{-6} (1-177.5 \times 10^{-6} \times 273) \times 101.2 \times 10^{6}$$

$$= -2776 + 1415$$

$$= -1361 \text{ J/mol.}$$

Example 4. One gm mole of nitrogen is contained in a constant volume tank at low pressure. Assuming nitrogen to be an ideal gas, calculate the amount of heat required to raise its temperature from 3000 to 1000 K. For nitrogen.

$$C_p' \text{ (J/mol K)} = 27.336 + 5.233 \times 10^{-3} \times \text{(J/K)}$$

Solution. At constant volume dE = Q

Also $\quad C_v = \dfrac{dE}{dT}$

or $\quad dE = C_v dT$

$\therefore \quad Q = \int_{T_1}^{T_2} C_v dT$

But $\quad C_p - C_v = R$

$\quad C_v = C_p - R$

Hence $\quad Q = \int_{T_1}^{T_2} (C_p - R)\, dT$

$$Q = \int_{300}^{1000} (19.022 + 5.223 \times 10^{-3} \times T) dT$$

$$[\because R = 8.314 \text{ J/Mol K}]$$

$$Q = 19.022 \times 700 + \frac{5.223 \times 10^{-3}}{2} \times 91 \times 10^{4}$$

$$= 15696 \text{ J.}$$

Example 5. The specific heat C of a certain substance at very low temperature is found to depend upon the absolute temperature T according to the relation

$$C = AT^3$$

where A is a constant

(i) If units are calories/gram and kelvin degrees, what units must the constant A have?

(ii) How much heat must be added to raise the temperature of a mass *m* of the material from T_1 to T_2? *(Agra 1972 Suppl.)*

Solution. Given $C = AT^3$

Thus $A = \dfrac{C}{T^3}$

(i) *To calculate the unit of A*

$$\text{Units of A} = \frac{\text{Cal. per gram per degree kelvin}}{(\text{degree kelvin})^3}$$

$$= \text{cal. per gram per } (^\circ\text{k})^4.$$

or

$$\text{Cal gm}^{-1}\ (^\circ\text{K})^{-4}$$

(ii) *To calculate the heat required to raisee the temperature of mass m by dT.*

$$\text{Heat required} = \int_{T_1}^{T_2} nCdT = \int_{T_1}^{T_2} mAT^3 dT$$

$$= mA\left[\frac{T^4}{4}\right]_{T_1}^{T_2}$$

$$= \frac{1}{4}\,mA\,(T_2^4 - T_1^4)$$

Specific Heat of Gases

When we heat a gas, there is in general an increase in volume as well as pressure. Generally either pressure or volume is kept constant and so there are two specific heats:

(i) Specific heat at constant volume (C_v)

(ii) Specific heat at constant pressure (C_p)

(i) **Specific heat of a gas at constant volume** is the quantity of heeat needed to raise the temperature of 1 g of gas through 1°C, the volume being kept constant.

(ii) **Specific heat of a gas at constant pressure** is the quantity of heat required to raise the temperature of 1 g of gas through 1°C, the pressure being kept constant.

The specific heat at constant pressure (C_p) is always greater than the specific heat at constant volume (C_v) by an amount equal to the thermal equivalent of the external work done by the gas. Thus

$$C_p - C_v = Q = W$$

where Q is the amount of heat used to do extra work and W is the external work done in the same units as Q. Difference between two specific heats is given by

$$C_p - C_v = R$$

where C_p, C_v and R are all expressed in the same units.

Example 1. Find the value of gas constant R for one gram molecule of a gas.

Solution. We know

1 mole of any gas = 22400 ml of gas at N.T.P

For one mole of gas the gas equation is

$$PV = RT$$

or $$R = \frac{PV}{T}$$

$$= \frac{76 \times 13.6 \times 981 \times 22400}{273}$$

$$= 8.31 \times 10^7$$

$$= 8.31 \text{ Joules/mole°K.}$$

Variation of Specific Heat of Diatomic Gases with Temperature

It has been observed experimentally that the specific heat of diatomic gases like H_2, N_2, Cl_2 and HCl etc. increases with increase of temperature and decreases with decrease of temperature finally approaching the limiting value of $\frac{3}{2}$ (= 2.98 calories) which is the value of specific heat of monoatomic gases at all temperatures. Thus it leads us to conclude that at very low temperatures all diatomic gases have same value of specific heat of constant volume i.e., as that of monoatomic gas. Euken found experimentally that the specific heat of hydrogen at constant volume approaches $\frac{3}{2}$ R at about 60°K.

Example 1. Assuming the ideal gas molecules as rigid, but capable of free rotation about either of two mutually perpendicular axis which are at right angles to the line joining the two atoms of the molecule. Find C_v, C_p and γ and compare γ with the experimental value of γ for a diatomic gas.

Solution. If we assume that the molecules of a diatomic gas are rigid, then it can have only five degrees of freedom (three of translational and two rotational).

The kinetic energy pere degree of freedom according to equipartitional principle of energy, is equal to $\frac{1}{2}kT$.

Therefore, kinetic energy associated with five degrees of freedom $= \frac{5}{2}kT$.

Thus K.E. per gram molecule $= \frac{5}{2}$ k.T.

and the K.E. per gram molecule $= \frac{5}{2}$ N.k.T.

or $$W = \frac{5}{2}RT$$

Hence

$$\text{Specific heat } C_v = \frac{dW}{dT} = \frac{5}{2}R$$

$$C_p = C_v = R$$

$$= \frac{5}{2}R + R = \frac{7}{2}R$$

and $$\gamma = \frac{C_p}{C_v} = \frac{\frac{7}{2}R}{\frac{5}{2}R}$$

$$= \frac{7}{5}$$

$$= 1.40$$

The experimentally observed value of g at 15°C are given in the following table:

Gas	Value of γ
H_2	1.41
N_2	1.404
O_2	1.401
HCl	1.41
CO	1.404

These values of γ agree with those calculated theoretically.